The Control Revolution

The Control Revolution

Technological and Economic Origins of the Information Society

James R. Beniger

Harvard University Press
Cambridge, Massachusetts, and London, England

Library of Congress Cataloging-in-Publication Data

Beniger, James R. (James Ralph), 1946–
    The control revolution.

    Bibliography: p.
    Includes index.
    1. Communication—Social aspects—United States.
2.   Information storage and retrieval systems—
Technological innovations.   3.   Computers and civilization.
I. Title.   II. Title: Information society.
HM258.B459   1986   302.2   85-31743
ISBN 0-674-16985-9 (alk. paper) (cloth)
ISBN 0-674-16986-7 (paper)

Why do we find ourselves living in an Information Society? How did the collection, processing, and communication of information come to play an increasingly important role in advanced industrial countries relative to the roles of matter and energy? And why is this change recent — or is it?

James Beniger traces the origin of the Information Society to major economic and business crises of the past century. In the United States, applications of steam power in the early 1800s brought a dramatic rise in the speed, volume, and complexity of industrial processes, making them difficult to control. Scores of problems arose: fatal train wrecks, misplacement of freight cars for months at a time, loss of shipments, inability to maintain high rates of inventory turnover. Inevitably the Industrial Revolution, with its ballooning use of energy to drive material processes, required a corresponding growth in the exploitation of information: the Control Revolution.

Between the 1840s and the 1920s came most of the important information-processing and communication technologies still in use today: telegraphy, modern bureaucracy, rotary power printing, the postage stamp, paper money, typewriter, telephone, punch-card processing, motion pictures, radio, and television. Beniger shows that more recent developments in microprocessors, computers, and telecommunications are only a smooth continuation of this Control Revolution. Along the way he touches on many fascinating topics: why breakfast was invented, how trademarks came to be worth more than the companies that own them, why some employees wear uniforms, and whether time zones will always be necessary.

The book is impressive not only for the breadth of its scholarship but also for the subtlety and force of its argument. It will be welcomed by sociologists, economists, historians of science and technology, and all curious general readers.

James R. Beniger is Associate Professor at the Annenberg School of Communications, University of Southern California, Los Angeles.

Preface

To SAY that the advanced industrial world is rapidly becoming an Information Society may already be a cliché. In the United States, Canada, Western Europe, and Japan, the bulk of the labor force now works primarily at informational tasks such as systems analysis and computer programming, while wealth comes increasingly from informational goods such as microprocessors and from informational services such as data processing. For the economies of at least a half-dozen countries, the processing of information has begun to overshadow the processing of matter and energy.

But why? Among the multitude of things that human beings value, why should it be information, embracing both goods and services, that has come to dominate the world's largest and most advanced economies? Despite scores of books and articles proclaiming the advent of the Information Society, no one seems to have even raised—much less answered—this important question.

My own desire to understand the new centrality of information began in the summer of 1963, before my junior year in high school, when the National Science Foundation sponsored my participation in an eight-week program in mathematics and computer science at Oregon State University. At a time when no teenage hacker culture had yet emerged, living with thirty students from around the country while learning to program proved to be the next best thing, my personal windfall from Sputnik (I still delight in being one of the youngest people to have run a program on vacuum tubes). Why have computers become so central to modern society, I wondered that summer, when all they can do is to transform information from one form to another? How could our entire era, popularly described even in the early 1960s as the "Computer Age," be evoked by so modest an activity as information processing?

Even if we could explain the growing importance of information and its processing in modern economies, I realized, we would immediately confront a second question: Why now? Because information plays an important role in all human societies, we would also have to explain why it has only *recently* emerged as a distinct and critical commodity. Material culture has also been crucial throughout human history, after all, and yet capital did not displace land as the major economic base until the Industrial Revolution. To what comparable technological and economic "revolution" might we attribute the emergence of the Information Society?

My answer, as the title of this book indicates, is what I call the Control Revolution, a complex of rapid changes in the technological and economic arrangements by which information is collected, stored, processed, and communicated, and through which formal or programmed decisions might effect societal control. From its origins in the last decades of the nineteenth century, the Control Revolution has continued unabated, and recently it has been accelerated by the development of microprocessing technologies. In terms of the magnitude and pervasiveness of its impact upon society, intellectual and cultural no less than material, the Control Revolution already appears to be as important to the history of this century as the Industrial Revolution was to the last.

But history alone cannot explain why it is information that increasingly plays the crucial role in economy and society. The answer must be sought in the nature of *all* living systems—ultimately in the relationship between information and control. Life itself implies control, after all, in individual cells and organisms no less than in national economies or any other purposive system.

My interest in such systems developed from the first course I attended as a Harvard freshman, Soc Sci 8, taught in the fall of 1965 by the cognitive scientist George A. Miller. Although I had the great pleasure, fifteen years later, of being George Miller's colleague at Princeton, I doubt that he can ever know how much his early teaching on information processing and communication inspired at least one undergraduate to view things social as interacting processing systems—and to appreciate the importance of communication and control in all such systems.

Once we view national economies as concrete processing systems engaged in the continuous extraction, reorganization, and distribution of environmental inputs to final consumption, the impact of industrialization takes on new meaning. Until the Industrial Revolution, even the largest and most developed economies ran literally at a human pace, with processing speeds enhanced only slightly by draft animals and by wind and water power, and with system control increased correspondingly by modest bureaucratic structures. By far the greatest effect of industrialization, from this perspective, was to speed up a society's entire material processing system, thereby precipitating what I call a crisis of control, a period in which innovations in information-processing and communication technologies lagged behind those of energy and its application to manufacturing and transportation.

Identifying the crisis of control and the resulting Control Revolution has helped me to answer another question that has nagged me since my days as an American history major, namely, why the period 1870–1910 is so interesting to modern students and seems so decisive for society as we know it today. Here my thinking has been most influenced by Alfred Chandler of the Harvard Business School, one of the few historians to exploit the view of societies as material processing systems. Chandler's 1977 book, *The Visible Hand: The Managerial Revolution in American Business*, first suggested to me the possibility that the American economy had become a distinctively more *purposive* system during those decades.

The Information Society, I have concluded, is not so much the result of any recent social change as of increases begun more than a century ago in the speed of material processing. Microprocessor and computer technologies, contrary to currently fashionable opinion, are not new forces only recently unleashed upon an unprepared society, but merely the latest installment in the continuing development of the Control Revolution. This explains why so many of the computer's major contributions were anticipated along with the first signs of a control crisis in the mid-nineteenth century.

Although some readers may see this as a "multidisciplinary" approach to history, my goal has been to understand not multiple subjects but only one: the origin of the Information Society. If the world economy uses information for the same general purpose as does a single organism, if economic changes influence theoretical work on information processing, and if the resulting technological breakthroughs increase our material control, as I will argue in the following chapters, then it seems a shame to leave this interesting phenomenon of information processing and control divided up—like a secret treasure map among conspirators—among biologists, economists, historians, and engineers. We segment experience only to make it easier to understand, after all, and although the various academic disciplines have beyond question proved themselves good means toward that end, they are surely not ends in themselves.

I am not advocating that social scientists regularly try to elucidate the subject matter of many specialties. Indeed, this book could never have been written had not generations of scholars devoted themselves to narrower and more manageable topics. But their contributions will not be complete unless we occasionally attempt to bring their separate truths together into a larger one. From this perspective, my goal might seem to be narrow: to understand the expanding economy of information as a means of control.

My research and writing have profited from both the criticism and the support of many people. Among these, Alfred Chandler, Thomas Parke Hughes, and Tony Oettinger have my deepest gratitude and respect. Each generously read large sections of the manuscript, offered many useful suggestions, and—though not agreeing with everything I wrote—provided warm encouragement. Without people like these, scholarship would be just another job.

Other scholars who kindly gave of their time to comment on various sections include Will Baumol, Daniel Bell, Al Biderman, Robert Bierstedt, Lord Briggs, Claude Fischer, Alexander Leitch, Marion Levy, Niklas Luhmann, Allan Mazur, David Sills, Neil Smelser, and Art Stinchcombe. Susan Cotts Watkins not only carefully read my penultimate draft but also provided invaluable advice and encouragement over regular lunches during her year at the Institute for Advanced Study. Robert Wright, a former student of mine and now an accomplished science columnist and editor, managed to scribble many helpful comments on an early draft while commuting on the New York City subways.

Because questions of living systems took me furthest from my own formal academic training, I made a special effort to solicit the advice of biologists. Among those who generously responded with useful comments, encouragement, or both, I would like to thank A. G. Cairns-Smith, Manfred Eigen, Richard Keogh, Ernst Mayr, Claude Villee, Paul Weisz, and Ed Wilson. Through a faculty seminar and coteaching with several members of Princeton's biology department, I have also learned a great deal from John Bonner, Henry Horn, Bob May, and George Sugihara.

Additional stimulation came from invitations to test various ideas in this book on a range of audiences: the American Association for the Advancement of Science, the American Association for Public Opinion Research, the American Sociological Association, the Annenberg Schools' Washington Program, Harvard's Program on Information Resources Policy, the National Academy of Science, New York University's Media Ecology Conference, the Social Science History Association, and the University of Pennsylvania Department of the History and Sociology of Science, as well as a number of brownbag luncheon meetings sponsored by the graduate students in my department at Princeton. For invitations to make these various presentations, I am grateful to Michael Armer, Hamilton Cravens, Tom Hughes, Elizabeth Martin, Tony Oettinger, Neil Postman, Everett Rogers, Howard Schuman, David Sills, and Charles Turner.

For sustaining my morale throughout the conception, planning, and writing of this book, I am particularly indebted to Clifford Nass, who entered the process as a mathematics and computer science major in my undergraduate course on technology and social change, graduated to our doctoral program in sociology, and finished as my (prize-winning) teaching assistant, collaborator on several projects, and friend. Among Cliff's many contributions, in addition to making detailed comments on each draft of the manuscript, I must single out his patience in convincing me to take seriously the concept of *preprocessing*. Of the hundreds of other students with whom I have argued various of this book's ideas over the past ten years, seven undergraduates stand out in my mind as particularly influential: Paul Fernhout, Bob Giuffra, Howard Pearlmutter, Glenn Picher, Peter Swire, Nicholas Ulanov, and David Wonnacott.

Much credit for this book belongs to Harvard University Press. Michael Aronson spotted merit in my partial manuscript and enlisted wise reviewers to suggest improvements. I learned a great deal about writing from the spidery green line of my copy editor, Patricia Flaherty, a gracious diplomat who made the book read better.

Literally hundreds of employees of Princeton University and of the Annenberg School of Communications at the University of Southern California helped with the preparation of this book. Among these contributors, I must single out for special thanks the staffs of Princeton's

Firestone Library and University Computer Center, USC's Doheny Library, the Annenberg School's Learning and Production Centers, and my secretary, Rachel Osborn. I am particularly indebted to Peter Clarke and Susan Evans of the Annenberg School, among the first to appreciate *The Control Revolution* and among its most steadfast supporters, for providing me with a comfortable home in which to complete it.

Another friend, Kay Ferdinandsen, offered advice on all drafts of the manuscript and sustained my efforts in countless other ways. Sometime between the completion of Chapter 6 and the start of Chapter 7, we managed to get married.

I must acknowledge my other good fortune, during the most formative period of my thinking about the Information Society, in having become acquainted with two of the pioneering scholars of the subject. Although I would have relied heavily upon their published ideas in any case, getting to know them in person before their deaths provided a special inspiration in my life as well as in this work. It is to them, Fritz Machlup and Ithiel de Sola Pool, that I gratefully dedicate this book.

Contents

The Control Revolution

1

Introduction

Here have we war for war and blood for blood,
controlment for controlment.

—King of England to the French
ambassador (Shakespeare, *King John*)

ONE TRAGEDY of the human condition is that each of us lives and dies
with little hint of even the most profound transformations of our society
and our species that play themselves out in some small part through
our own existence. When the earliest *Homo sapiens* encountered *Homo
erectus*, or whatever species was our immediate forebear, it is unlikely
that the two saw in their differences a major turning point in the
development of our race. If they did, this knowledge did not survive
to be recorded, at least not in the ancient writings now extant. Indeed,
some fifty thousand years passed before Darwin and Wallace rediscovered the secret—proof of the difficulty of grasping even the most
essential dynamics of our lives and our society.

Much the same conclusion could be drawn from any of a succession
of revolutionary societal transformations: the cultivation of plants and
the domestication of animals, the growth of permanent settlements,
the development of metal tools and writing, urbanization, the invention
of wheeled vehicles and the plow, the rise of market economies, social
classes, a world commerce. The origins and early histories of these
and many other developments of comparable significance went unnoticed or at least unrecorded by contemporary observers. Today we are
hard pressed to associate specific dates, places, or names with many
major societal transformations, even though similar details abound for
much lesser events and trends that occurred at the same times.

This condition holds for even that most significant of modern societal
transformations, the so-called Industrial Revolution. Although it is
generally conceded to have begun by mid-eighteenth century, at least
in England, the idea of its revolutionary impact does not appear until

the 1830s in pioneering histories like those of Wade (1833) and Blanqui (1837). Widespread acceptance by historians that the Industrial Revolution constituted a major transformation of society did not come until Arnold Toynbee, Sr., popularized the term in a series of public lectures in 1881 (Toynbee 1884). This was well over a century after the changes he described had first begun to gain momentum in his native England and at least a generation after the more important ones are now generally considered to have run their course. Although several earlier observers had described one or another of the same changes, few before Toynbee had begun to reflect upon the more profound transformation that signaled the end—after some ten thousand years—of predominantly agricultural society.

Two explanations of this chronic inability to grasp even the most essential dynamics of an age come readily to mind. First, important transformations of society rarely result from single discrete events, despite the best efforts of later historians to associate the changes with such events. Human society seems rather to evolve largely through changes so gradual as to be all but imperceptible, at least compared to the generational cycles of the individuals through whose lives they unfold. Second, contemporaries of major societal transformations are frequently distracted by events and trends more dramatic in immediate impact but less lasting in significance. Few who lived through the early 1940s were unaware that the world was at war, for example, but the much less noticed scientific and technological by-products of the conflict are more likely to lend their names to the era, whether it comes to be remembered as the Nuclear Age, the Computer Age, or the Space Age.

Regardless of how we explain the recurrent failure of past generations to appreciate the major societal transformations of their own eras, we might expect that their record would at least chasten students of contemporary social change. In fact, just the opposite appears to be the case. Much as if historical myopia could somehow be overcome by confronting the problem head-on, a steadily mounting number of social scientists, popular writers, and critics have discovered that one or another revolutionary societal transformation is now in progress. The succession of such transformations identified since the late 1950s includes the rise of a new social class (Djilas 1957; Gouldner 1979), a meritocracy (Young 1958), postcapitalist society (Dahrendorf 1959), a global village (McLuhan 1964), the new industrial state (Galbraith 1967), a scientific-technological revolution (Richta 1967; Daglish 1972; Prague Academy 1973), a technetronic era (Brzezinski 1970), postindustrial society (Touraine 1971; Bell 1973), an information economy (Porat 1977), and the micro millennium (Evans 1979), to name only a few. A more complete catalog of these and similar transformations, listed by year of first exposition in a major work, is given in Table 1.1.

The writer who first identified each of the transformations listed in Table 1.1 usually found the brunt of the change to be—coincidentally enough—either in progress or imminent. A recent best-seller, for example, surveys the sweep of human history, notes the central importance of the agricultural and industrial revolutions, and then finds in contemporary society the seeds of a third revolution—the impending "Third Wave":

> Humanity faces a quantum leap forward. It faces the deepest social upheaval and creative restructuring of all time. Without clearly recognizing it, we are engaged in building a remarkable new civilization from the ground up. This is the meaning of the Third Wave . . . It is likely that the Third Wave will sweep across history and complete itself in a few decades. We, who happen to share the planet at this explosive moment, will therefore feel the full impact of the Third Wave in our own lifetimes. Tearing our families apart, rocking our economy, paralyzing our political systems, shattering our values, the Third Wave affects everyone. (Toffler 1980, p. 26)

Even less breathless assessments of contemporary change have been no less optimistic about the prospect of placing developing events and trends in the broadest historical context. Daniel Bell, for example, after acknowledging the counterevidence of Toynbee and the Industrial Revolution, nevertheless concludes, "Today, with our greater sensitivity to social consequences and to the future . . . we are more alert to the possible imports of technological and organizational change, and this is all to the good" (1980, pp. x–xi).

The number of major societal transformations listed in Table 1.1 indicates that Bell appears to be correct; we do seem more alert than previous generations to the possible importance of change. The wide variety of transformations identified, however, suggests that, like the generations before us, we may be preoccupied with specific and possibly ephemeral events and trends, at the risk of overlooking what only many years from now will be seen as the fundamental dynamic of our age.

Because the failures of past generations bespeak the difficulties of overcoming this problem, the temptation is great not to try. This reluctance might be overcome if we recognize that understanding our-

Table 1.1. Modern societal transformations identified since 1950

Year	Transformation	Sources
1950	Lonely crowd	Riesman 1950
	Posthistoric man	Seidenberg 1950
1953	Organizational revolution	Boulding 1953
1956	Organization man	Whyte 1956
1957	New social class	Djilas 1957; Gouldner 1979
1958	Meritocracy	Young 1958
1959	Educational revolution	Drucker 1959
	Postcapitalist society	Dahrendorf 1959
1960	End of ideology	Bell 1960
	Postmaturity economy	Rostow 1960
1961	Industrial society	Aron 1961; 1966
1962	Computer revolution	Berkeley 1962; Tomeski 1970; Hawkes 1971
	Knowledge economy	Machlup 1962; 1980; Drucker 1969
1963	New working class	Mallet 1963; Gintis 1970; Gallie 1978
	Postbourgeois society	Lichtheim 1963
1964	Global village	McLuhan 1964
	Managerial capitalism	Marris 1964
	One-dimensional man	Marcuse 1964
	Postcivilized era	Boulding 1964
	Service class society	Dahrendorf 1964
	Technological society	Ellul 1964
1967	New industrial state	Galbraith 1967
	Scientific-technological revolution	Richta 1967; Daglish 1972; Prague Academy 1973
1968	Dual economy	Averitt 1968
	Neocapitalism	Gorz 1968
	Postmodern society	Etzioni 1968; Breed 1971
	Technocracy	Meynaud 1968
	Unprepared society	Michael 1968
1969	Age of discontinuity	Drucker 1969
	Postcollectivist society	Beer 1969
	Postideological society	Feuer 1969
1970	Computerized society	Martin and Norman 1970
	Personal society	Halmos 1970
	Posteconomic society	Kahn 1970
	Postliberal age	Vickers 1970
	Prefigurative culture	Mead 1970
	Technetronic era	Brzezinski 1970
1971	Age of information	Helvey 1971
	Compunications	Oettinger 1971

Year	Transformation	Sources
1971	Postindustrial society	Touraine 1971; Bell 1973
	Self-guiding society	Breed 1971
	Superindustrial society	Toffler 1971
1972	Limits to growth	Meadows 1972; Cole 1973
	Posttraditional society	Eisenstadt 1972
	World without borders	Brown 1972
1973	New service society	Lewis 1973
	Stalled society	Crozier 1973
1974	Consumer vanguard	Gartner and Riessman 1974
	Information revolution	Lamberton 1974
1975	Communications age	Phillips 1975
	Mediacracy	Phillips 1975
	Third industrial revolution	Stine 1975; Stonier 1979
1976	Industrial-technological society	Ionescu 1976
	Megacorp	Eichner 1976
1977	Electronics revolution	Evans 1977
	Information economy	Porat 1977
1978	Anticipatory democracy	Bezold 1978
	Network nation	Hiltz and Turoff 1978
	Republic of technology	Boorstin 1978
	Telematic society	Nora and Minc 1978; Martin 1981
	Wired society	Martin 1978
1979	Collapse of work	Jenkins and Sherman 1979
	Computer age	Dertouzos and Moses 1979
	Credential society	Collins 1979
	Micro millennium	Evans 1979
1980	Micro revolution	Large 1980, 1984; Laurie 1981
	Microelectronics revolution	Forester 1980
	Third wave	Toffler 1980
1981	Information society	Martin and Butler 1981
	Network marketplace	Dordick 1981
1982	Communications revolution	Williams 1982
	Information age	Dizard 1982
1983	Computer state	Burnham 1983
	Gene age	Sylvester and Klotz 1983
1984	Second industrial divide	Piore and Sabel 1984

selves in our own particular moment in history will enable us to shape and guide that history. As Bell goes on to say, "to the extent that we are sensitive [to the possible importance of technological and social change], we can try to estimate the consequences and decide which policies we should choose, consonant with the values we have, in order to shape, accept, or even reject the alternative futures that are available to us" (1980, p. xi).

Much the same purpose motivates—and I hope justifies—the pages that follow. In them I argue, like many of the writers whose names appear in Table 1.1, that society is currently experiencing a revolutionary transformation on a global scale. Unlike most of the other writers, however, I do not conclude that the crest of change is either recent, current, or imminent. Instead, I trace the causes of change back to the middle and late nineteenth century, to a set of problems—in effect a crisis of control—generated by the industrial revolution in manufacturing and transportation. The response to this crisis, at least in technological innovation and restructuring of the economy, occurred most rapidly around the turn of the century and amounted to nothing less, I argue, than a revolution in societal control.

The Control Revolution

Few turn-of-the-century observers understood even isolated aspects of the societal transformation—what I shall call the "Control Revolution"—then gathering momentum in the United States, England, France, and Germany. Notable among those who did was Max Weber (1864–1920), the German sociologist and political economist who directed social analysis to the most important control technology of his age: bureaucracy. Although bureaucracy had developed several times independently in ancient civilizations, Weber was the first to see it as the critical new machinery—new, at least, in its generality and pervasiveness—for control of the societal forces unleashed by the Industrial Revolution.

For a half-century after Weber's initial analysis bureaucracy continued to reign as the single most important technology of the Control Revolution. After World War II, however, generalized control began to shift slowly to computer technology. If social change has seemed to accelerate in recent years (as argued, for example, by Toffler 1971), this has been due in large part to a spate of new information-processing, communication, and control technologies like the computer, most notably the microprocessors that have proliferated since the early 1970s.

Such technologies are more properly seen, however, not as causes but as consequences of societal change, as natural extensions of the Control Revolution already in progress for more than a century.

Revolution, a term borrowed from astronomy, first appeared in political discourse in seventeenth-century England, where it described the restoration of a previous form of government. Not until the French Revolution did the word acquire its currently popular and opposite meaning, that of abrupt and often violent change. As used here in Control Revolution, the term is intended to have both of these opposite connotations.

Beginning most noticeably in the United States in the late nineteenth century, the Control Revolution was certainly a dramatic if not abrupt discontinuity in technological advance. Indeed, even the word *revolution* seems barely adequate to describe the development, within the span of a single lifetime, of virtually all of the basic communication technologies still in use a century later: photography and telegraphy (1830s), rotary power printing (1840s), the typewriter (1860s), transatlantic cable (1866), telephone (1876), motion pictures (1894), wireless telegraphy (1895), magnetic tape recording (1899), radio (1906), and television (1923).

Along with these rapid changes in mass media and telecommunications technologies, the Control Revolution also represented the beginning of a restoration—although with increasing centralization—of the economic and political control that was lost at more local levels of society during the Industrial Revolution. Before this time, control of government and markets had depended on personal relationships and face-to-face interactions; now control came to be reestablished by means of bureaucratic organization, the new infrastructures of transportation and telecommunications, and system-wide communication via the new mass media. By both of the opposite definitions of *revolution*, therefore, the new societal transformations—rapid innovation in information and control technology, to regain control of functions once contained at much lower and more diffuse levels of society—constituted a true revolution in societal control.

Here the word *control* represents its most general definition, purposive influence toward a predetermined goal. Most dictionary definitions imply these same two essential elements: *influence* of one agent over another, meaning that the former causes changes in the behavior of the latter; and *purpose*, in the sense that influence is directed toward some prior goal of the controlling agent. If the definition used here differs at all from colloquial ones, it is only because many people reserve the word *control* for its more determinate manifestations, what I shall call "strong control." Dictionaries, for example, often include in their definitions of control concepts like direction, guidance, regulation, command, and domination, approximate synonyms of *influence* that vary mainly in increasing determination. As a more general concept, however, *control* encompasses the entire range from absolute control to the weakest and most probabilistic form, that is, any purposive influence on behavior, *however slight.* Economists say that television advertising serves to control specific demand, for example, and political scientists say that direct mail campaigns can help to control issue-voting, even though only a small fraction of the intended audience may be influenced in either case.

Inseparable from the concept of control are the twin activities of information processing and reciprocal communication, complementary factors in any form of control. Information processing is essential to all purposive activity, which is by definition goal directed and must therefore involve the continual comparison of current states to future goals, a basic problem of information processing. So integral to control is this comparison of inputs to stored programs that the word *control* itself derives from the medieval Latin verb *contrarotulare*, to compare something "against the rolls," the cylinders of paper that served as official records in ancient times.

Simultaneously with the comparison of inputs to goals, two-way interaction between controller and controlled must also occur, not only to communicate influence from the former to the latter, but also to communicate back the results of this action (hence the term *feedback* for this reciprocal flow of information back to a controller). So central is communication to the process of control that the two have become the joint subject of the modern science of cybernetics, defined by one of its founders as "the entire field of control and communication theory, whether in the machine or in the animal" (Wiener 1948, p. 11). Similarly, the pioneers of mathematical communication theory have defined the object of their study as purposive control in the broadest sense: communication, according to Shannon and Weaver (1949, pp. 3–5), includes "all of the procedures by which one mind may affect another"; they note that "communication either affects conduct or is without any discernible and probable effect at all."

Because both the activities of information processing and communication are inseparable components of the control function, a society's ability to maintain control—at all levels from interpersonal to international relations—will be directly proportional to the development of its information technologies. Here the term *technology* is intended not in the narrow sense of practical or applied science but in the more general sense of any intentional extension of a natural process, that is, of the processing of matter, energy, and information that characterizes all living systems. Respiration is a wholly natural life function, for example, and is therefore not a technology; the human ability to breathe under water, by contrast, implies some technological extension. Similarly, voting is one general technology for achieving collective decisions in the control of social aggregates; the Australian ballot is a particular innovation in this technology.

Technology may therefore be considered as roughly equivalent to that which can be done, excluding only those capabilities that occur naturally in living systems. This distinction is usually although not always clear. One ambiguous case is language, which may have developed at least in part through purposive innovation but which now appears to be a mostly innate capability of the human brain. The brain itself represents another ambiguous case: it probably developed in interaction with purposive tool use and may therefore be included among human technologies.

Because technology defines the limits on what a society *can* do, technological innovation might be expected to be a major impetus to social change in the Control Revolution no less than in the earlier societal transformations accorded the status of revolutions. The Neolithic Revolution, for example, which brought the first permanent settlements, owed its origin to the refinement of stone tools and the domestication of plants and animals. The Commercial Revolution, following exploration of Africa, Asia, and the New World, resulted directly from technical improvements in seafaring and navigational equipment. The Industrial Revolution, which eventually brought the nineteenth-century crisis of control, began a century earlier with greatly increased use of coal and steam power and a spate of new machinery for the manufacture of cotton textiles. Like these earlier revolutions in matter and energy processing, the Control Revolution resulted from innovation at a most fundamental level of technology—that of information processing.

Information processing may be more difficult to appreciate than matter or energy processing because information is epiphenomenal: it derives from the *organization* of the material world on which it is wholly dependent for its existence. Despite being in this way higher order or derivative of matter and energy, information is no less critical to society. All living systems must process matter and energy to maintain themselves counter to entropy, the universal tendency of organization toward breakdown and randomization. Because control is necessary for such processing, and information, as we have seen, is essential to control, both information processing and communication, insofar as they distinguish living systems from the inorganic universe, might be said to define life itself—except for a few recent artifacts of our own species.

Each new technological innovation extends the processes that sustain life, thereby increasing the need for control and hence for improved control technology. This explains why technology appears autonomously to beget technology in general (Winner 1977), and why, as argued here, innovations in matter and energy processing create the need for further innovation in information-processing and communication technologies. Because technological innovation is increasingly a collective, cumulative effort, one whose results must be taught and diffused, it also generates an increased need for technologies of information storage and retrieval—as well as for their elaboration in systems of technical education and communication—quite independently of the particular need for control.

As in the earlier revolutions in matter and energy technologies, the nineteenth-century revolution in information technology was predicated on, if not directly caused by, social changes associated with earlier innovations. Just as the Commercial Revolution depended on capital and labor freed by advanced agriculture, for example, and the Industrial Revolution presupposed a commercial system for capital allocations and the distribution of goods, the most recent technological revolution developed in response to problems arising out of advanced industrialization—an ever-mounting crisis of control.

Crisis of Control

The later Industrial Revolution constituted, in effect, a consolidation of earlier technological revolutions and the resulting transformations of society. Especially during the late nineteenth and early twentieth centuries industrialization extended to progressively earlier technological revolutions: manufacturing, energy production, transportation, agriculture—the last a transformation of what had once been seen as the extreme opposite of industrial production. In each area industrialization meant heavy infusions of capital for the exploitation of fossil fuels, wage labor, and machine technology and resulted in larger and more complex systems—systems characterized by increasing differentiation and interdependence at all levels.

One of the earliest and most astute observers of this phenomenon was Emile Durkheim (1858–1917), the great French sociologist who examined many of its social ramifications in his *Division of Labor in Society* (1893). As Durkheim noted, industrialization tends to break down the barriers to transportation and communication that isolate local markets (what he called the "segmental" type), thereby extending distribution of goods and services to national and even global markets (the "organized" type). This, in turn, disrupts the market equilibrium under which production is regulated by means of direct communication between producer and consumer:

> Insofar as the segmental type is strongly marked, there are nearly as many economic markets as there are different segments. Consequently, each of them is very limited. Producers, being near consumers, can easily reckon the extent of the needs to be satisfied. Equilibrium is established without any trouble and production regulates itself. On the contrary, as the organized type develops, the fusion of different segments draws the markets together into one which embraces almost all society . . . The result is that each industry produces for consumers spread over the whole surface of the country or even of the entire world. Contact is then no longer sufficient. The producer can no longer embrace the market in a glance, nor even in thought. He can no longer see limits, since it is, so to speak, limitless. Accordingly, production becomes unbridled and unregulated. It can only trust to chance . . . From this come the crises which periodically disturb economic functions.   (1893, pp. 369–370)

What Durkheim describes here is nothing less than a crisis of control at the most aggregate level of a national system—a level that had had little practical relevance before the mass production and distribution of factory goods. Resolution of the crisis demanded new means of communication, as Durkheim perceived, to control an economy shifting from local segmented markets to higher levels of organization—what might be seen as the growing "systemness" of society. This capacity to communicate and process information is one component of what structural-functionalists following Durkheim have called the problem of *integration*, the growing need for coordination of functions that accompanies differentiation and specialization in any system.

Increasingly confounding the need for integration of the structural division of labor were corresponding increases in commodity flows through the system—flows driven by steam-powered factory production and mass distribution via national rail networks. Never before had the processing of material flows threatened to exceed, in both volume and speed, the capacity of technology to contain them. For centuries most goods had moved with the speed of draft animals down roadway and canal, weather permitting. This infrastructure, controlled by small organizations of only a few hierarchial levels, supported even national economies. Suddenly—owing to the harnessing of steam power—goods could be moved at the full speed of industrial production, night and day and under virtually any conditions, not only from town to town but across entire continents and around the world.

To do this, however, required an increasingly complex system of manufacturers and distributers, central and branch offices, transportation lines and terminals, containers and cars. Even the logistics of nineteenth-century armies, then the most difficult problem in processing and control, came to be dwarfed in complexity by the material economy just emerging as Durkheim worked on his famous study.

What Durkheim described as a crisis of control on the societal level he also managed to relate to the level of individual psychology. Here he found a more personal but directly related problem, what he called *anomie*, the breakdown of norms governing individual and group behavior. Anomie is an "abnormal" and even "pathological" result, according to Durkheim (1893, p. 353), an exception to his more general finding that increasing division of labor directly increases normative integration and, with it, social solidarity. As Durkheim argued, anomie results not from the structural division of labor into what he called distinct societal "organs" but rather from the breakdown in communication among these increasingly isolated sectors, so that individuals employed in them lose sight of the larger purpose of their separate efforts:

> The state of anomie is impossible wherever solidary organs are sufficiently in contact or sufficiently prolonged. In effect, being continguous, they are quickly warned, in each circumstance, of the need which they have of one another, and, consequently, they have a lively and continuous sentiment of their mutual dependence . . . But, on the contrary, if some opaque environment is interposed, then only stimuli of a certain intensity can be communicated from one organ to another. Relations, being rare, are not repeated enough to be determined; each time there ensues new groping. The lines of passage taken by the streams of movement cannot deepen

because the streams themselves are too intermittent. If some rules do come to constitute them, they are, however, general and vague. (1893, pp. 368–369)

Like the problem of economic integration, anomie also resulted—in Durkheim's view—from inadequate means of communication. Both problems were thus manifestations, at opposite extremes of aggregation, of the nineteenth-century control crisis.

Unlike Durkheim's analysis, which was largely confined to the extremes of individual and society, this book will concentrate on intervening levels, especially on technology and its role in the processing of matter, energy, and information—what might be called the *material economy* (as opposed to the abstract ones that seem to captivate most modern economists). Chapter 6 includes separate sections on the production, distribution, and consumption of goods and services in the industrializing economy of the United States in the nineteenth century and on the new information-processing and communication technologies—just emerging during Durkheim's lifetime—that served to control the increasing volume and speed of these activities. We will find that, just as the problem of control threatened to reach crisis proportions late in the century, a series of new technological and social solutions began to contain the problem. This was the opening stage of the Control Revolution.

Rationalization and Bureaucracy

Foremost among all the technological solutions to the crisis of control—in that it served to control most other technologies—was the rapid growth of formal bureaucracy first analyzed by Max Weber at the turn of the century. Bureaucratic organization was not new to Weber's time, as we have noted; bureaucracies had arisen in the first nation-states with centralized administrations, most significantly in Mesopotamia and ancient Egypt, and had reached a high level of sophistication in the preindustrial empires of Rome, China, and Byzantium. Indeed, bureaucratic organization tends to appear wherever a collective activity needs to be coordinated by several people toward explicit and impersonal goals, that is, to be *controlled*. Bureaucracy has served as the generalized means to control any large social system in most institutional areas and in most cultures since the emergence of such systems by about 3000 B.C.

Because of the venerable history and pervasiveness of bureaucracy, historians have tended to overlook its role in the late nineteenth century as a major new control technology. Nevertheless, bureaucratic administration did not begin to achieve anything approximating its modern form until the late Industrial Revolution. As late as the 1830s, for example, the Bank of the United States, then the nation's largest and most complex institution with twenty-two branch offices and profits fifty times those of the largest mercantile house, was managed by just three people: Nicholas Biddle and two assistants (Redlich 1951, pp. 113–124). In 1831 President Andrew Jackson and 665 other civilians ran all three branches of the federal government in Washington, an increase of sixty-three employees over the previous ten years. The Post Office Department, for example, had been administered for thirty years as the personal domain of two brothers, Albert and Phineas Bradley (Pred 1973, chap. 3). Fifty years later, in the aftermath of rapid industrialization, Washington's bureaucracy included some thirteen thousand civilian employees, more than double the total—already swelled by the American Civil War—only ten years earlier (U.S. Bureau of the Census 1975, p. 1103).

Further evidence that bureaucracy developed in response to the Industrial Revolution is the timing of concern about bureaucratization as a pressing social problem. The word *bureaucracy* did not even appear in English until the early nineteenth century, yet within a generation it became a major topic of political and philosophical discussion. As early as 1837, for example, John Stuart Mill wrote of a "vast network of administrative tyranny . . . that system of *bureaucracy*, which leaves no free agent in all France, except the man at Paris who pulls the wires" (Burchfield 1972, p. 391); a decade later Mill warned more generally of the "inexpediency of concentrating in a dominant bureaucracy . . . all power of organized action . . . in the community" (1848, p. 529). Thomas Carlyle, in his *Latter-Day Pamphlets* published two years later, complained of "the Continental nuisance called 'Bureaucracy' " (1850, p. 121). The word *bureaucratic* had also appeared by the 1830s, followed by *bureaucrat* in the 1840s and *bureaucratize* by the 1890s.

That bureaucracy is in essence a control technology was first established by Weber, most notably in his *Economy and Society* (1922). Weber included among the defining characteristics of bureaucracy several important aspects of any control system: impersonal orientation of structure to the information that it processes, usually identified as "cases," with a predetermined formal set of rules governing all decisions and responses. Any tendency to humanize this bureaucratic machinery, Weber argued, would be minimized through clear-cut division of labor and definition of responsibilities, hierarchical authority, and specialized decision and communication functions. The stability and permanence of bureaucracy, he noted, are assured through regular promotion of career employees based on objective criteria like seniority.

Weber identified another related control technology, what he called *rationalization*. Although the term has a variety of meanings, both in Weber's writings and in the elaborations of his work by others, most definitions are subsumed by one essential idea: control can be increased not only by increasing the capability to process information but also by decreasing the amount of information to be processed. The former approach to control was realized in Weber's day through bureaucratization and today increasingly through computerization; the latter approach was then realized through rationalization, what computer scientists now call *preprocessing*. Rationalization must therefore be seen, following Weber, as a complement to bureaucratization, one that served control in his day much as the preprocessing of information prior to its processing by computer serves control today.

Perhaps most pervasive of all rationalization is the increasing tendency of modern society to regulate interpersonal relationships in terms of a formal set of impersonal and objective criteria. The early technocrat Claude Henri Comte de Saint-Simon (1760–1825), who lived through only the first stages of industrialization, saw such rationalization as a move "from the government of men to the administration of things" (Taylor 1975, pt. 3). The reason why people can be governed more readily *qua* things is that the amount of information about them that needs to be processed is thereby greatly reduced and hence the degree of control—for any constant capacity to process information—is greatly enhanced. By means of rationalization, therefore, it is possible to maintain large-scale, complex social systems that would be overwhelmed by a rising tide of information they could not process were it necessary to govern by the particularistic considerations of family and kin that characterize preindustrial societies.

In short, rationalization might be defined as the destruction or ignoring of information in order to facilitate its processing. This, too, has a direct analog in living systems, as we shall see in the next chapter. One example from within bureaucracy is the development of standardized paper forms. This might at first seem a contradiction, in that the proliferation of paperwork is usually associated with a growth in information to be processed, not with its reduction. Imagine how much more processing would be required, however, if each new case were recorded in an unstructured way, including every nuance and in full detail, rather than by checking boxes, filling blanks, or in some other way reducing the burdens of the bureaucratic system to only the limited range of formal, objective, and impersonal information required by standardized forms.

Equally important to the rationalization of industrial society, at the most macro level, were the division of North America into five standardized time zones in 1883 and the establishment the following year of the Greenwich meridian and International Date Line, which organized world time into twenty-four zones. What was formerly a problem of information overload and hence control for railroads and other organizations that sustained the social system at its most macro level was solved by simply ignoring much of the information, namely that solar time is different at each node of a transportation or communication system. A more convincing demonstration of the power of rationalization or preprocessing as a control technology would be difficult to imagine.

So commonplace has such preprocessing become that today we dismiss the alternative—that each node in a system might keep a slightly different time—as hopelessly cumbersome and primitive. With the continued proliferation of distributed computing, ironically enough, it might soon become feasible to return to a system based on local solar time, thereby shifting control from preprocessing back to processing—where it resided for centuries of human history until steam power pushed transportation beyond the pace of the sun across the sky.

New Control Technology

The rapid development of rationalization and bureaucracy in the middle and late nineteenth century led to a succession of dramatic new information-processing and communication technologies. These innovations served to contain the control crisis of industrial society in what can be treated as three distinct areas of economic activity: production, distribution, and consumption of goods and services.

Control of production was facilitated by the continuing organization and preprocessing of industrial operations. Machinery itself came increasingly to be controlled by two new information-processing technologies: closed-loop feedback devices like James Watt's steam governor (1788) and preprogrammed open-loop controllers like those of the Jacquard loom (1801). By 1890 Herman Hollerith had extended Jacquard's punch cards to tabulation of U.S. census data. This information-processing technology survives to this day—if just barely—owing largely to the corporation to which Hollerith's innovation gave life, International Business Machines (IBM). Further rationalization and control of production advanced through an accumulation of other industrial innovations: interchangeable parts (after 1800), integration of production within factories (1820s and 1830s), the development of modern accounting techniques (1850s and 1860s), professional managers (1860s and 1870s), continuous-process production (late 1870s and early 1880s), the "scientific management" of Frederick Winslow Taylor (1911), Henry Ford's modern assembly line (after 1913), and statistical quality control (1920s), among many others.

The resulting flood of mass-produced goods demanded comparable innovation in control of a second area of the economy: distribution. Growing infrastructures of transportation, including rail networks, steamship lines, and urban traction systems, depended for control on a corresponding infrastructure of information processing and telecommunications. Within fifteen years after the opening of the pioneering Baltimore and Ohio Railroad in 1830, for example, Samuel F. B. Morse— with a congressional appropriation of $30,000—had linked Baltimore to Washington, D.C., by means of a telegraph. Eight years later, in 1852, thirteen thousand miles of railroad and twenty-three thousand miles of telegraph line were in operation (Thompson 1947; U.S. Bureau of the Census 1975, p. 731), and the two infrastructures continued to coevolve in a web of distribution and control that progressively bound the entire continent. In the words of business historian Alfred Chandler, "the railroad permitted a rapid increase in the speed and decrease in the cost of long-distance, written communication, while the invention of the telegraph created an even greater transformation by making possible almost instantaneous communication at great distances. The railroad and the telegraph marched across the continent in unison . . . The telegraph companies used the railroad for their rights-of-way, and the railroad used the services of the telegraph to coordinate the flow of trains and traffic" (1977, p. 195).

This coevolution of the railroad and telegraph systems fostered the development of another communication infrastructure for control of mass distribution and consumption: the postal system. Aided by the introduction in 1847 of the first federal postage stamp, itself an important innovation in control of the national system of distribution, the total distance mail moved more than doubled in the dozen years between Morse's first telegraph and 1857, when it reached 75 million miles—almost a third covered by rail (Chandler 1977, p. 195). Commercialization of the telephone in the 1880s, and especially the development of long-distance lines in the 1890s, added a third component to the national infrastructure of telecommunications.

Controlled by means of this infrastructure, an organizational system rapidly emerged for the distribution of mass production to national and world markets. Important innovations in the rationalization and control of this system included the commodity dealer and standardized grading of commodities (1850s), the department store, chain store, and wholesale jobber (1860s), monitoring of movements of inventory or "stock turn" (by 1870), the mail-order house (1870s), machine packaging (1890s), franchising (by 1911 the standard means of distributing automobiles), and the supermarket and mail-order chain (1920s). After World War I the instability in national and world markets that Durkheim had noted a quarter-century earlier came to be gradually controlled, largely because of the new telecommunications infrastructure and the reorganization of distribution on a societal scale.

Mass production and distribution cannot be completely controlled, however, without control of a third area of the economy: demand and consumption. Such control requires a means to communicate information about goods and services to national audiences in order to stimulate or reinforce demand for these products; at the same time, it requires a means to gather information on the preferences and behavior of this audience—reciprocal feedback to the controller from the controlled (although the consumer might justifiably see these relationships as reversed).

The mechanism for communicating information to a national audience of consumers developed with the first truly mass medium: power-driven, multiple-rotary printing and mass mailing by rail. At the outset of the Industrial Revolution, most printing was still done on wooden handpresses—using flat plates tightened by means of screws—that differed little from the one Gutenberg had used three centuries earlier. Steam power was first successfully applied to printing in Germany in 1810; by 1827 it was possible to print up to 2,500 pages in an hour. In 1893 the New York *World* printed 96,000 eight-page copies every hour—a 300-fold increase in speed in just seventy years.

The postal system, in addition to effecting and controlling distribution, also served, through bulk mailings of mass-produced publications, as a new medium of mass communication. By 1887 Montgomery Ward mailed throughout the continent a 540-page catalog listing more than 24,000 items. Circulation of the Sears and Roebuck catalog increased from 318,000 in 1897 (the first year for which figures are available) to more than 1 million in 1904, 2 million in 1905, 3 million in 1907, and 7 million by the late 1920s. In 1927 alone, Sears mailed 10 million circular letters, 15 million general catalogs (spring and fall editions), 23 million sales catalogs, plus other special catalogs—a total mailing of 75 million (Boorstin 1973, p. 128) or approximately one piece for every adult in the United States.

Throughout the late nineteenth and early twentieth centuries uncounted entrepreneurs and inventors struggled to extend the technologies of communication to mass audiences. Alexander Graham Bell, who patented the telephone in 1876, originally thought that his invention might be used as a broadcast medium to pipe public speeches, music, and news into private homes. Such systems were indeed begun in several countries—the one in Budapest had six thousand subscribers by the turn of the century and continued to operate through World War I (Briggs 1977). More extensive application of telephony to mass communication was undoubtedly stifled by the rapid development of broadcast media beginning with Guglielmo Marconi's demonstration of long-wave telegraphy in 1895. Transatlantic wireless communication followed in 1901, public radio broadcasting in 1906, and commercial radio by 1920; even television broadcasting, a medium not popular until after World War II, had begun by 1923.

Many other communication technologies that we do not today associate with advertising were tried out early in the Control Revolution as means to influence the consumption of mass audiences. Popular books like the novels of Charles Dickens contained special advertising sections. Mass telephone systems in Britain and Hungary carried advertisements interspersed among music and news. The phonograph, patented by Thomas Edison in 1877 and greatly improved by the 1890s in Hans Berliner's "gramophone," became another means by which a sponsor's message could be distributed to households: "Nobody would refuse," the United States Gramaphone Company claimed, "to listen to a fine song or concert piece or an oration—even if it is interrupted by a modest remark, 'Tartar's Baking Powder is Best' " (Abbot and Rider 1957, p. 387). With the development by Edison of the "motion

picture" after 1891, advertising had a new medium, first in the ki-
netoscope (1893) and cinematograph (1895), which sponsors located in
busy public places, and then in the 1900s in films projected in "movie
houses." Although advertisers were initially wary of broadcasting because audiences could not be easily identified, by 1930 sponsors were
spending $60 million annually on radio in the United States alone (Boor-
stin 1973, p. 392).

These mass media were not sufficient to effect true control, however,
without a means of feedback from potential consumers to advertisers,
thereby restoring to the emerging national and world markets what
Durkheim had seen as an essential relationship of the earlier segmental
markets: communication from consumer to producer to assure that the
latter "can easily reckon the extent of the needs to be satisfied" (1893,
p. 369). Simultaneously with the development of mass communication
by the turn of the century came what might be called *mass feedback*
technologies: market research (the idea first appeared as "commercial
research" in 1911), including questionnaire surveys of magazine readership, the Audit Bureau of Circulation (1914), house-to-house interviewing (1916), attitudinal and opinion surveys (a U.S. bibliography
lists nearly three thousand by 1928), a Census of Distribution (1929),
large-scale statistical sampling theory (1930), indices of retail sales
(1933), A. C. Nielsen's audimeter monitoring of broadcast audiences
(1935), and statistical-sample surveys like the Gallup Poll (1936), to
mention just a few of the many new technologies for monitoring consumer behavior.

Although most of the new information technologies originated in the
private sector, where they were used to control production, distribution, and consumption of goods and services, their potential for
controlling systems at the national and world level was not overlooked
by government. Since at least the Roman Empire, where an extensive
road system proved equally suited for moving either commerce or
troops, communications infrastructures have served to control both
economy and polity. As corporate bureaucracy came to control increasingly wider markets by the turn of this century, its power was
increasingly checked by a parallel growth in state bureaucracy. Both
bureaucracies found useful what Bell has called "intellectual technology":

The major intellectual and sociological problems of the post-industrial
society are . . . those of "organized complexity"—the management of large-

scale systems, with large numbers of interacting variables, which have to be coordinated to achieve specific goals . . . An *intellectual technology* is the substitution of algorithms (problem-solving rules) for intuitive judgments. These algorithms may be embodied in an automatic machine or a computer program or a set of instructions based on some statistical or mathematical formula; the statistical and logical techniques that are used in dealing with "organized complexity" are efforts to formalize a set of decision rules.   (1973, pp. 29–30)

Seen in this way, intellectual technology is another manifestation of bureaucratic rationality, an extension of what Saint-Simon described as a shift from the government of men to the administration of things, that is, a further move to administration based not on intuitive judgments but on logical and statistical rules and algorithms. Although Bell sees intellectual technology as arising after 1940, state bureaucracies had begun earlier in this century to appropriate many key elements: central economic planning (Soviet Union after 1920), the state fiscal policies of Lord Keynes (late 1920s), national income accounting (after 1933), econometrics (mid-1930s), input-output analysis (after 1936), linear programming and statistical decision theory (late 1930s), and operations research and systems analysis (early in World War II).

In the modern state the latest technologies of mass communication, persuasion, and market research are also used to stimulate and control demand for governmental services. The U.S. government, for example, currently spends about $150 million a year on advertising, which places it among the top thirty advertisers in the country; were the approximately 70 percent of its ads that are presented free as a public service also included, it would rank second—just behind Proctor and Gamble (Porat 1977, p. 137). Increasing business and governmental use of control technologies and their recent proliferation in forms like data services and home computers for use by consumers have become dominant features of the Control Revolution.

The Information Society

One major result of the Control Revolution had been the emergence of the so-called Information Society. The concept dates from the late 1950s and the pioneering work of an economist, Fritz Machlup, who first measured that sector of the U.S. economy associated with what he called "the production and distribution of knowledge" (Machlup 1962). Under this classification Machlup grouped thirty industries into five major categories: education, research and development, communications media, information machines (like computers), and information services (finance, insurance, real estate). He then estimated from national accounts data for 1958 (the most recent year available) that the information sector accounted for 29 percent of gross national product (GNP) and 31 percent of the labor force. He also estimated that between 1947 and 1958 the information sector had expanded at a compound growth rate double that of GNP. In sum, it appeared that the United States was rapidly becoming an Information Society.

Over the intervening twenty years several other analyses have substantiated and updated the original estimates of Machlup (1980, pp. xxvi–xxviii): Burck (1964) calculated that the information sector had reached 33 percent of GNP by 1963; Marschak (1968) predicted that the sector would approach 40 percent of GNP in the 1970s. By far the most ambitious effort to date has been the innovative work of Marc Uri Porat for the Office of Telecommunications in the U.S. Department of Commerce (1977). In 1967, according to Porat, information activities (defined differently from those of Machlup) accounted for 46.2 percent of GNP—25.1 percent in a "primary information" sector (which produces information goods and services as final output) and 21.1 percent in a "secondary information" sector (the bureaucracies of noninformation enterprises).

The impact of the Information Society is perhaps best captured by trends in labor force composition. As can be seen in Figure 1.1 and the corresponding data in Table 1.2, at the end of the eighteenth century the U.S. labor force was concentrated overwhelmingly in agriculture, the location of nearly 90 percent of its workers. The majority of U.S. labor continued to work in this sector until about 1850, and agriculture remained the largest single sector until the first decade of the twentieth century. Rapidly emerging, meanwhile, was a new industrial sector, one that continuously employed at least a quarter of U.S. workers between the 1840s and 1970s, reaching a peak of about 40 percent during World War II. Today, just forty years later, the industrial sector is close to half that percentage and declining steadily; it might well fall below 15 percent in the next decade. Meanwhile, the information sector, by 1960 already larger (at more than 40 percent) than industry had ever been, today approaches half of the U.S. labor force.

At least in the timing of this new sector's rise and development, the data in Figure 1.1 and Table 1.2 are compatible with the hypothesis

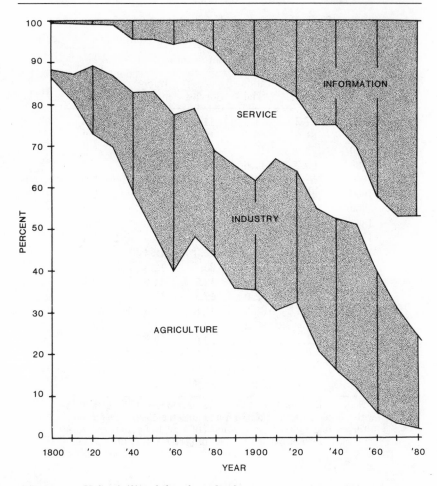

Figure 1.1. U.S. civilian labor force by four sectors, 1800–1980.

that the Information Society emerged in response to the nineteenth-century crisis of control. When the first railroads were built in the early 1830s, the information sector employed considerably less than 1 percent of the U.S. labor force; by the end of the decade it employed more than 4 percent. Not until the rapid bureaucratization of the 1870s and 1880s, the period that—as I argue on independent grounds in Chapter 6—marked the consolidation of control, did the percentage employed in the information sector more than double to about one-eighth of the civilian work force. With the exception of these two great discontinuities, one occurring with the advent of railroads and the crisis of control in the 1830s, the other accompanying the consolidation of

Table 1.2. U.S. experienced civilian labor force by four
sectors, 1800–1980

| | Sector's percent of total | | | | Total |
Year	Agri-cultural	Indus-trial	Service	Infor-mation	labor force (in millions)
1800	87.2	1.4	11.3	0.2	1.5
1810	81.0	6.5	12.2	0.3	2.2
1820	73.0	16.0	10.7	0.4	3.0
1830	69.7	17.6	12.2	0.4	3.7
1840	58.8	24.4	12.7	4.1	5.2
1850	49.5	33.8	12.5	4.2	7.4
1860	40.6	37.0	16.6	5.8	8.3
1870	47.0	32.0	16.2	4.8	12.5
1880	43.7	25.2	24.6	6.5	17.4
1890	37.2	28.1	22.3	12.4	22.8
1900	35.3	26.8	25.1	12.8	29.2
1910	31.1	36.3	17.7	14.9	39.8
1920	32.5	32.0	17.8	17.7	45.3
1930	20.4	35.3	19.8	24.5	51.1
1940	15.4	37.2	22.5	24.9	53.6
1950	11.9	38.3	19.0	30.8	57.8
1960	6.0	34.8	17.2	42.0	67.8
1970	3.1	28.6	21.9	46.4	80.1
1980	2.1	22.5	28.8	46.6	95.8

Sources: Data for 1800–1850 are estimated from Lebergott (1964)
with missing data interpolated from Fabricant (1949); data for 1860–
1970 are taken directly from Porat (1977); data for 1980 are based
on U.S. Bureau of Labor Statistics projections (Bell 1979, p. 185).

control in the 1870s and especially the 1880s, the information sector
has grown steadily but only modestly over the past two centuries.

Temporal correlation alone, of course, does not prove causation.
With the exception of the two discontinuities, however, growth in the
information sector has tended to be most rapid in periods of economic
upturn, most notably in the postwar booms of the 1920s and 1950s, as
can be seen in Table 1.2. Significantly, the two periods of discontinuity
were punctuated by economic depressions, the first by the Panic of
1837, the second by financial crisis in Europe and the Panic of 1873.
In other words, the technological origins of both the control crisis and
the consolidation of control occurred in periods when the information
sector would not have been expected on other economic grounds to
have expanded rapidly if at all. There is therefore no reason to reject

the hypothesis that the Information Society developed as a result of the crisis of control created by railroads and other steam-powered transportation in the 1840s.

A wholly new stage in the development of the Information Society has arisen, since the early 1970s, from the continuing proliferation of microprocessing technology. Most important in social implications has been the progressive convergence of all information technologies— mass media, telecommunications, and computing—in a single infrastructure of control at the most macro level. A 1978 report commissioned by the President of France—an instant best-seller in that country and abroad—likened the growing interconnection of information-processing, communication, and control technologies throughout the world to an alteration in "the entire nervous system of social organization" (Nora and Minc 1978, p. 3). The same report introduced the neologism *telematics* for this most recent stage of the Information Society, although similar words had been suggested earlier—for example, *compunications* (for "computing + communications") by Anthony Oettinger and his colleagues at Harvard's Program on Information Resources Policy (Oettinger 1971; Berman and Oettinger 1975; Oettinger, Berman, and Read 1977).

Crucial to telematics, compunications, or whatever word comes to be used for this convergence of information-processing and communications technologies is increasing digitalization: coding into discontinuous values—usually two-valued or binary—of what even a few years ago would have been an analog signal varying continuously in time, whether a telephone conversation, a radio broadcast, or a television picture. Because most modern computers process digital information, the progressive digitalization of mass media and telecommunications content begins to blur earlier distinctions between the communication of information and its processing (as implied by the term *compunications*), as well as between people and machines. Digitalization makes communication from persons to machines, between machines, and even from machines to persons as easy as it is between persons. Also blurred are the distinctions among information types: numbers, words, pictures, and sounds, and eventually tastes, odors, and possibly even sensations, all might one day be stored, processed, and communicated in the same digital form.

In this way digitalization promises to transform currently diverse forms of information into a generalized medium for processing and exchange by the social system, much as, centuries ago, the institution of common currencies and exchange rates began to transform local markets into a single world economy. We might therefore expect the implications of digitalization to be as profound for macrosociology as the institution of money was for macroeconomics. Indeed, digitalized electronic systems have already begun to replace money itself in many informational functions, only the most recent stage in a growing systemness of world society dating back at least to the Commercial Revolution of the fifteenth century.

Societal Dynamics Reconsidered

Despite the chronic historical myopia that characterizes the human condition as documented in the opening pages of this chapter, it is unlikely that the more astute observers of our era would fail to glimpse—however dimly—even a single aspect of its essential social dynamic. For this reason the ability of a conceptual framework to subsume social changes noted by previous observers might be taken as one criterion for judging its claim to portray a more fundamental societal transformation. We shall see that the various transformations identified by contemporary observers as listed in Table 1.1 can be readily subsumed by the major implications of the Control Revolution: the growing importance of information technology, as in Richta's scientific-technological revolution (1967) or Brzezinski's technetronic era (1970); the parallel growth of an information economy (Machlup 1962, 1980; Porat 1977) and its growing control by business and the state (Galbraith 1967); the organizational basis of this control (Boulding 1953; Whyte 1956) and its implications for social structure, whether a meritocracy (Young 1958) or a new social class (Djilas 1957; Gouldner 1979); the centrality of information processing and communication, as in McLuhan's global village (1964), Phillips's communications age (1975), or Evans's micro millennium (1979); the information basis of postindustrial society (Touraine 1971; Bell 1973); and the growing importance of information and knowledge in modern culture (Mead 1970).

In short, the argument that motivates our investigation of the nineteenth-century crisis of control and the resulting Control Revolution is that particular attention to the material aspects of information processing, communication, and control makes possible the synthesis of a large proportion of the literature on contemporary social change. It will be useful, however, to consider first the broader theoretical and historical context of industrialization and technological change. A more detailed review of the literature on contemporary social change will therefore be postponed until the nineteenth-century developments, our own Information Society, and the emerging world system have been examined in greater detail. We now turn to the analytic groundwork for this larger task, consideration of information processing and control at the more general level of living systems.

I

Living Systems, Technology, and the Evolution of Control

2

Programming and Control: The Essential Life Process

The most humble organism is something much higher than the inorganic dust under our feet, and no one with an unbiased mind can study any living creature, however humble, without being struck with enthusiasm at its marvellous structure and properties.

—Darwin, *Descent of Man*

THE CONTROL REVOLUTION and the resulting Information Society now emerging in the United States, Canada, Western Europe, and Japan leave us with certain nagging questions. Why has information, among the multitude of commodities, come to dominate economic statistics— come indeed to replace industry as the sector best reflecting the extent of a nation's development? Among the diversity of technologies, why the growing importance of microprocessors and computers, devices that can do nothing more than convert information from one form to another?

No study of technological innovation or economic history can possibly answer such questions—no more than, say, the history of organic evolution could explain the importance of information to all living things. In both cases explanation lies not in the particulars of evolutionary or human history but in the nature of the physical universe. Historical detail can only obscure the more fundamental laws that govern energy conversion and material processing, for example, in human societies no less than in other living systems. To ignore these laws in attempting to account for the history of the past century would be to forgo answers to the more pressing questions of our current age: Why *information*, and Why *now?*

This is not to imply that the Control Revolution can be explained as the inevitable result of some autonomous dynamic of the material universe. Social change results from the purposive behavior of people acting from individual and often idiosyncratic motives in pursuit of real goals, justification enough to study the political and economic history of technological innovation. No amount of innovation can ever free us from the physical laws that constrain all technological development, however, which explains why we will also find meaning in questions posed at the suprahistorical level. For example, could a control revolution have come *before* an industrial one? (The answer, as we shall see, is clearly *no*.)

An earlier materialist historian, Karl Marx, made the same point: "Men make their own history, but they do not make it just as they please; they do not make it under circumstances chosen by themselves, but under circumstances directly encountered, given, and transmitted from the past" (1852, p. 15). That these circumstances have shifted from land and capital to information—that information has emerged as the material base of modern economies—challenges the social theory we have inherited from the nineteenth century, much as the Industrial Revolution challenged Marx and other thinkers of that era to reconsider preindustrial theories. Then rapid industrialization forced theoretical reconstructions in terms of capital, energy, and material processing; today the Information Society demands similar reanalysis according to the physical relationships governing information storage, processing, communication, and control.

Prerequisite for any such reanalysis is a more general understanding of information's role in the production, distribution, and consumption of material goods, processes whose increasing volume and speed brought the nineteenth-century control crisis and resulting revolution in information technology. Such understanding may not require anything grandiose enough to be called a new theory, perhaps, but it would be advanced by a model of society more general than the changes we intend to explain.

Society as Processor

One such model, suggested in the previous chapter, is that of society as a *processing system*, one that sustains itself by extracting matter and energy from the environment and distributing them among its members. The science of economics has for centuries been grounded in the study of such material flows (Quesnay 1758; Walras 1874; Leontief 1941), while energy flows have provided the foundation for ecology since the late 1920s (Transeau 1926; Elton 1927; Tansley 1935). One behavioral scientist describes the processing role of living systems more generally: "They are open systems with significant inputs, throughputs, and outputs of various sorts of matter-energy and information. Processing these is all they do—a deceptively simple fact not widely recognized by the scientists who study them" (Miller 1978, p. 1027).

One of the first social scientists to elaborate this view of human society—with far-reaching implications—was the Australian economist Colin Clark in his path-breaking *Conditions of Economic Progress* (1940). Clark divided economic activity into what he called the *primary* (extractive), *secondary* (manufacturing), and *tertiary* (service) sectors of the economy. The relative importance of each sector, Clark argued, is determined by its relative productivity. With increasing industrialization, labor shifts from the primary to the secondary sector and then, with the resulting increase in demand for services, to the tertiary sector. The rate of transfer between sectors is a function of the differential value of their outputs per worker.

Using this conceptual apparatus, Clark measured economic progress as the rate of labor transfer to higher sectors and predicted transitional stages in the development of national economies. Following his lead, Hatt and Foote (1953) split out from the tertiary sector two new ones: a white-collar or *quaternary* sector including services like banking, insurance, and real estate, and an intellectual knowledge or *quinary* sector involving medicine, education, and research. According to their analysis, social mobility, including the increasing professionalization of work, is another major cause of labor shifts among sectors.

Despite the simplicity of these analytic frameworks, they have enabled economists to transcend the detail of a particular economy or historical period. In answer to our question about the timing of industrial and control revolutions, for example, Clark's work and its elaboration by Hatt and Foote suggest that information processing develops subsequent to the extraction and processing of matter and energy, with industrialization of the latter sectors a necessary precondition for the sustained growth of the former. Processing and distribution require control, under this model, and control depends in turn on information services. This would explain why, as we saw in Chapter 1, rapid increases in the volume and speed of throughput processing were accompanied by a spate of innovations in information technology during the late nineteenth century.

Such explanations, derived from the gross properties and physical exigencies shared by all concrete open systems, constitute an important step toward understanding the Control Revolution and resulting Information Society. For a social or behavioral scientist, however, the explanations raise as many questions as they answer. What precisely are the relationships among information, processing, and control and how are these concepts related to other aspects of human society and culture?

Organization for Control

Because the Control Revolution and its aftermath have so forcefully impressed on us the importance of information, we might be tempted to embrace that concept alone as the ultimate means to understand technological development and societal change. As we have seen, however, information is an epiphenomenon of the physical world, one that appears in even inanimate matter and energy when they are ordered, for example, into comets and crystals. To reduce a living system to the level of information, therefore, would be to study the ordering of its matter and energy, that is, to analyze it in terms of chemistry and physics—hardly a new or inspiring idea. If we wish to exploit the higher-order or derivative aspects of living systems, we must instead determine how they differ from the inanimate matter studied by physical scientists; in this sense only will we eschew reductionism.

When we compare even the simplest living systems to the most complex inorganic materials, one difference stands out: the much greater *organization* found in things organic. Living things require many pages to describe the organization of their physical structures, while the structure of an inorganic compound can always be captured in a relatively short string of symbols. Quartz, for example, which accounts for much of the earth's crust, is well described (well enough to replicate most experiments using it, for example) by the chemical symbols SiO_2 or the words *silicon dioxide*. Crystals can be uniquely described by the combination of their chemistry and atomic arrangement: only thirty-two different types of symmetry and seven systems of relationships among axes are possible; angles between corresponding faces must be constant. In other words, the complexity of crystals derives not from organization but from regularity and repetition, that is, from *order*.

Compared to organization, order contains relatively little information. A simple organism like the amoeba, for example, is not at all well ordered; it is a formless bag full of sticky fluid in which irregularly shaped molecules float haphazardly. In stark contrast to even the most complex crystalline structure, however, the amoeba is highly organized, and indeed its description requires several hundred large volumes, the information storage capacity of the DNA in which the structure is in fact recorded. That living material has greater complexity of organization holds even at the molecular level, which explains why students normally learn physical before organic chemistry.

John von Neumann (1903–1957), mathematician and computer pioneer, was one of the first to argue that the difference between order and organization is that the latter "always involves 'purposive' or end-directedness." The quotation comes from a conversation recalled by the biologist Colin Pittendrigh, who has applied the insight in defending teleology in the life sciences. In Pittendrigh's words, "What it was the biologist could not escape was the plain fact—or rather the fundamental fact—which he must (as scientist) explain: that the objects of biological analysis are organizations (he calls them organisms) and, as such, are end-directed. Organization is more than mere order; order lacks end-directedness: organization *is* end-directed" (1970, p. 392).

What we recognize in the end-directedness or purpose of organization is the essential property of *control*, already defined as purposive influence toward a predetermined goal. Control accounts for the difference between even the most complex inorganic crystal and simple organisms like the amoeba: the amoeba controls both itself and its environment; the crystal does not. As noted in the previous chapter, everything living processes information to effect control; nothing that is not alive can do so—nothing, that is, except certain artifacts of our own invention, artifacts that proliferated with the Control Revolution.

Purposive organization and control, in other words, define the tangible discontinuity that distinguishes life from the inorganic universe. On one side, the exclusive province of the physical sciences, we find only matter, energy, and their ordering in the epiphenomenon we call *information*. On the other side, our own side in that we ourselves are living systems, we find structures *purposively organized* (in von Neumann's sense) for information processing, communication, and control, the special subject matter of the behavioral and life sciences.

Social scientists have long recognized that society implies organization, a distinction that extends downward through organisms and organic material to the molecular level. As biochemist Albert Lehninger puts it, "In living organisms it is quite legitimate to ask what the function of a given molecule is. However, to ask such questions about molecules in collections of inanimate matter is irrelevant and meaningless" (1975, p. 3). This same distinction, between "the most humble organism" and "the inorganic dust under our feet," is what Darwin (1871, p. 169) celebrated—without benefit of information theory or modern genetics—in the passage that opens this chapter.

Here, then, is the most fundamental reason why the Control Revolution has been so profound in its impact on human society: it transformed no less than the essential life function itself. Rapid technological expansion of what Darwin called life's "marvellous structure and properties" and what we now see to include organization, information processing, and communication to effect control constitute a change unprecedented in recorded history. We would have to go back at least to the emergence of the vertebrate brain if not to the first replicating molecule—marking the origin of life on earth—to find a leap in the capability to process information comparable to that of the Control Revolution. But what precisely is it that living systems are organized to control, and why is this particular function so important to life?

Control and Energy

Similar questions raised about the organization of nonliving matter early in the nineteenth century culminated in *thermodynamics*, the science of heat and its relationship to other energy forms. Although thermodynamics developed out of steam power engineering following the Industrial Revolution, many scientists still consider its laws—whose progressive elaboration spanned the Control Revolution—to rank among the greatest achievements of Western thought.

According to the second law of thermodynamics, the so-called principle of the degradation of energy, a system's energy cannot be converted from one form to another—including work—without decreasing its organization and hence ability to do further work. A steam engine, for example, can work only so long as it is organized into relatively more and less heated parts, the organization known as a *heat gradient*. Because degradation of organization can only increase, according to the second law, the energy available to do work can only decrease.

In closed systems defined by impermeable boundaries, energy must remain constant in keeping with the first law of thermodynamics, the so-called principle of the conservation of energy, that matter and energy can neither be created nor destroyed. Because this total energy must remain constant, a closed system can only *lose* its ability to do work. Inevitably, by the second law, such systems must lapse into the state of totally unorganized, randomly distributed, inconvertible particles, the state of totally bound energy known—appropriately enough—as *heat death*.

When thermodynamic theory first developed, many scientists thought it to be violated by organisms that sustain and even increase their organization, continuing to do work for many years in apparent contradiction of the second law. Resolution of the contradiction occurred as living organisms came to be seen, like the steam engines whose coal smoke blackened the skies of nineteenth-century industrial towns, not as closed systems but as open ones, that is, systems with continuous inputs and outputs of matter and energy. Just as steam engines run off organic energy stored in wood or coal, vegetation also gets its energy from outside sources, usually directly from the sun through photosynthesis, while animals steal energy from other organisms by eating them. Like steam engines, living things are open systems that continuously lose energy to their environments—witness the heat of the cushion beneath the cat. Even the wastes excreted by living systems retain some energy in their molecules and this, too, is lost.

Thermodynamics thus explains what it is that all living systems must control, and why such control is essential to life itself. All open systems, if they are to postpone for a time their inevitable heat death, must control the extraction and processing of matter, its internal distribution and storage, continuous conversion into energy, and elimination as by-product wastes. Living organisms, for example, convert energy originally from the sun into forms more useful to the processes of life: body heat, chemical energy for metabolism, electric energy to fire nerve impulses, mechanical energy to contract muscles and move about (luminescent organisms can even convert the energy back into light).

Control and Society

Because societies must also be concrete open systems if they are to sustain their organization against the progressive degrading of their collective energy, the view of organisms as concrete open processing systems applies equally to their social aggregates. The essence of human society, in other words, is its continuous processing of physical throughputs, from their input to the concrete social system to their final consumption and output as waste. The major sectors of an economy as first delineated by Colin Clark reflect the same essential life process: agriculture and mining, the extraction and breakdown of matter to produce energy; manufacturing, the synthesis of matter and energy into more organized forms; and services, the organization and support of these processes in well-integrated systems.

Unlike living organisms, however, *social* systems are made up of relatively autonomous components—individuals, families, groups, organizations—that can act for different and even cross-purposes. Because system processing must depend on exchanges among these individual components, the need for their coordination and control means that information processing and communication will account for a relatively greater proportion of matter and energy flows than they do in single organisms. The actual proportion will depend on several factors, including size of the population and its spatial dispersion, complexity of organization, and volume and speed of processing, among others.

This conclusion suggests that the proper subject matter of the social and behavioral sciences, if they are to complement studies of the flows of matter (input-output economics) and energy (ecology), ought to be information: its generation, storage, processing, and communication to effect control. Much the same vision of social science has already been expressed in a variety of disciplines. As early as 1950, for example, Norbert Wiener, a pioneer of cybernetic theory, argued that "society can only be understood through a study of the messages and the communication facilities which belong to it" (1950, p. 9). Zoologist E. O. Wilson, after surveying the thousands of social species from colonial jellyfish and corals to the primates, including *Homo sapiens*, declared "reciprocal communication of a cooperative nature" to be what he called the "diagnostic criterion" of society as most generally defined (1975, p. 595). Niklas Luhmann, a German sociologist, proclaims: "The system of society consists of communications. There are no other elements, no further substance but communications" (1984, pp. 1–2).

Living as we do in the Information Society, surrounded by microprocessing and telecommunications technologies, we might suppose that social theorists have always appreciated the importance of information and communication to social organization in general and to human society in particular. What could be more obvious than that, as Wiener put it in his first book on cybernetics, "the social system is an organization, like the individual, that is bound together by a system of communication" (1948, p. 24). With the continuing development of global computer networks, telecommunications, and mass media, it grows increasingly difficult to appreciate how relatively recent and different is the information-processing view of human organization and society.

One indication is provided by the Encyclopedia Britannica's fifty-four-volume set, *Great Books of the Western World*, published in 1952. Not one of three key concepts—information, communication, and control—made the list of 102 "great ideas" used to organize 520 classic works from Homer to Freud, a list that did include *form, matter, mechanics*, and *physics*. The three informational concepts did not make Britannica's penultimate list of 115 ideas, nor even the 88 additional ones ranked "among the most likely candidates for inclusion" (Adler 1952, pp. 1223–1224). Even the publication's 1,798-item "inventory of terms" included only *communication*, which turned up twenty-four times among 163,000 citations. Not until the rise of the Information Society, it appears, did concepts like information processing, communication, and control first surface in social theoretical discourse.

Most of the conceptual apparatus we need to understand the Control Revolution, it turns out, was directly inspired by the Control Revolution itself. The new ideas followed major technological advances, with some lag, during especially the period from the 1870s through the 1930s. This means that, although our interest here lies primarily in understanding the Control Revolution, we can also chronicle its impact on the history of ideas even as we develop the same ideas to help account for the larger societal change. For those who consider intellectual history to have a material basis, we could hardly do otherwise.

Control through Programming

One idea directly inspired by the Control Revolution in technology is the concept of a *program*, a word that first appeared in the seventeenth century for public notices but which in the past 150 years has spread through other organizational and informational technologies: plans for formal proceedings (1837), political platforms (1895), broadcast presentations (1923), electronic signals (1935), computer instructions (1945), educational procedures (1950), and training (1963). In general, *program* has come to mean any prearranged information that guides subsequent behavior.

We have already defined *control* as purposive influence toward a predetermined goal. If purpose is to be explained in other than metaphysical terms, however, its goal must exist prior to the behavior that it influences in some material form, on the government rolls, for example, that motivated the original Latin *contrarotulare*. All control is thus *programmed:* it depends on physically encoded information, which must include both the goals toward which a process is to be influenced and the procedures for processing additional information toward that end.

Often such programming is built directly into the physical structure of a purposive mechanism as, for example, information about the length of a minute and the number of minutes in an hour is built directly into the parts of a clock. Genetic programming is of this nature, built into our material essence at the moment of conception. Other programming, like that of an alarm clock, can be modified by the environment. All living systems can be reprogrammed in at least this way—Pavlov's dogs were just one example of the universal rule. Still other information-processing systems, because of their highly generalized capability for purposive action, might be controlled by a wide variety of programming: microprocessors and computers, for example, or bureaucracy, a controlling structure that routinely serves even a radically different government after a revolution or coup. By far the most generally programmable structures to be found in any living system are the brains evolved by the vertebrates, especially the human brain.

To account for purposive processes and behavior in terms of programming might seem merely to shift the problem of explanation from one set of concepts to another—from material action to information—with no real gain in understanding. This characterizes causal explanations in general, however, not just those based on the concept of programming. Without attempting to review the many controversies surrounding scientific explanation, we might simply note that, in seeking to account for purpose in terms of programming, we at least avoid two major conceptual problems, those of *vitalism* and *teleology*.

Unlike other attempts to account for the goal-directed behavior of living organisms, including postulated forces like *animus* (Aristotle 1931), *élan vital* (Bergson 1907), and *Entelechie* (Driesch 1908), programming must at least exist in a *physical* form, thereby eliminating vitalist and other metaphysical baggage. Because programs must exist prior to the phenomena they explain, moreover, they circumvent a major objection to teleological, functional, or "in order to" explanations—that their effects precede rather than follow their causes (Nagel 1961, chap. 12; Wright 1976).

Programmed behavior is not teleological but *teleonomic*, a term introduced by the biologists Colin Pittendrigh (1958), Julian Huxley (1960), and Ernst Mayr (1961, 1974b), among others, in an effort to rid their discipline of both teleological explanation and a longstanding contradiction: insistence that all natural processes have mechanistic interpretations, on the one hand, in the face of seemingly purposive sequences like organic growth, reproduction, and animal behavior on the other. In Mayr's words, "a teleonomic process or behavior is one that owes its goal directedness to the operation of a program" (1976, p. 389); a program is "coded or prearranged information that controls a process (or behavior) leading it toward a given end" (pp. 393–394).

Teleonomy, in other words, is equivalent to explanation in terms of programming and control. Although grounded in these concepts of modern information theory, teleonomic explanations are compatible with theories arising as early as the secularization of social thought in the seventeenth century. The view that society emerges from the interaction of goal-directed behavior controlled at various levels of aggregation, for example, dates back to the seventeenth-century political philosophers and eighteenth-century economists. The positivist version, stressing that individual goals are conditioned on objective facts like the behavior of other actors, first developed in the writing of Hobbes (1651), Locke (1690), and the classical economists (Smith 1776; Ricardo 1817). The idealist version, emphasizing the normative and subjective programming of behavior, can be traced back to the later writings of Kant, especially to the *Critiques* of what he called *Praktischen Vernunft* or "practical reason" (1788) and *Urtheilskraft* or "judgment" (1790, pt. 2).

In contrast to these classical approaches, however, teleonomic explanations obviate the need to attribute consciousness, planning, purpose, or any other anthropomorphic qualities to aggregate levels, the special problem of *reification*. Adam Smith's "invisible hand" of market forces (1776, p. 423), for example, can be seen to result from the interconnected programming of individuals and their organizations, including individual tactics, strategies, and utilities, on the one hand, and organizational procedures, written law, and cultural norms on the other. This list of programming is more than an analogy. Each item describes a programmable form of control—including both goals and procedures—that might be encoded in some physical form, whether electrochemically in the human brain, in inked patterns on paper, or in charged particles in computer memory. Each program might therefore be a contributing cause—temporally prior and material—of aggregate market behavior.

To specify the example as a modern engineering problem that, like Smith's invisible hand, combines individual action with certain purposive organization, consider the task of controlling rush-hour traffic. In the classical tradition traffic patterns might be seen to result from the interaction of goal-directed behavior by individual commuters, the "invisible hand" of rush-hour traffic. Behavior might also be ignored altogether, however, in favor of the various levels of programming involved: *genetic programming*, encoded in each cell of each commuter and determining the distributions of reaction times and stress levels; *cultural programming*, encoded in neural structures of the brain and defining certain norms and etiquette of the commute; *organizational programming*, encoded in traffic law and employer regulations and determining patterns of carpooling and parking; and *mechanical programming*, encoded in the timing devices of traffic lights and helping to maintain the larger patterns planned by traffic control engineers.

Because we do not normally shift our attention from a manifest social behavior like rush-hour traffic to the various programs that might interact to control it, such an approach might at first seem awkward—a needlessly complex way to account for a relatively simple social phenomenon. Like most macrosocietal processes, however, rush-hour traffic patterns are in fact complex for a simple reason: they are controlled by multiple, multilevel, and densely interconnected programs. In concentrating our analysis on these programs, we at least address their complexity directly, rather than through reification of planning and purpose to the aggregate level.

Programming and Social Theory

Teleonomic explanations are also useful in reconciling theoretical approaches that might seem mutually exclusive but which in fact are merely pitched at different levels of control. Consider, for example, the normatively controlled, goal-directed behavior that sociologists call *social action*. The most influential synthesis of action theory, Talcott Parsons's *Structure of Social Action* (1937), opens with a question borrowed from the historian Crane Brinton (1933, p. 226): "Who now reads Spencer?" Like Brinton, Parsons rejected Herbert Spencer's positivist-utilitarian view of society as an integrated system, subject to the laws of natural selection and evolution, in favor of social analysis grounded in the voluntaristic action of individuals.

The difference reflects the progress of the Control Revolution which, as already noted, resulted in a fundamental change in human thought between the 1870s and 1930s. At the turn of the century most intellectuals still accepted the view expressed in Spencer's *Principles of Biology* (1864–1867) and *Principles of Sociology* (1876–1896) that social organization and control might be treated as autonomous processes like those studied by physical scientists. By the 1930s, however, social theorists like Brinton and Parsons were beginning to see organization as purposively constructed and effecting control through goal-directed behavior. This new world view, compatible with concepts like information processing, programming, and decision, was at the same time gaining adherents in a wide variety of other disciplines, from mathematical logic, philosophy, and psychology to the founding work in computer theory.

Despite Parsons's early involvement in the change, however, and the great success of his work on action theory, he drifted back to the study of autonomous social systems (1951, 1960, 1971) and even to evolution (1966), eventually becoming what one sociologist ironically called "a Spencerian of sorts" (Bierstedt 1975, p. 155). Meanwhile, according to the same source, Spencer's work has enjoyed "a remarkable resurrection, one that may require the restoration of his statue in the pantheon of sociological heroes" (p. 154). Obviously the distinction that Parsons insisted separated his work from Spencer's was in fact overdrawn. The two theorists merely pitched their analyses at different levels of control: Spencer closer to organization and its processes, Parsons closer to individual programs, their goals, and resultant behavior.

Had Parsons managed to develop the concept of programming, just emerging as he wrote his first book, he would have had the perfect means to reconcile his action theory with the theories of Spencer. Because programming must exist in some material form, it is tractable by positivist-utilitarian methods like those championed by Spencer. Because both genetic and cultural programming can be communicated to new individuals, whether through biological reproduction or through socialization and cultural diffusion, both do indeed evolve much as Spencer argued, whether through so-called natural (nonpurposive) or purposive selection. Parsons's emphasis on normative programming remains compatible with this view, however, as does the importance he placed on psychological and other subjective influences on behavior, which he understood cannot be reduced in the positivist sense of "location in space" (1937, p. 86). This is a fair statement of the fact that, for example, the "style" of a computer chess program is not reducible to any particular part of its listing and indeed might appear to change as the program is run on different types of machines.

The concept of programming would also have clarified the ontology of what Parsons called a "social system," helping to distinguish it from the concrete manifestation that we might also describe with that term. As Parsons was repeatedly forced to stress, "In the limiting conception a society is composed of human individuals, organisms, but a social system is not . . . It is an organized set of behaviors of persons interacting with each other: a pattern of roles" (Grinker 1967, p. 328). We might accept this position without further agreeing that roles are merely "abstracted" by reflective observers from concrete behavior. On the contrary, what Parsons calls a "pattern of roles" exists concretely in the programming—operational definitions, goals, and processing rules— by which it is taught, diffused, and renegotiated by each new generation.

Sociology is not the only discipline that might have benefited from an earlier appreciation of programming as the cause of purposive behavior. Perhaps because information is higher-order or derivative of matter and energy on which it depends for its existence, physical scientists also overlooked the role of programming until well into this century. Nothing illustrates this better than the case of Maxwell's demon, a hypothetical being who bedeviled theoretical physics between 1871 and the 1920s and whose progressive demystification reflects the crucial transition in thought over the period of the Control Revolution.

Lessons of Maxwell's Demon

James Clerk Maxwell (1831–1879), the Scottish physicist, was without question one of the most brilliant scientists of the nineteenth century. His theory of electromagnetic radiation, published in 1865, occupies a position in modern physics comparable to Newton's classic work on mechanics. Many historians of engineering consider control theory to date from the 1868 paper by Maxwell on Watt's flyball governor (Evans

1977). Three years later, in *The Theory of Heat*, Maxwell introduced a demon

> whose faculties are so sharpened that he can follow every molecule in its course. Such a being, whose attributes are still as essentially finite as our own, would be able to do what is at present impossible to us. For we have seen that the molecules in a vessel full of air at uniform temperature are moving with velocities by no means uniform . . . Now let us suppose that such a vessel is divided into two portions, A and B, by a division in which there is a small hole, and that a being, who can see the individual molecules, opens and closes this hole, so as to allow only the swifter molecules to pass from A to B, and only the slower ones to pass from B to A. He will thus, without expenditure of work, raise the temperature of B and lower that of A, in contradiction to the second law of thermodynamics.   (1871, pp. 308–309)

According to the second law of thermodynamics, as we have seen, the organization of a closed system like the vessel can only decrease and the process is irreversible, Maxwell's demon notwithstanding. Relative disorganization or randomness of a system is important enough to have been given the special name *entropy*, a term coined by the German physicist Rudolf Clausius in 1865. So basic and pervasive is the tendency toward increased entropy in both nonliving and living systems that the British astronomer and physicist Sir Arthur Eddington declared, "The law that entropy always increases—the second law of thermodynamics—holds, I think, the supreme position among the laws of Nature" (1928, p. 74).

Here, then, was the challenge of Maxwell's demon, which seemed to overturn—by merely opening and closing a small hole—no less than the supreme law of nature. The fact that even the great Maxwell failed to resolve the paradox, not to mention similar failures by three generations of scientists, illustrates how recent and nonobvious are concepts like information, programming, decision, and control—precisely the concepts that Maxwell unwittingly ascribed to his hypothetical demon.

That Maxwell's demon is in fact a program is demonstrated by Figure 2.1, which gives the flow chart of an algorithm that perfectly duplicates the demon's behavior. For each molecule in the vessel, the demon must continually determine two pieces of information—velocity and trajectory—for the program's essential inputs (represented by the parallelogram near the top of Figure 2.1). Based on only these two

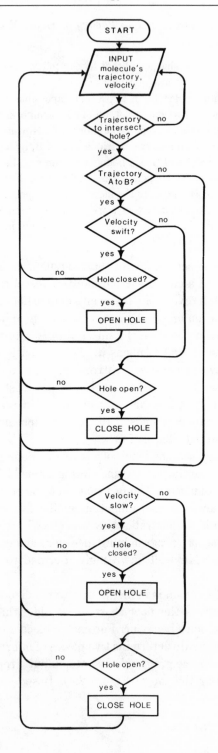

inputs, the algorithm determines whether or not the demon ought to behave in the only two ways it can: by opening or closing the small hole in the partition separating portions A and B of the vessel (outputs are represented in the flow chart as rectangles).

With the same simple sequence of decisions, some mechanism might in fact be programmed to act in the way described by Maxwell, namely, "to allow only the swifter molecules to pass from A to B, and only the slower ones to pass from B to A." Would such a mechanism overturn the second law of thermodynamics—the supreme law of nature—by reversing the progressive degradation of energy in the vessel and thereby decreasing its entropy?

Because Maxwell took information for granted, he did not consider the costs of providing his demon with a continual flow of inputs to its program. Information always requires at least an expenditure of energy sufficient to transmit it—with a resulting increase in entropy. Unless there is light, for example, Maxwell's demon cannot "see" the molecules in the vessel. Once light is introduced, however, the entropy of the vessel can only increase—despite the demon's best efforts—because the free energy lost in transmitting information exceeds the energy gained by his purposive sorting. Even if the demon does not "see" the molecules, that is, follow them by means of reflected light, he will need some alternative means of acquiring information. Because information can only be communicated in some form of matter or energy, the results are always the same: the demon's energy needs will exceed the energy freed by his sorting and his operation must inevitably succumb to heat death.

This argument was foreshadowed in work on statistical physics that the Austrian Ludwig Boltzmann published in 1894. Boltzmann observed that entropy is related to "missing information," that is, to the number of alternatives that remain to a physical system after all the macroscopically observable information about it has been recorded (Shannon and Weaver 1949, p. 3). Extending this insight, Leo Szilard, the Hungarian physicist who would later be instrumental in building the first atomic bomb, used the statistics of quantum mechanics to calculate the minimum amount of energy required to transmit *any* information. Whatever means of information processing Maxwell's demon might employ, according to Szilard's calculations, the heat gradient would be exceeded by the energy costs of information necessary to sustain it

Figure 2.1. Flow chart of an algorithm for Maxwell's demon.

(Szilard 1929). In other words, the demon is no threat to nature's supreme law—Maxwell's paradox is resolved.

His demon continues to interest us, however, because its sixty-year career so precisely mirrors the intellectual development of the Control Revolution. Born just three years after Maxwell's seminal paper on control theory, in the same decade that brought technological innovations like the telephone, phonograph, and microphone, the demon was not retired until the understanding of information had reached a high level of quantitative sophistication—owing largely to the work of new communications engineers like Nyquist (1924, 1928) and Hartley (1928) and culminating in Shannon's mathematical theory after World War II.

The basic lessons of Maxwell's demon—that control involves programming, that programs require inputs of information, that information does not exist independent of matter and energy and therefore must incur costs in terms of increased entropy—all seem commonplace today. One evidence is that grade-school children around the world can now program simple algorithms like that used by Maxwell's demon.

Programming and Decision

A second basic concept that, like the idea of a program, was profoundly changed by the Control Revolution in technology is that of *decision*. Although programs are responsible for all control and hence all purposive action, this alone does not explain how such control is possible. How do inert progams manage to influence concrete technologies, not to mention living organisms, guiding both toward predetermined goals? Work in a wide range of disciplines converged by the 1930s on a single answer: programs control by determining decisions.

As we saw in the algorithm for Maxwell's demon, the process of control involves comparison of new information (inputs) to stored patterns and instructions (programming) to decide among a predetermined set of contingent behaviors (possible outputs). As we have seen, this describes the use of official government rolls as commemorated by the medieval Latin verb *contrarotulare*. Contingent decision: If a man's property is listed on the rolls, tax accordingly; if not, assess his property and fine or imprison him. To decide is to control, in short, and to control is to decide; information processing and programming make both possible.

Once again the crucial aspect of a definition is exposed by etymology, by the origin of *decision* in the Latin verb *decidere*, to cut off from the source, that is, purposely to determine the ultimate function—to influence toward a desired goal. As late as the seventeenth century, *decide* was still used as a transitive verb meaning to cut or separate off as did the sea, for example, for Thomas Fuller: "Again, our seat denies us traffick here, the sea too near decides us from the rest" (1642, p. 122).

This early usage suggests modern applications like *decision trees*. The number of alternatives or *degrees of freedom* available to Fuller as he moved from his coastal position up the River Tyne, for example, was progressively reduced as he passed each of the Tyne's tributaries. The amount of information contained in his eventual destination (output), once he passed even a single branch, was greater than that in any point of embarkation (input) downstream.

As we have already seen for programming, such decision implies neither voluntarism nor free will; it carries no vitalist or other metaphysical implications. All decisions are determined in the sense that, for a given program, the same input will always result in the same decision or output. Decisions may be probabilistic, of course, as in deciding by the flip of a coin, but the decision to flip it must be strictly determined at some higher level.

Any decision tree of finite length can be duplicated by a finite automaton, thereby equating the question of *decidability* with that of *computability*, a connection that has brought considerable convergence of mathematics, logic, and computer science with traditionally humanistic disciplines like philosophy, linguistics, and cognitive psychology. As with the debunking of Maxwell's demon, the progressive development of computability theory, known in mathematics as recursive function theory and in computer science as the theory of algorithms, reflects a crucial transition in human thought about programming and control between the 1870s and 1930s, the first stage of the revolution in control technology.

The mid-nineteenth century saw a critical reexamination of the foundations of mathematics, inspired largely by the discovery of the first non-Euclidean geometries under the influence of Karl Friedrich Gauss (1777–1855), who coined the term. While geometers worked at revising and perfecting their axioms, an effort that culminated in David Hilbert's *Grundlagen der Geometrie* in 1899, logicians were similarly inspired. Modern logic is generally considered to have begun with publication in 1879 of Gottlob Frege's *Begriffsschrift*, which boldly attempted to formalize various parts of mathematics (van Heijenoort 1967, p. vi). In contrast to axiomization, which uses ordinary language and logic, Frege introduced a formal system based on abstract signs, formulas, and explicit rules for their manipulation. The monumental *Principia Mathematica* of A. N. Whitehead and Bertrand Russell, published in three volumes between 1910 and 1913, helped to popularize Frege's notion of a formal system.

Hilbert, in particular, insisted that all mathematics could be reduced to an inventory of formulas, rules, and proofs, the subject matter of a new science of proof theory or metamathematics. When the International Congress of Mathematicians met in Paris in 1900, Hilbert presented twenty-three of what he considered the most important problems—including several involving questions of decidability—to occupy mathematicians during the coming century. Hilbert himself had already set out to find an algorithm to decide the truth or falsity of *any* mathematical proposition, what came to be known as the *Entscheidungsproblem*. This program occupied many of the world's leading mathematicians, including Wilhelm Ackermann (1924), John von Neumann (1927), and Jacques Herbrand (1930, 1931).

Hopes for the program were dashed in 1931, however, when Kurt Gödel published his famous incompleteness theorem, one of the great intellectual achievements of the century. The theorem proved, in effect, that the procedure sought by Hilbert, including the one suggested by Whitehead and Russell in the *Principia*, could not exist. In any formal system adequate for number theory, Gödel showed, there exists an *undecidable* formula for which neither itself nor its negation can be proved; a major corollary establishes that the consistency of any such system is also undecidable. The revolution in logic begun by Frege a half-century earlier had at last found its dialectical resolution: intuitionism remained dead, on the one hand, but mathematicians were now forced to couch their new formalisms in semiformal language.

Of the many effects of Gödel's theorem, none was more important to the intellectual development of the Control Revolution than the resulting clarification of *decision procedures*, the mechanical rules that determine, in finitely many steps, whether a combination of symbols is decidable. Today the accepted formalization of such procedures is the *Turing machine*, a simple mathematical model of a general-purpose computer introduced in 1936 by the British logician Alan Turing, who evidently took the idea of a "mechanical" rule quite literally.

On the basis of Turing's work and several alternative formulations published in the same year (Church 1936a,b; Kleene 1936; Post 1936; Rosser 1936; all reprinted in Davis 1965), many problems and functions of longstanding importance to logic and related disciplines have been found to be undecidable. One noted example is the theorem of Alonzo Church (1936a) establishing that no decision procedure exists for the predicate calculus beyond the monadic. Hilbert's best-known problem, the tenth, provides another classic example: it was finally shown to be undecidable, owing largely to work on the behavior of Turing machines, by a twenty-two-year-old Russian mathematician, Yuri Matyasevich, in 1970 (Davis and Hersh 1973).

Links between abstract questions of decidability and the development of control technology were established by von Neumann, Turing, and the many other mathematicians who helped build computers during World War II. Although formal proof cannot be made, there are good arguments to accept the Turing machine as equivalent to a real computer. Perhaps most compelling is Church's hypothesis (1936b), known also as the Church-Turing thesis, that all decidable number theoretic functions are mathematically recursive (computable) and vice versa. Because all functions that operate on strings of symbols—and indeed from any countable set to any other—can be represented by natural numbers, the Church-Turing thesis (now generally adopted by other mathematicians) has wide application in computer science, artificial intelligence, and cognitive theory.

One computer scientist (Hofstadter 1979, p. 429) restates the Church-Turing thesis as "What is human-computable is machine-computable," calling the idea "one of the most important concepts in the philosophy of mathematics, brains, and thinking" (p. 561). Certainly the modern revolution in logic, in which the thesis played a central role, will remain a basic foundation for the Control Revolution as it continues to unfold in new industrial applications like those of artificial intelligence and expert systems.

Social Technologies of Decision

Neologisms reflect the continued diffusion of new thinking about decision into the popular ethos. The logician Willard Quine introduced the term *decidability* in 1940 (Burchfield 1972, p. 750), noting in a later work that "elementary algebra is completable and mechanically decidable while elementary number theory is not" (1950, p. 247). The term *decision procedure*, also introduced by Quine (1945, p. 3), was by the late 1950s applied to popular games in the semipopular press; according to the July 1957 *Technology* magazine, "any game of a finite kind which can be completed in a finite number of moves . . . must have a decision procedure, even though we may not know for any particular game what this procedure is" (p. 182). Three years later the London *Times* discussed "decision-making machines" (March 24, 1960, p. 2); the image of the "decision-taking mechanism of the human brain" dates from 1964 (Young 1964, p. 28).

Games also provided the inspiration for a separate application of decision to control theory, namely statistical decision theory, a formalization of rational choice among alternatives with different probabilities and utilities. Drawing on von Neumann's 1928 paper establishing what has come to be known as *game theory*, the mathematical statistician Abraham Wald in 1939 introduced a more general *decision theory* incorporating classic statistical work on hypothesis testing and estimation. Wald's generalization and synthesis of statistics, including his application of multiple decision spaces, weight and risk functions, and minimization of maximum risks (so-called minimax solutions), provided the basis for much of the intellectual technology of control that would emerge during and after World War II: quality control, linear programming, operations research, systems analysis, and even game theory, which found wide practical application after von Neumann's text with the economist Oskar Morgenstern appeared in 1944. Sociologist Daniel Bell assesses the role of decision theory as follows:

> Any single choice may be as unpredictable as the quantum atom responding erratically to the measuring instrument, yet the aggregate patterns can be charted as neatly as the geometer triangulates the height and the horizon. If the computer is the tool, then decision theory is its master. Just as Pascal sought to play dice with God, and the physiocrats attempted to draw an economic grid that would array all exhanges among men, so the decision theorists seek their own *tableau entier*—the compass of rationality, the "best" solution to the choices perplexing men.   (1973, p. 33)

Logical and statistical theories of decision were not the only ones to emerge in the 1920s and 1930s, a period that saw development of the concept in two other areas: normative decision theory and deontic logic. Both were influenced by the work of Frank P. Ramsey, a mathematician who in 1926 stated simple consistency postulates implying the existence of both personal probabilities and utilities.

Psychologist Kurt Lewin, influenced by lectures on decision theory that von Neumann gave in Berlin in 1928, formulated the first treatment of utility (what he called *Aufforderungscharakter* or "valence") and related concepts of subjective probability in the early 1930s (1931a,b,c). These and similar formulations gained mathematical rigor in work by Bruno de Finetti published in 1937. Two years later, using Milton's "Reason is also choice" as a motivation, John Hicks applied the developing ideas to economic decision under uncertainty, work for which he received the Nobel Prize in 1972. Following World War II, the ideas of Ramsey, von Neumann, de Finetti, and Hicks found synthesis in work by Kenneth Arrow (1951a,b), corecipient of the 1972 Nobel Prize in economics, and statistician Leonard Savage (1951, 1954); both treatments converged with statistical decision theory. A few years later, Patrick Suppes (1956, 1957) and R. Duncan Luce (1959) began to incorporate many of the same normative decision concepts into psychology (Luce and Suppes 1965).

Ramsey had initially called his theory "a branch of logic" (1926, p. 157), and indeed that idea developed into a fourth distinct approach to decision. Until the late 1920s principles of *deontic* logic, the logic of obligation, tended to be stated as footnotes to work on moral philosophy or *modal* logic, the logic of the possible and necessary. Even modal logic, extensively developed since its origins in ancient Greece, had fallen into neglect during the Renaissance and found little place in modern mathematical logic following Frege. Revival of interest in especially the systematic treatment of modal logic in the 1920s, however, led to attempts to treat deontic logic in a similar fashion.

The first such attempt was that of Ernst Mally (1926), who introduced the modern usage of "Deontik" by adding a modal operator denoting *ought* to the formal language of propositional logic. Despite continuations of Mally's effort by Karl Menger (1934, 1939), Kurt Grelling (1939), and Karl Reach (1939), widespread interest did not come until the end of World War II and publication of the classic paper by Georg Henrik von Wright (1951). Von Wright's essential insight, that there exists a significant analogy between the deontic *obligation* (ought) and *permission* and the modal *necessity* and *possibility*, has influenced most subsequent work in the field.

Thus it was that the simple idea of decision, combined with the concepts of information processing and programming, stimulated considerable intellectual activity during the 1920s and 1930s in four interrelated yet distinct areas: recursive function theory or the theory of algorithms, statistical and normative decision theory, and deontic logic. Practical applications of this work in computer control technologies as well as in a wide range of academic disciplines have continued to multiply. Like programming, the concept of decision holds considerable promise for social and behavioral science, especially one grounded in principles of information processing and communication.

As we have already seen, for example, Talcott Parsons might have benefited from explicit use of the concept of programming in his early action theory, a formal model that did include a half-dozen other logical components: an actor, a situation involving both conditions and means, a normative orientation, and an end (Parsons 1937, p. 44). When we compare this schema with the postcomputer interest of philosophers in normative action theory as derived from deontic logic (for example, Rescher 1966), one major addition stands out: explicit attention to decision. Similar integration of deontic logic and other abstract formalizations of decision making has already developed in organizational studies, computer science, psychology, and social theory in general, most notably in work by Herbert Simon (1965, 1966).

With the growing understanding of programming and decision, the intellectual development of the Control Revolution may at last have come full circle from the 1870s and the work of Maxwell and Frege. Consider, for example, the discussion by cognitive psychologists of a "decision demon" (Selfridge 1959), including a lengthy treatment in a textbook (Lindsay and Norman 1977, chap. 7) suggestively titled *Human Information Processing: An Introduction to Psychology*. Could Maxwell have possibly guessed that, in the instant he conceived his own mythical demon, he glimpsed a complex tangle of ideas that would remain at the center of intellectual discourse, technological development, and social change for more than a century?

Life's Decision Demon: DNA

What Maxwell could not have suspected when he introduced his demon, thirty years before Mendel's work on heredity became known and seventy years before the first published hint of a genetic code, is that the heart of his hypothetical creature—that of programming, decision, and control—does in fact beat in every cell of every living thing on earth. This programmable control structure, which first appeared in simple form at the origin of life perhaps four billion years ago, is the complex macromolecule of deoxyribonucleic acid or *DNA*. DNA does not sort molecules by opening and closing a small hole, of course, but it does achieve the same result, that of maintaining and even increasing the organization of living materials counter to entropy, in apparent violation of thermodynamics' second law.

This fact is commemorated by the common origin of *organization* and *organism* in the Indo-European *worg*, meaning *work*. A system can sustain work, we have seen, only if its internal energy is purposively organized in a heat gradient, as, for example, in the steam engine, which inspired early work on thermodynamics. Only living systems can maintain and even increase such organization—to work as if guided by some vitalist equivalent of Maxwell's demon. This does not mean that life decreases entropy in the universe, however, but only within its own systems and only by increasing entropy in the matter it consumes. Hence all living systems, including human societies, must be seen as eddies in the entropic stream—as countercurrents resisting for a time the rush of the universe toward final heat death before they too are caught up in the flow.

The continuous processing of matter and energy that makes possible this miracle, apparently unique in at least our own solar system and essential to societies no less than to individual cells, resides ultimately in DNA, the decision demon that organizes matter and energy at the most fundamental level of control. DNA is thus not only the most basic of all control technologies, in the figurative sense, but also one whose capabilities are unlikely to be rivaled by technologies of our own making for many generations to come.

Consider that a one-inch strand of genetic particles can store the amount of information contained in up to twelve thousand typed manuscript pages, the equivalent of about twenty books as long as the one you are reading now. Because the nucleus of a single human cell contains roughly five feet of genetic material, it might store the equivalent of nearly two thousand such books. The same amount of information would fill three to eight thousand floppy disks, depending on the type, or nearly sixty 24-thousand-foot reels of computer tape packed as densely as possible. Some eleven hours would be needed to play as much information on a high-fidelity system or over FM radio, more than forty-six hours to send it via telephone. The programming contained in a single one of our cells, in other words, might easily exceed in information our personal libraries, record or tape collections, and video game cartridges combined.

Even more impressive by engineering standards, all of the cell's information is stored in a structure just barely visible with the most powerful electron microscope. George Beadle, the Nobel Prize–winning geneticist, calculated that the complete genetic coding for every person on earth could be wound back and forth into a cube only one-eighth of an inch thick. This tiny stack, by Beadle's estimate, would contain roughly sixty thousand times as much information as in all the books ever published (1964, p. 6). "The deciphering of the DNA code," Beadle concludes, "has revealed our possession of a language much older than hieroglyphics, a language as old as life itself, a language that is the most living language of all—even if its letters are invisible and its words are buried deep in the cells of our bodies" (Beadle and Beadle 1966, p. 207).

It is hardly surprising that, with the growing understanding of programmed control by computers after World War II, scientists would begin to decipher the ancient language of DNA and to exploit its programming in new technologies. More surprising is that the essential insight of modern genetics that DNA can be treated as programming quite apart from the biochemistry of the amino acids that it controls owed little to contemporaneous development of information theory, cybernetics, or even von Neumann's seminal work (1966) on self-reproducing automata (Judson 1979, pp. 242–244).

Perhaps the earliest mention of genes as physically encoded information can be found was *What Is Life?*, a popular book published in 1944 by Erwin Schrödinger, the Austrian-born theoretical physicist who shared the 1933 Nobel Prize for his formulation of wave mechanics, including the most widely used mathematical tools of modern quantum mechanics. Twenty years before the decipherment of the genetic code, Schrödinger—who had little knowledge of biochemistry—wrote:

It has often been asked how this tiny speck of material, the nucleus of the fertilized egg, could contain an elaborate code-script involving all the future development of the organism . . . the number of atoms in such a structure need not be very large to produce an almost unlimited number of possible arrangements. For illustration, think of the Morse code. The two different signs of dot and dash in well-ordered groups of not more than four allow of thirty different specifications . . . with five signs and groups up to 25, the number is 372,529,029,846,191,405 . . . with the molecular picture of the gene it is no longer inconceivable that the miniature code should precisely correspond with a highly complicated and specific plan of development and should somehow contain the means to put it into operation.   (1944, pp. 61–62)

As happened so often since the advent of the Control Revolution, concepts from information and communication technology—here Morse's binary telegraph code—helped scientists to reconceptualize traditional subjects like cellular biology. Because information and control are so basic to living systems in general, nonspecialists who understood these concepts have managed to contribute to a wide range of fields. Modern biology, for example, developed through important work by a large number of theoretical physicists and physical chemists: Max Delbruck, Leo Szilard, Francis Crick, Maurice Wilkins, George Gamow, and Linus Pauling, among others (Judson 1979, p. 605).

With the first postulation of DNA structure by Watson and Crick in 1953, literally hundreds of theorists—working more as cryptographers than as biochemists—set out to crack the code by which the four different genetic particles (called *nucleotides*) represent the various amino acids used in protein synthesis. As if to establish the role that informational concepts were to play in modern biology, the first suggestion for a code came not from a biologist but from George Gamow, a theoretical physicist. The importance of this contribution (Gamow 1954), Crick later recalled, "was that it was really an abstract theory of coding and was not cluttered up with a lot of unnecessary chemical details" (1966, p. 4).

Nor could it have been, since Gamow had only sparse knowledge of the relevant biochemistry. He did get the number of basic amino acids correct at twenty, however, which prompted Crick and Watson—immediately on receipt of Gamow's code—to list the correct acids, even though the assumption that so few compounds could be the basis of so many others was, according to one biochemist, "for its time, quite audacious and wholly unsupported" (Stent 1971, p. 35). As one history of modern biology concludes, "Gamow disentangled the problem, stating it for the first time in its modern form" (Judson 1979, p. 250).

Technologies of Genetic Programming

Since perfection of the recombinant DNA process in 1973, modern biology has rapidly spawned *biotechnology*, an application of several academic disciplines—including microbiology, genetics, molecular biology, biochemistry, and chemical engineering—which is already a multibillion-dollar-a-year industry with some one hundred fifty firms (Abelson 1983b). By 1982 the United States was issuing more than eleven hundred patents annually in biotechnology, about 2 percent of all patents granted (Zaborsky 1983).

The term *biotechnology* has now lent itself to an eight-volume reference set (Rehm and Reed 1981), to at least two textbooks (Smith 1982; Wiseman 1982), to its own patent digest (begun March 15, 1982), and to a glossy monthly trade journal (begun March 1983). Today one finds in mathematical puzzles (Hofstadter 1982) and popular books (Hofstadter 1979; Eigen and Winkler 1981) what theorists like Schrödinger and Gamow first advanced as one of the more important insights in the history of science: that genetics may be treated as an informational problem, that is, one depending for its solution on the interrelationships among programming, information processing, decision, and control.

Biotechnology continues to exploit DNA *qua* programming, information that controls the physical structures and processes of living systems. Recombinant DNA, or "gene splicing," for example, involves reprogramming a living organism by inserting into its DNA a foreign gene for a product with commercial value. Industry is just beginning to realize the possibility of engineering proteins—which approach infinity in their potential numbers—by using either genes from existing organisms or synthetic genetic programs created by stringing together chains of nucleotides (Abelson 1983a). Because such technologies are the functional equivalent for living organisms of what more traditional programming is for computers, genetic engineering must itself be viewed as an information technology—another logical outgrowth of the Control Revolution.

The Control Revolution Reconsidered

We began this chapter with certain nagging questions concerning the Control Revolution and resulting Information Society: Why has information—of all commodities—come to dominate the economies of at least a half-dozen advanced industrial nations? Why the growing importance of microprocessors and computers, devices that can do nothing more than convert information from one form to another? How are information processing and programming related to communication and control, and why have these simple functions become so central to human society and culture?

Ultimate answers to these questions, we have found, lie at the heart of physical existence. In order to oppose entropy and put off for a time the inevitable heat death, every living system must maintain its organization by processing matter and energy. Information processing and programmed decision are the means by which such material processing is controlled in living systems, from macromolecules of DNA to the global economy. The Industrial Revolution sharply increased the volume and speed of energy conversion and material processing and thereby precipitated the various technological responses we call the Control Revolution. Even if industrialization had been more gradual, however, the ultimate result would have been much the same.

To understand the basis of human society in information processing, communication, and control, moreover, is to appreciate the profound irony in popular sentiment against technology that has persisted over the past century. Of all the revolutionary innovations in technology since the Industrial Revolution, few have aroused more widespread suspicion, resentment, and even open hostility than have the various capabilities for information processing. Since the mid-nineteenth century, as we saw in the first chapter, increasing bureaucratization has been opposed as somehow dehumanizing, a sentiment that persisted into the 1950s in popular works like William H. Whyte's *Organization Man* (1956). Since then, many of the same feelings have begun to shift to newer information-processing technologies: to the computer in the 1960s and 1970s and more recently to microprocessors, especially as manifest in devices like industrial robots, word processors, and video games, all of which were routinely accused in the early 1980s of dehumanizing influences. Just as the terms *bureaucracy* and *bureaucrat* long ago came to suggest narrow outlook, lack of humanity, and otherwise vague reprobation, similar connotations have been attached to computer technology and personnel, although less frequently as a laity begins to take responsibilities from the priesthood.

The irony, of course, is that information processing might be more properly seen as the most natural of functions performed by human technologies, at least in that it is shared by every cell of every living thing on earth. In view of the fact that information processing distinguishes all living things and a few of their artifacts from the rest of the universe, moreover, the ability must by definition be as old as life itself—on this planet or any other. Information processing is also arguably the most human of life functions in that particular capabilities of our brains to process information best distinguish us from all other species.

No human technology has more in common with all living things than do our various capabilities to process information, whether they be institutionalized in the formal structures and procedures of bureaucracy, input electronically to computer memory, or photolithographed into the silicon wafers of microprocessors. It is through the understanding of these capabilities, the essential life processes of organization, programming, and decision to effect control, that we can best hope to answer the many challenging questions raised by the Control Revolution.

3

Evolution of Control: Culture and Society

Social existence is controlled existence . . . Without some constraint
of individual leanings the coordination of action and regularity of conduct which turn a human aggregation into a society could not mater-
ialize . . . The concept of social control brings us to the focus of sociology
and its perpetual problem—the relation of the social order and the
individual being, the relation of the unit and the whole . . . Control is
simply coterminous with society, and in examining the former we simply describe the latter.

—S. F. Nadel, "Social Control and Self-Regulation"

To APPLY the model of the previous chapter to the Control Revolution
directly without first considering the earlier evolution of culture and
society would be to risk underestimating the importance of control to
all social structures, those of other species no less than those of our
own. It would also be to risk failure in grasping the full impact of the
Control Revolution itself. Only through appreciation of even the simplest animal societies as complex systems of information processing
and communication—systems that maintain an often fragile balance of
control, both internal and external—can we begin to understand the
particular way in which industrialization disrupted the material economy of the nineteenth century and why the modern information society
emerged as a direct result.

The Industrial Revolution was not the first time that external pressures on processing systems produced a sharp increase in their capability to control throughputs, a change abrupt enough to merit
classification as a revolution in control. As we saw in Chapter 2, in the
example of rush-hour traffic, living systems have thus far evolved four
levels of programmable structures and programs: DNA molecules encoded with genetic programming, the brain with cultural programming, organizations with formal processing and decision rules, and
mechanical and electronic processors with algorithms. This succession

of four levels of programming, each one of which appears to have complemented and extended more than superseded already existing levels, constitutes the total history of control as we know it—a relatively smooth development punctuated by only these four major revolutions in control.

Table 3.1 places each of the four control revolutions in historical perspective. The first, an all but imperceptible molecular change that we revere as the origin of life, occurred approximately four billion years ago, surprisingly soon after the formation of the earth itself. As can be seen in Table 3.1, it took another 3.9 billion years for information processing to transcend the genetic level; this second control revolution occurred about 100 million years ago as certain vertebrates first began to learn through imitation, the earliest and most primitive form of cultural programming. The third control revolution, one that expanded information processing from neural to social structures, came almost one hundred million years later—about 3000 B.C.—in the bureaucratic organizations of Mesopotamia and ancient Egypt. As we saw in Chapter 1, bureaucracy did not become pervasive until the nineteenth-century control crisis and the appearance of a complementary fourth level of control: mechanical, electric, and electronic technologies for information storage, processing, and communication.

Because the first level of control characterizes all living things, the second level the higher vertebrates (birds and mammals), the third and fourth levels only *Homo sapiens*, it might be tempting to stake claim to the superiority of our own species based on this progression. All such claims have been rejected, however—Aristotle's eleven-point *scala naturae* (1912, pp. 732a–733b), his hierarchy of "psychic powers" (1931, pp. 414a–415b), and all forms of the "Great Chain of Being" that dominated science and philosophy until the late eighteenth century (Lovejoy 1936). Even with no knowledge of genetics, Darwin could appreciate the "marvellous structures and properties"—which we now know derive from purposive information processing and control—that characterize "the most humble organism."

Indeed, equally plausible arguments can be made for the superiority of even the simple prokaryotes, which include the single-celled bacteria and blue-green algae. Consider their case: Prokaryotes ruled earth for some three billion years, fully three-fourths of the total tenure of life on the planet. They still rule in that they are found inside all more complex organisms and virtually everywhere else: in the depths of oceans, in arctic glaciers, in hot springs—even at the top of the strat-

Table 3.1. Time line of the development of four levels of control

Years ago (logarithmic)		
	-	Formation of the Earth
	-	*Level One: Life (Molecular Programming)*
	-	Motility
1 billion -	-	Organs specialized in control
	-	True nervous systems
	-	Vertebrate brain
100 million -	-	*Level Two: Culture (Learning by Imitation)*
	-	Learning by teaching
10 million -		
	-	Divergence of hominids and apes
	-	Rapid increase in hominid brain size
	-	Genus *Homo*
1 million -		
100 thousand -		
	-	*Homo sapiens*
10 thousand -		
	-	Neolithic or Agricultural Revolution
	-	*Level Three: Bureaucracy*
B.C.–A.D.		
1 thousand -		
	-	Commercial Revolution
	-	Industrial Revolution
100 years -	-	*Level Four: Technology (Control Revolution)*
	-	Computers
	-	Microprocessors
10 years -		
	-	Genetic programming technology

osphere. There are more bacteria, as separate individuals, than any other type of organism—as many as one hundred million in a single gram of fertile soil. Earth's entire ecosystem depends on the algae, primary producers, through photosynthesis, of the energy that powers all forms of life.

It can even be argued that the prokaryotes, as the earliest of life forms still existing on earth, grabbed up all the planet's choicest ecological niches and thereby forced subsequent species to evolve at greater and more inferior levels of size and complexity. By virtue of having only a single cell, after all, the prokaryotes live free of the nagging burden of gravity, one of the so-called weak forces of nature that affect only objects above a certain ratio of surface area to volume. This argument is facetious, of course, but not in its moral: success in control cannot be judged by the nature of its programming or the complexity of the results, only by the persistence of its organization counter to entropy—the one true test of all living systems. By this standard the prokaryotes—with neither brain nor technology—are at least holding their own.

That the prokaryotes have survived virtually unchanged over several billions of years ought to be evidence enough that material processing systems do not necessarily respond to external changes by evolving greater capacities to control their environments—let alone wholly new levels of programmable structures. If we hope to understand the emergence of the most recent control level, that of programmable technology, we might consider this as another instance of the dynamic at work at earlier levels. If the revolution in control technology did indeed come in response to a crisis in the economy's capability to process throughputs, as suggested in the first chapter, then we must seek to understand this in terms of the more general conditions that have constituted the causes of such change. That the change obviously came not through natural but purposive selection by governments and markets, a direct legacy of the third (bureaucratic) control revolution five millennia earlier, ought not to obscure what might be learned from more general features of control common to all living systems.

Three Problems for Control

Undoubtedly the most difficult aspect of control to appreciate, especially in technologies of our own making, is its evolution across generations of programs. This evolution spans both human generations and, in the contemporary use of the term, generations of technologies like computers as well. As argued in Chapter 1, human society and culture seem often to evolve through changes so gradual as to be all but imperceptible compared to the generational cycles of the individuals through whose lives they unfold.

Like the concepts of programming and decision, the idea of evolution through differential selection was profoundly affected by the growing understanding of information during the Control Revolution, particularly in the critical period from the mid-nineteenth century to the 1920s. Sixty years bridge the first edition of Darwin's *Origin of Species* (1859) and the so-called Modern Synthesis, the work of Alfred Lotka, Vito Volterra, and other mathematical ecologists and geneticists that culminated in Ronald Fisher's *Genetical Theory of Natural Selection* (1930), Sewall Wright's "Evolution in Mendelian Populations" (1931), and J. B. S. Haldane's *Causes of Evolution* (1932). According to E. O. Wilson, "very little theory in the strict sense was created between 1930 and 1960 beyond that already laid down in the 1920s" (1975, p. 63).

In human culture as in all programming, long-term maintenance of control requires the *replication* of programs in new entities: by establishing a new branch office of an organization, for example, or through socialization of our children. If programming is to *evolve* through such replication, however, there must be both a relative *fidelity* to the transference of information and some type of *selection*, whether purposive or nonpurposive. These two factors, fidelity of communication and differential selection of messages, have increasingly drawn the attention of social scientists since World War II in, for example, studies of the diffusion of innovations, a subject that has itself diffused through a wide variety of academic disciplines: geography (Hägerstrand 1952, 1953; Brown 1965), economics (Griliches 1957; Mansfield 1961, 1963), sociology (Coleman et al. 1957), and communications studies (Katz 1957; Rogers 1962, 1971). The most recent interest in cultural diffusion can be found in biology, especially in the work of Cavalli-Sforza, Feldman, and colleagues (1981, 1982) on cultural versus gene-based selection in the quantitative theory of population genetics.

That processes like cultural diffusion and change are too rarely seen in terms of programming is largely the fault of cybernetics, the formal study of control that has unfortunately concentrated almost exclusively on behavior. The example cyberneticists cite most frequently, for example, is the behavior of a simple thermostat, one that can turn a furnace on and off and thereby control temperature by means of feedback from the environment heated. The major shortcoming of such examples is that they draw attention to the control behavior itself—especially to feedback—and therefore away from the more fundamental aspect of control, namely programming. As Ernst Mayr put it:

> It is not the thermostat that determines the temperature of a house, but the person who sets the thermostat . . . Negative feedbacks improve the precision of goal seeking, but they do not determine it. Feedback devices are only executive mechanisms that operate during the translation of a program. Therefore, it places the emphasis on the wrong point to define teleonomic processes in terms of the presence of feedback devices. They are mediators of the program, but as far as the basic principle of goal achievement is concerned, they are of minor consequence. (1976, p. 391)

Overemphasis on control behavior and feedback may be even more misleading than Mayr suggests. By concentrating on behavior per se, cyberneticists tend to overlook two other aspects of control perhaps most obvious in genetic programming but which must prevail in any control system. At a more fundamental level than behavior itself is maintenance of the *mechanism* on which control depends; at a higher level is the programming process on which depends the continuing *evolution* of control. In a thermostatic system, for example, thermostats must be distributed, positioned, and maintained according to some program; only a still higher-level program can decide that such control is desirable and, if so, whether some other technology might not better fulfill the same function.

We encounter these three levels of control involving mechanism, behavior, and the process of programming in all control systems, from DNA to human organization and technology. The levels might be seen to involve three different temporal dimensions or to address three distinct problems for control:

Existence or being, the problem of maintaining organization—even in the absence of external change—counter to entropy

Experience or behaving, the problem of adapting goal-directed processes to variation and change in external conditions

Evolution or becoming, the problem of reprogramming less successful goals and procedures while at the same time preserving more successful ones

To have some mechanism that might be programmed and the matter and energy in which to store its programming is the problem of existence or being. Experience or behaving involves the problem of maintaining control regardless of the inputs to an existing program. To be able to reprogram and thereby adapt to new contingencies is the problem of evolution or becoming. Taken together, the three problems subsume all those of programmed control because they exhaust the three different temporal dimensions of one universal problem: how to maintain organization counter to entropy. Existence solves the problem at any one instant in the sense of differential calculus, that is, as time goes to zero as a limit; behavior solves the problem over the lifetime of programs in a single system; and evolution becomes the essential solution across generations of both programs and systems.

With computers, for example, we recognize the fragility of existence in system crashes, power failures, and accidental erasures of memory, the problems of behavior in exceeding parameters or in infinite looping, and the failure of evolution in the impossibilty of reprogramming read-only memory (ROM) to new specifications. Reprogramming is the general solution to problems of both behavior and evolution in all living systems. At the organic level reprogramming comes in response to changes in selection pressure on the replicating programs (genes) *qua* species that both exist and behave, that is, on the class of all entities that might be controlled by the program (Hamilton 1964; Williams 1966; Dawkins 1976, 1982).

Similar categories have already been applied to human societies, for example, to distinguish structure, process, and history in modern structural-functionalist social theory (Merton 1968). According to the venerable campus cliché, a university has precisely the same three functions: to create knowledge (its evolution or becoming in research), to preserve knowledge (the archival function of existence or being), and to disseminate knowledge (the experience or behavior we know as teaching and learning).

A neurophysiologist (Gerard 1960, p. 255) has suggested the terms *being, behaving,* and *becoming* for virtually the same three aspects of material systems, an application similar although not identical to the one here. We emphasize not the temporal aspects of the three problems but rather their functional implications as essential elements of control, the solution to any one of which can have unanticipated consequences for the other two. Proliferation of a new control technology, for example, can have major effects on the evolution of programming. As we shall see in Chapter 6, the development of mass media and national advertising had precisely such implications for cultural programming by the turn of the century.

An appendix to this chapter, "What Is Life? An Information Perspective," presents my argument that the properties most basic to life itself, in an abstract sense, are subsumed by static and dynamic aspects of the three control functions presented here. Evidence comes from properties of living systems cited by biologists in a half-dozen recent textbooks. The same three functions are shown to correspond to the three fundamental genetic processes of replication, regulation, and reproduction, thereby establishing the control model on the level of programming itself.

In adapting this model to an analysis of the Control Revolution and resulting Information Society, I want to be clear that I do not consider society actually to be an organism, "alive" in some sense apart from the life of its individual members. Considering the misleading inferences that have plagued organismic models since Aristotle's *Politica*, from the pre-Darwinian treatments by Hegel and Comte through those of Herbert Spencer, Oswald Spengler, and the twentieth-century functional theorists, we would do well to avoid such models entirely.

My own analysis of the Control Revolution will be motivated not by any organismic model but by the more modest conviction that all concrete control systems—both individual and collective—must overcome the same fundamental problems if they are to maintain themselves counter to entropy. Living organisms can provide an independent test of our three-dimensional model, by this reasoning, and at the same time afford new insights into the manifest properties of concrete control. This will be indirectly tested in a rapid journey through the first three control revolutions, those that breathed life into an inorganic universe, brought cultural programming to specialized nervous systems, and built formal organizational control in culture-based social structures.

The Origin of Life

We may never know the precise details of what at least for us must be the most sublime of all control revolutions: the origin of life—and of control—on the third planet from the sun, a relatively solitary star on the outskirts of a typical spiral cluster of perhaps a half-trillion stars that we happen to know as the Milky Way, which is itself one of a hundred billion such galaxies. Despite the lack of details, however, we can infer what must have happened, albeit at a level of abstraction above that of particular forms of matter and energy, namely at the level of their organization as *information*.

Because even the simplest organic compounds contain much greater energy than the most complex inorganic ones, one obvious place to seek life's origin is in some naturally occurring energy source. Lightning, volcanoes, and of course the sun's ultraviolet rays have all been cited in various speculations. Whatever the source, little energy would have been needed to knock apart particles in the hydrogen-rich atmosphere that must have formed as the earth condensed out of interstellar gas and dust some 4.6 billion years ago. Laboratory tests using electric sparks discharged in containers of methane, ammonia, and other simple gases have produced several amino acids, basic constituents of the proteins found in all living things. Sometimes the amino acids have actually come chained together like simple proteins and even nucleotides, the ringed molecules that constitute the coded instructions of DNA.

Regardless of how it might have happened, more complex molecules must have arisen from the infusion of energy into the early atmosphere. These molecules would have inevitably mixed and combined with other molecules and atoms, the products dissolving in the oceans to form a kind of organic soup of increasing complexity. No teleonomic entity, not life nor any other, could arise from the haphazard combination, breakdown, and recombination of simple compounds, of course; happenstance alone does not yield purpose. Given sufficient time, however, at least a simple molecule would certainly have appeared whose various parts had the particular property of affinity for elements of the same kind (as in crystals) or for their chemical "negatives" (the case, as we have seen, with the DNA molecule). In either event the new molecule—call it *Replicator A*—would have immediately begun to populate the primordial soup with exact replicas of itself, directly in the case of a crystal, alternately with its negative for a molecule like DNA.

Though not itself alive, Replicator A was nevertheless the earliest ancestor of all life on earth, unless, of course, it has had multiple origins. Replicator A was also the first concrete processing system, one capable of taking simple chemical particles and combining them into exact copies of itself. As for the information processing necessary to *control* such a system, Replicator A had the unprecedented ability to copy the instructions for its own composition, what might be called its own structurally coded *program*. This it replicated over and over again in new molecules, each of which—by virtue of being an exact copy—could replicate *itself* many times in still more copies.

First there was Replicator A, then two such molecules, then four, eight, sixteen, and thirty-two. Because no predators as yet existed, growth of the Replicator A population was likely to approach exponential, limited only by the number of free-floating constituents and perhaps by the disintegrating effects of external energy. After ten such "generations" there would have been 512 Replicator A–type molecules in the primordial ooze, by the twenty-first generation well over a million, by the twenty-fifth almost seventeen billion (assuming, of course, that no molecules broke down in the process).

No known means of copying and recopying can go on indefinitely without error, however. Sooner or later, given sufficient time, generations, and replications, something would have gone awry in the combination of chemical particles by Replicator A–type molecules. Because any resultant mutation would consist mostly if not entirely of parts of proven replicability, it is likely that one of the first such aberrant molecules would also be a replicator, capable of making exact copies of itself from free-floating particles. At this point the earth had two different types of replicating molecules, a population of Replicator A and one variant and potential competitor.

If Replicator A had the superior capability for reproduction, its mutant would not have gained much of a foothold in the primordial ooze and indeed might well have disappeared altogether. In either event, neither Replicator A nor any of its less successful variants could be said to have possessed life, although they stood only a single step from crossing into the realm of the living. This required the additional ability to *reprogram* in response to environmental and selection pressures, which the previous chapter showed to be necessary for the higher control levels of experience and evolution.

Sooner or later, given enough time, a mutant replicator would have appeared that had an edge on Replicator A in the competition for free-floating particles. This could have happened in either of two ways: through a copying error that had immediate natural advantages over Replicator A or as a result of changing conditions on earth, which might have eventually swung selection pressures in favor of an earlier, less successful mutation still surviving in the organic soup. In either case, the relatively more successful type of molecule, call it *Replicator B*, would have begun to spread its program at the expense of Replicator A.

Although no trumpets could have sounded when Replicator B first started to compete successfully with Replicator A, life must be said to have begun at precisely this instant. This was the moment when the total programming for constructing molecules on earth was upgraded the first notch *in the direction of functional advantage*. The teleonomy of evolution was born and with it the final step in the goal-directed programming process we know as *life*.

Several points are instructive here. First, Replicators A and B must be said to be alive not because they manifest all of the properties that biologists usually attribute to life (certainly not metabolism or responsiveness) but rather because they achieved all three levels of control: existence, experience, and evolution. Second, life is not a property of individual molecules, which are themselves unchanged from their preliving condition, but rather of the *population* of all replicating molecules. This is contrary to Helena Curtis's conclusion, as noted in the appendix, that "life" itself does not exist, only individual living organisms. Third, and perhaps most important for our investigation here, life would have begun as the response of concrete open systems to external pressures on their capabilities to process throughputs, in this case the manufacture of replica molecules using free-floating chemical constituents.

It may seem that this explanation is too simple to account for the origin of a process that has culminated four billion years later in the wide diversity of complex organisms that we know today. The argument does gain support from the survival of *viruses*, organisms that are nothing more than replicator molecules encased in protein. Fossil evidence suggests that the appearance of such simple life forms may be relatively straightforward, at least compared to that of more complex forms. Although life itself arose only 200 to 600 million years after the origin of the earth, about as soon as the new planet had cooled sufficiently for life as we know it to survive, another three *billion* years passed before the appearance of even rudimentary specialized organs. This suggests that the origin of life may be less difficult to explain than the second control revolution, one that produced the most generalized living organ of all: the vertebrate brain.

Origin of the Brain

The brain is, like the genome, a generalized information processor, controlled by programming and capable of all of the functions that characterize concrete open systems: input, storage, comparison, decision, output. The evolutionary importance of the brain is that it enabled organisms to "short circuit"—in effect—control by the genome. Compared to genetic control, control by the brain has certain obvious advantages: it can be based on a much greater amount of programming, respond much more quickly to environmental conditions, control social communication with other organisms, and be reprogrammed continuously. Genetical evolution continues in slow response to selection pressures, but the evolution of culture—programming of the brain— has been relatively quicker and more flexible, often changing many times within the lifetime of a single organism.

As biologist John Bonner (1980) argues, the first branch in the evolutionary process leading eventually to the brain occurred in the bacteria, the earliest known living organisms. Several billion years ago, before the appearance of other specialized organs, some bacteria evolved *motility*, the ability to move about by rotating their flagella at the base (Berg and Anderson 1973; Silverman and Simon 1974; Berg 1975). Once motile, bacteria evolved *chemotaxis*, attraction to some substances and avoidance of others.

That motility and chemotaxis characterize a rudimentary information-processing system has become clear now that bacteriologists have begun to isolate the various components of such systems. Specific proteins at the cell surface of a bacterium cell, for example, serve as simple but quite effective *input devices*. Twenty such receptor proteins have been identified for one species; they are specific for twelve chemotactic attractants and eight repellents (Adler 1976). Bacteria have even been found to possess short-term *memory*, to be able to *compare* inputs at different times, and to make *decisions* based on the integration of simultaneous inputs from attractants and repellents (MacNab and Koshland 1972).

As with the origin of DNA, molecular programming and indeed life itself, in other words, the generalized information processor that became the brain also arose as the response of a concrete open system— the genome—to external pressures on its capabilities to process throughputs. This pressure first appeared as the relative advantage of responding to environmental conditions more quickly than was possible through differential reproduction across generations of organisms. Once this lower level of response had produced motility, which it did almost immediately, selection pressures shifted from the replication of genetic programming to information processing, reprograming, and communication by organisms during their own lifetimes. In

Bonner's words, "The brain, the whole neuromuscular system, and therefore the capacity for culture are foreshadowed in the most rudimentary organisms and known to have originated in the earliest of all living forms" (1980, p. 56).

True nervous systems appeared about six hundred million years ago; virtually all multicelled animals have at least rudimentary forms. They are built up of *neurons*, specialized cells that transmit information by means of electrical impulses. Although chemicals still play a major role in neural communication, especially at the junctions or *synapses* at which one neuron influences another, nervous systems can be described quantitatively by means of standard variables from electronics (McLennan 1970). The parallels between this evolutionary development and the emergence of electronics as the basis of the nineteenth-century Control Revolution have not been lost on scientists themselves. Consider astronomer Sir Fred Hoyle, for example, who writes of evolution:

> I am overwhelmingly impressed by the way in which chemistry has gradually given way to electronics. It is not unreasonable to describe the first living creatures as entirely chemical in character. Although electrochemical processes are important in plants, organized electronics, in the sense of data processing, does not enter or operate in the plant world. But primitive electronics begins to assume importance as soon as we have a creature that moves around . . . The first electronic systems possessed by primitive animals were essentially guidance systems, analogous logically to sonar or radar.   (1964, pp. 24–25)

The beginning of increasingly centralized control of the actions and reactions of animals in response to their environments marked the second evolutionary branch leading eventually to the brain. Why should such centralization have been evolutionarily advantageous for some creatures? One possibility is pressure to sense and respond to the environment in quicker or more complex ways; another quite different explanation is that centralization results only incidentally from the increasing size of organisms, which requires a corresponding increase in the number of neurons needed to control them. In either case, centralization is relatively more efficient not only because less connecting fiber is necessary but also because nerve impulses travel over shorter routes and thereby the speed of responses is increased.

For the lower invertebrates, in which learning is rudimentary at best, centralization of control seems little more than an extension of motility functions that arose in the bacteria. The more primitive nervous systems appear wholly occupied by the processing of sensory inputs from the environment and their translation into the appropriate outputs to control muscular movements. Higher invertebrates like the insects, however, have both short-term and long-term memory, a fact well documented for fruit flies (Quinn and Dudai 1976) and honey bees (Menzel, Erber, and Masuhr 1974), among others. Social insects have remarkable abilities to memorize mazes, the rate of learning in ants being only slightly inferior to that of rats (Schneirla 1953).

Evolution of the larger and more complex brain and spinal cord of the vertebrates made possible an entirely new form of learning—the purposive imitation of other individuals that we call *cultural* programming. Learning by imitation arose perhaps one hundred million years ago and today distinguishes higher vertebrates, the birds and mammals, some twelve thousand species. Learning through intentional *teaching* began roughly fifty million years ago and can today be found in about two dozen species: wolves and dogs, lions, elephants, and the anthropoid apes, including, of course, *Homo sapiens* (Lumsden and Wilson 1981, p. 4).

At some point in vertebrate evolution, the brain surpassed even the remarkable capability of DNA to store and process information (Bonner 1980, p. 164). Brains reigned as the highest-level information processors on earth for about one hundred million years, then began themselves to be challenged by formal bureaucracy. We are just now approaching the time when the largest bureaucracies will be exceeded in their information-processing capabilities by the fourth level of control: microprocessing and computing systems. Computers would not exist independently of social organization, of course, no more than bureaucracy would apart from individual brains, or brains from genomes. Each new level of control presupposes all previous ones. We might therefore extend to the levels of organizational and electronic processing the symbiotic relationship that Bonner finds between genome and brain:

> The genome has produced a monster [the brain] that can respond to selective forces with lightning speed; yet it is theoretically a benevolent monster in that it responds to the same selective forces as the genome, but does so more rapidly and efficiently. It is in this sense that I would call the genome and the brain symbiants. Each depends upon the other for benefits . . . The brain needs the genome, and the genome gains by the agility of the brain to adapt quickly . . . The brain makes for a par-

ticularly efficient survival machine, and in this way helps in perpetuating the genes.   (1980, p. 34)

The survival of countless organisms without brains indicates that the organ is hardly necessary—and probably not even advantageous—for the exploitation of many ecological niches. Like all organs, brains have their costs, and these may outweigh their benefits in many environments. For niches in which spatial location and other environmental variables are important, however, the capacity to process information within an organism's lifetime has certain advantages over processing across generations. Not the least of such selection pressures leading eventually to the brain favored the reciprocal communication of a cooperative nature that distinguishes society itself.

Brain, Culture, and Society

Not all species that are social have brains and, conversely, not all species with brains are social. The former include many species of so-called *colonial* microorganisms and invertebrates in which intercommunication is so well developed that the distinction between multicellularity and society breaks down. Some species of sponges, for example, can be disaggregated by passage through a fine mesh, after which the separated cells move independently, regroup in small clusters, differentiate, and regenerate a new sponge in its original form; neighboring sponges sometimes fuse and reorganize as a new colony (Hartman and Reiswig 1973; Simpson 1973).

Examples of relatively big-brained animals that are not social occur throughout the class Mammalia. Within the same family or even genus, species often range from solitary to fully social. Of the forty-seven species of bats, for example, seven are solitary except for copulation and mother-offspring associations, and three live in truly social groups—multimale and multifemale—throughout the year (Bradbury 1977, pp. 22–26). Even the eleven species of anthropoid apes span the entire range from the solitary orangutan, through parental families (gibbon, siamang) and age-graded troops (gorilla), to the fully social chimpanzees and, of course, *Homo sapiens* (Eisenberg, Muckenhirn, and Rudran 1972). Sometimes wide variation can be found within a single species as, for example, in the gray whale, for which sociality changes with the seasons.

The brain seems to have evolved, in other words, for reasons that had nothing to do with sociality per se, and society emerged for reasons that had nothing to do with a brain: many species have one or the other but not both. Of the million or more animal species that currently exist, several tens of thousands are social (Boorman and Levitt 1980, pp. 2–3) and about twelve thousand (the birds and mammals) have relatively well-developed brains. Brains and sociality coincide in a few thousand species, however, perhaps ten times the overlap of the two factors expected by chance alone.

Despite imperfect correlation, a combination of both sociality and brain *does* appear to be essential to culture. All protocultural organisms capable of learning through imitation have relatively well-developed brains, and true culture is confined to the bigger-brained mammals. Bonner's flat assertion that "social existence is a necessary but not sufficient basis for culture" (1980, p. 76) appears to have but partial exceptions: the orangutan only occasionally meets to form secondary groupings (Rodman 1973), and the jackal, siamang, and gibbons (six species), all of which teach their young, are social only to the extent of forming parental families. On the other hand, numerous species of cold-blooded vertebrates—including reptiles, frogs, and fish—have evolved both brains and social organization but not even the most rudimentary culture. The only clear relationship among the three variables, therefore, appears to be that relatively big brains and sociality are *both* necessary but not sufficient—even in combination—for culture based on purposive teaching and learning to emerge.

Because brains and sociality are the two most obvious variables associated with culture, it seems plausible that cultural programming arose out of the need to process specifically *social* information. This would suggest one reason for the rise of cultural over genetic programming: coevolution of the complexity of inputs and processors. Social existence, depending as it does on reciprocal communication of a cooperative nature, exerts pressure for greater information-processing capabilities. Better processors, in turn, permit social programming, processing, and communication of greater complexity, which sustain the pressure for still greater processing capabilities. Only occasionally has this process produced true culture, perhaps, but cultural programming has never emerged in its absence.

Indeed, the concept of culture subsumes all of the complex programming, decision making, and communication necessary for societies composed of big-brained and hence relatively autonomous individuals. The term has been variously applied to processing structures (social organization and technology), to programs (specific learned rules of behavior), to preprocessing of inputs and outputs (selective perception and symbolization), and to other informational activities (including, as we have seen, imitation, teaching, and learning). These various meanings reinforce the conclusion that culture involves the processing of specifically *social* information—information about other individuals of the same species.

If so, several conclusions might be drawn about the nature of this information: it would be associated with *particular* individuals, advantage would lie in processing *quickly*, and outputs would be relatively *flexible*. Why? Because these are characteristics that, if adaptive, would cause culture to be selected over genetic programming, the relative disadvantages of which increase directly with pressure for speed and flexibility and with the utility of information—like past experiences with particular individuals—generated during a lifetime.

As we have already seen, for example, it is apparently adaptive for the young of some species to learn to distinguish their own parents from other adults, particularly among birds that nest close together. The best solution available through genetic programming is that of *imprinting*, learning once and irreversibly during a critical period of infancy, usually during the first five days after hatching (shortly before hatching in some species). Such open genetic programming serves well for even relatively complex functions, as in the learning of patterns of star rotations about the North Star by nestlings of the indigo bunting (Emlen 1975).

When it becomes adaptive to learn to recognize many individuals over the course of a lifetime, however, and to process, store, and update information about them, genetic programming cannot serve as well as cultural programming—if at all. Certainly cultural programming is essential to society based on cooperation among many individuals *qua* individuals. Although only a few cases of truly personal recognition have been documented in the invertebrates it is widespread and possibly universal among the birds and mammals (Wilson 1975, p. 382), the two vertebrate orders with the most developed forms of social organization.

This raises the question of how cooperation among individuals might be explained within the paradigm of evolution through natural selection. As Darwin originally formulated the theory, evolution proceeds through the competition of individual organisms and the greater chance that the fitter will survive to reproduction. What, then, is the basis of cooperation among these same organisms? Here evolutionary theory confronts the Hobbesian question: how could society have ever emerged from the so-called "state of nature"—characterized by Hobbes (1651, pt. I) as "a condition of war of all against all"?

Until the mid-1960s few biologists attempted to characterize the circumstances under which organic evolution might favor sociality; most dismissed cooperative behavior as adaptive at the higher levels of population or species. Even more recent writers have resorted to non-Darwinian explanations: individual sacrifice for the sake of the group (Wynne-Edwards 1962; Ardrey 1970), relationships that exist for the good of the species (Lorenz 1963), adaptation of the entire earth as a single organism (Lovelock 1979). More systematic examinations (Williams 1966; Hamilton 1975), however, have found little basis for selection at these higher levels; the process tends to be weak even at the group level. Today most biologists accept cooperation and other social behavior—on terms wholly compatible with Darwinian theory— as derived from combinations of three general types of selection:

1. *Group selection*. In those relatively rare instances when a population is divided into noncommunicating "islands" or *demes*, the survival of which depends on each deme's genetic or cultural programming, selection will favor any programs, including those for cooperation, that increase a deme's chances—even at the expense of individual members (Levins 1970). Here selection results not from competition among individuals but among demes.

2. *Kin selection*. Because relatives share some fraction of identical genes, programs that have greater benefits for kin than costs to the individual will tend to spread through a population. Programs for cooperation will be selected, but only if the behavior remains relatively confined within groups that share many genes. Unlike group-selected cooperative behavior, kin-selected cooperation depends on the information-processing capabilities of individual organisms: discrimination of relatives is essential. This need not involve a processor as elaborate as a brain, however; in species with little dispersal, for example, cooperation might simply be directed toward *all* others with whom an individual has been associated from birth, as in litters of puppies (Scott and Fuller 1965).

3. *Reciprocity selection*. Programming for altruistic behavior that benefits the recipient more than it costs the donor will be selected, if the same behavior is likely to be reciprocated, because the *average* fitness of all individuals with the program is thereby increased. Because it will be to each individual's advantage to "cheat" by accepting benefits

without incurring the costs of reciprocity, however, this form of cooperation requires much greater information-processing capability than does behavior associated with kin selection. Individuals must be recognized *qua* individuals, not merely as members of some group—like birth associates—with relatively high proportions of kin. Unlike kin-based cooperation, reciprocity also requires individual long-term memory or some social form of information storage and retrieval (branding of transgressors, credit ratings, etc.).

Numerous behavioral studies suggest that this latter type of selection may have determined the social evolution of large carnivores and primates, precisely the species distinguished by culture. Likely products of reciprocity selection include cooperative hunting (Scott 1967; Schaller 1972), multimale defense (Kummer 1971), and intragroup coalitions (Packer 1977). Here, then, in the relatively complex forms of organization that characterize the social primates and early hominids, lies the likely pressure for genetic programming to be supplanted by culture—the second level of control to appear on earth.

The Threshold of Sociality

One lingering problem with explanations of sociality based on selection for reciprocity is that, below a certain level of frequency in a population—because the behavior is not reciprocated often enough—programming for cooperation remains disadvantageous and hence will not spread. To restate the problem more formally, the trade-off between benefits of mutual cooperation and costs to the initiator when cooperative acts are not reciprocated might be seen as a *threshold frequency*, above which programming for cooperation spreads throughout a population but below which it will be eliminated. No matter how great the advantages of cooperation for society at large, it is difficult to see how such programming, either genetic or cultural, could have spread much beyond its earliest possessors, to whom it incurred no benefits—in an otherwise asocial society—and considerable costs.

The problem of how cultural programming might break through this "threshold of sociality," what for two centuries before Darwin philosphers had felt compelled to explain with some form of *social contract* (Hobbes 1651; Locke 1690; Rousseau 1762), remains for some social biologists an unresolved question (Wilson 1975, p. 120; Elster 1979). Despite several thousand empirical studies of the diffusion of innovations (Rogers 1962, 1971, 1983), no social scientist seems even to have attempted to address the difficult problem: adoption of innovations that are not beneficial until possessed by a certain minimal proportion (threshold frequency) of the population. As vaudeville comedians have posed the question, "What fool bought the first telephone when there was no one to call?"

A general set of methods for addressing such questions and thereby attempting to account for the origin of cooperative behavior and society based on culture derives from one of the more important applications of mathematics to decision developed during the Control Revolution, namely *game theory* (von Neumann 1928; von Neumann and Morgenstern 1944). Extension of this work to evolutionary theory, especially in the concept of the *evolutionarily stable strategy* or ESS, has been pioneered over the past decade by the British mathematicial biologist John Maynard Smith (1972, 1974; Maynard Smith and Price 1973), although the essential idea appears in earlier work (MacArthur 1965; Hamilton 1967). In the opinion of ethologist Richard Dawkins (1976, p. 90), "We may come to look back on the invention of the ESS concept as one of the most important advances in evolutionary theory since Darwin."

In the first application of game theory to the evolution of behavior, Maynard Smith postulated five alternative strategies for conflict between organisms (1972, p. 19; Maynard Smith and Price 1973, p. 16), which for each participant must involve either conventional fighting, escalation, or retreat. The five formal strategies can be expressed in words, although in somewhat simplified form:

1. *Hawk*. Begin at escalated level, continue until victorious; retreat only if injured.

2. *Bully*. Begin at escalated level, continue as in Hawk except that if opponent escalates, retreat.

3. *Retaliator*. Begin conventionally, escalate only if opponent escalates; otherwise continue for a predetermined time or until injured.

4. *Prober*. Begin conventionally, escalate only if opponent also fights conventionally; if opponent escalates, retreat.

5. *Mouse*. Begin conventionally, continue—never escalating—for a predetermined time, then retreat; retreat immediately if opponent escalates or appears stronger.

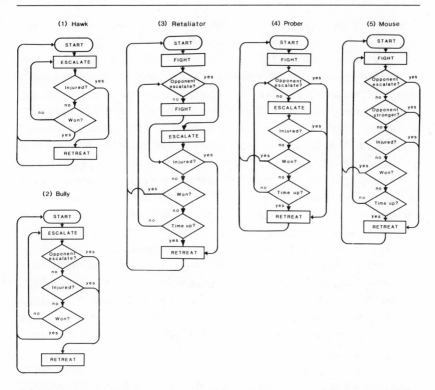

Figure 3.1. Flow charts of algorithms for five alternative strategies for conflict behavior. (Strategies adapted from Maynard Smith 1972, p. 19, and Maynard Smith and Price 1973, p. 26.)

Not surprisingly, considering that game theory has intellectual roots in the concepts of decision and control (as we saw in the previous chapter), what Maynard Smith means by *strategy* might be subsumed by the concept of *program*. The latter is the better term in the view of Dawkins (1982, p. 118), who finds *strategy* likely to be misunderstood for several reasons, not the least of which is the suggestion "that the animal is making rational decisions, attempting to outwit an equally rational opponent" (1980, p. 332). That Maynard Smith's five strategies for conflict are in fact programs—without assumptions of rationality— is demonstrated by Figure 3.1, which gives a simplified flow chart for each (input stages have been omitted to streamline the diagrams and facilitate their comparison). As in the flow chart for Maxwell's Demon (Figure 2.1), rectangles represent outputs, here one of the three available behaviors: conventional fighting, escalation, or retreat.

Although game strategies cannot be isolated in material programs in the way that, say, genetic programs can be traced to specific sections of chromosome, such strategies must be considered no less real—unless we are to embrace vitalism or some other metaphysical explanation of control. The Bully, Retaliator, or any other game strategy that we might devise, in other words, has the same relationship to concrete genetic or cultural programming as, for example, the programs that Mendel postulated for yellow hue and wrinkled skin have to the strings of nucleotides that actually encode such programming in garden peas. Just as Mendel could make correct inferences about actual programs in total ignorance of their material basis, mathematical biologists might do the same—ignorance of the concrete nature of an ESS in no way diminishes the possiblity of its existence. What cannot be determined by game theory alone, of course, is the relative degree to which a strategy is genetically, culturally, or structurally programmed. Game theory remains invaluable to the study of cultural programming and control, however, for roughly the same reason that classical genetic theory was valuable in the development of molecular biology: inferences from inputs and outputs usually aid in the unraveling of unknown programming.

With the same simple sequence of decisions, any programmable information processor might be made to behave—if only through simulation—as if controlled by one or more of the five strategies; pairs of the strategies could thus be made to "fight" each other to a decision or draw. As genetic or cultural programs, the same strategies might control actual animals that engage in conflict, in which case they would compete not only in individual contests but for limited space in the gene pool (in the case of genetic programming) or for space in the population of brains (in the case of cultural programming). Using computer simulation, Maynard Smith addressed the question of which program would win this competition, that is, spread throughout a population at the expense of other programs.

Because successful programs will become increasingly plentiful, either through genetic selection or adoption as cultural innovations, the one that prevails (if indeed there is only one) must be especially good at competing with many copies of itself, at least relative to other types of programs—including any new mutations or innovations. Such a program is the evolutionarily stable strategy or ESS, rigorously defined by Maynard Smith (1974) but essentially the program that, when controlling most members of a population, cannot be supplanted by any alternative program. In the case of the five fighting strategies, for example, Maynard Smith found that the third—the so-called Retaliator strategy—is an ESS: once predominant in a population, none of the other four strategies can do better. In some applications the ESS has turned out to be not a pure strategy but a ratio or relative frequency of two or more, what might itself be considered a still more complex program.

As applied to the second level of control, that of brains and culture, ESS methods might be used to address three general questions concerning the nature of society:

Its origin, the Hobbesian question, what we have already seen to involve the initial viability in an asocial population of programming for cooperative behavior

Its spread, which depends on the competitiveness of cooperative programs relative to a wide variety of noncooperative ones

Its stability once established, the question of whether the successful program or programs for cooperative behavior constitute an ESS

Political scientist Robert Axelrod, in collaboration with biologist William Hamilton, has made several recent contributions to our understanding of cooperation and its evolution, building on the earlier work of Trivers (1971), Boorman and Levitt (1980), Chase (1980), and Fagen (1980). To see what type of program might thrive in an environment dominated by noncooperating individuals, Axelrod (1980a) conducted a computer tournament for the Prisoner's Dilemma, an elegant application of game theory to the problem of achieving mutual cooperation (Rapoport and Chammah 1965). In the two-player Prisoner's Dilemma game, either player can cooperate or defect, with the selfish choice yielding the higher payoff (here in fitness) for both players. If both do defect, however, both do worse than if they had cooperated—hence the "dilemma" for each player.

Of the fourteen strategies (programs) game theorists in five disciplines submitted to Axelrod's tournament, the highest average score went to the simplest strategy of all: *Tit-for-Tat.* This program cooperated on the first move, and on all succeeding ones simply did whatever the other player had done—cooperate or defect—on the immediately prior move. In other words, *Tit-for-Tat* established cooperation—after the initial encounter—on the principle of reciprocity, an essential basis for human culture, as we shall see. After Axelrod circulated the tournament results and solicited entries for a second round, he received sixty-two programs—some quite intricate—from six countries and eight academic disciplines, from mathematics and computer science to economics, political science, and evolutionary biology. *Tit-for-Tat*, resubmitted by game theorist Anatol Rapoport, won again (Axelrod 1980b).

Through analysis of some three million moves made in the second round of the tournament, Axelrod determined three essential features of an evolutionarily successful program for cooperation: begin cooperatively, retaliate immediately, forgive just as fast. Over simulated evolutionary time, using these three rules, *Tit-for-Tat* continued to do well even as less successful programs became extinct, finally displacing all others and stabilizing in the population (Axelrod 1980b). Axelrod and Hamilton (1981) have also shown that a population programmed exclusively by *Tit-for-Tat* can resist invasion by any possible mutant program, given one essential feature that characterizes small societies but not necessarily large ones: that individuals who interact have sufficiently large probability of meeting again.

Further elaboration of these results (Axelrod and Hamilton 1981; Axelrod 1984) has begun to suggest how altruism and cooperative behavior might have evolved even at the microbial level, possibly explaining, for example, the existence of both chronic and acute phases in many diseases and of chromosomal nondisjunctions like Down's syndrome. Certainly findings at this level ought to dispel any lingering suspicion that cooperation derives from rational decisions or from the individual's attempt to outwit the collectivity. The line of investigation begun by von Neumann and running through the work of Maynard Smith, Axelrod, and their collaborators seems also to have gotten theory securely across the thresholds of sociality and programmed culture.

Sociality, Language, and Control

Upon crossing the threshold of sociality, as we have seen, a population encounters increasing dependence—if it is to maintain this relatively advantageous situation—on the reciprocal communication that characterizes all social structure. Reciprocal communication, in turn, gives increased advantage to the various information-processing capabilities that we generally subsume under culture: the preprocessing of selective perception and symbolization, the programming of learned rules and formal law, and the processing structures of formal organization.

Relative advantage alone cannot be the mother of invention in organic evolution, of course, but tens of thousands of animal species do manage to sustain some degree of sociality, and some twelve thousand receive cultural programming through imitation, about two dozen via intentional teaching, and one—*Homo sapiens*—through perhaps the most wondrous of all forms of communication: language. Although no claim to the superiority of our own species can be based on this capability (a million or more animal species alone appear to do quite well without it), language certainly does distinguish us most sharply from all other life, at least on earth, and enables us to communicate with one another in many ways unique among living things, for example, the way I am communicating with you now.

Language must be considered at least in part a technology, since it is an artifact that we can modify somewhat through our own innovation, but it also appears to reflect—and to be constrained by—processing capabilities innate to the human brain. Linguist Noam Chomsky (1965) finds the brain innately structured to string words together in certain arrangements and not others, for example, a "deep grammar" that permits far more rapid acquisition of language than would be possible by simple learning. Despite the faith of behaviorist psychologists like B. F. Skinner (1938) that all types of learning might be explained by a few simple reflexes, computer simulations have established that not nearly enough time exists in childhood to learn English sentences through such operant conditioning (Miller et al. 1960). Rather than passively being programmed to speak, young children seem actively to pursue language by incessantly babbling, inventing and testing new words, and acquiring grammatical rules rapidly and in predictable sequence (Brown 1973), basic steps of learning which Lenneberg (1967) has shown to be inherited.

Because the human brain developed at least in part in interaction with the use of tools and other technologies including language, the processor itself might be seen as an artifact of human invention or even of language. Whorf (1956) suggests that the artifactual languages used by any system determine many aspects of its structure and processes (Miller 1978, p. 34). If brain and language have indeed coevolved so that each continues to reflect essential features of the other, it would be impossible to distinguish them even as to the hardware versus software or the processor versus programming of human culture (much the same difficulty—as we shall see in the next section—surrounds the concept of law).

Language distinguishes human beings from all other living things in several ways. Among the diversity of what ethologists call "the behavior of communicating" (Smith 1977), language is unique in the amount of information that it can encode: the number of sentences in any language is infinite, and indeed—with the use of conjunctions— even a single sentence can be infinitely long. Language is also unique in the extent to which it must be learned, although a few species of birds must also learn certain features of their songs and other vocalizations (Hooker 1968; Konishi and Nottebohm 1969). Because such open programming requires greatly increased capacity for storing information as well as mechanisms—most commonly parental care—to ensure the appropriate inputs, it is obviously costly to evolve. Indeed, ethologist Martin Moynihan (1976) argues that the increased storage capacity required to learn social signaling may have been preadaptive in our own species for the evolution of language as a richly complex means of communication.

Unlike all other animal communication whose components differ from each other in message but not in type, all languages distinguish lexical categories: nouns and verbs (Sapir 1921), the "conjunctionalizations" that link them, and the "adnominal" and "adverbial" classes that modify them (Martin 1964). These unique distinctions of signals by type make possible the grammatical sequences that structure diverse kinds of relationships among them. Because all other animal communication lacks such distinctions, none could possibly be as complex as language. Although natural human grammars can also be complex, grammatical sequencing can develop without lexical distinctions, even if the rules remain simple. This suggests to Smith (1977, p. 421) that "some formalization of sequences may have been a relatively early step in linguistic evolution." Given the limits of short-term memory processes (Miller 1956a), hierarchical structuring of grammatical formalizations probably reflects a need to store information in organized "chunks," compounding numerous small units in a few more inclusive ones. "From a recipient's perspective," according to Smith (1977, p. 421), "a formalized sequence probably serves to make the temporal and quantitative relationships among displays [discrete social signaling acts] maximally apparent."

Also unique among animal communication is language's capacity to refer to other times and places ("displacement") and to generate totally new sentences that can nevertheless be understood ("productivity"). Through *metonymy*, use of a part to represent a whole, and *metaphor*, use of a known to represent an unknown, language has even transcended the need for all referents to be known in advance to receivers of a communication. According to linguist Roman Jakobson (1960), metonymy and metaphor constitute the two fundamental means to communicate meaning, the former through substitution by contiguity ("syntagmatically," the mode of, say, a realistic novel), the latter through substitution by similarity ("paradigmatically," the mode characteristic of most poetry).

Because language evolved in a species that already had formalized signal behavior comparable to other primates, speech did not replace the other signal systems so much as it coevolved with them. Speech still depends heavily on nonlanguage communication to provide contextual information, a fact rediscovered many times for the telegraph, the telephone, and most recently computer networks (Hiltz and Turoff 1978). It also explains why speech still bears the mark of the simpler signaling systems, so that "how are you," for example, is not ordinarily a query about health but a greeting ritual (Goffman 1967, 1971; Kendon and Ferber 1973). At the same time, however, speech has evolved in ways that help to salvage social interactions in which other signal systems might be so ambiguous as to present what Goffman (1971) has termed "compromising situations."

More than merely a means of communication, language constitutes what Berger and Berger (1976, chap. 4) call "the social institution above all others." Like kinship, religion, law, and other highly systematized and enduring processes that distinguish human societies, language constrains and controls each individual, translating the collective flood of personal sensation and experience into a classified, ordered, shared reality. Although each member of a species with a partially open signaling system can construct what ethologist Julian Huxley (1966, p. 259) terms "private ritualizations," communication will be effective only to the extent that each approximates a "socially standardized concept" (Carroll 1964; Smith 1977, pp. 76–77). Language, although it does not function apart from other cultural systems, constitutes the primary institution through which collective reality is continuously renegotiated and reproduced and thereby makes possible the organization and mobilization of individual experience. Much as sociologist William Graham Sumner (1906) said of the institution of common law, language is not enacted but *crescive;* it develops unconsciously out of collective action.

As we have already seen for the difference between language and brain, the distinction between processor and programming is difficult to make at any level of cultural institutionalization. Indeed, American sociologists have argued for a half-century since Sumner about whether institutions include formally organized processors like the family or are restricted to more abstract systems of interacting roles (Parsons 1960, p. 171), to still more systematized and transcendent processes like language (MacIver 1937; Bierstedt 1963; Ginsberg 1965), or to more explicitly controlling programming like norms (Davis 1949). "Always in human society there is what may be called a double reality," Kingsley Davis (1949, p. 52) notes, "on the one hand a normative system embodying what *ought* to be, and on the other a factual order embodying what *is*" (emphasis in original).

Bridging this gap between *is* and *ought* and reconciling the various approaches to institutions is the fact that language and derivative cultural systems exist apart from any one individual and in that sense, at least, serve to constrain and control all individuals. Such control, like any other, will necessarily involve all levels and components of concrete open processing systems: preprocessing (the Whorf-Sapir view of language), programs (norms), the process of programming (socialization), and processors themselves (formal social structures). This range is perhaps best covered, among American sociologists, by Erving Goffman's view of institutions as extending from what he calls "total institutions" (like prisons and mental hospitals) to the seemingly "natural" and timeless institutions of language, ritualized behavior, and similar cultural programming. Although total institutions appear to exert greatest control over the individual, Goffman (1961) notes that "every institution has encompassing tendencies."

As we might expect, the gap between the social *is* and the cultural *ought*—between process and program—is best understood at the highest level: language. During the first stirrings of the Control Revolution in the 1870s and early 1880s a number of scattered investigators and theorists in both mathematics and linguistics discovered "the conjugate notions of invariance and variation," in the words of Jakobson (1972, p. 39), what he calls "a remarkable parallelism in time and essence with the ideas that underlie the development of modern mathematics and physics." In the United States the ideas were best represented by the work of Charles Peirce (1839–1914), whose 1878 argument that the meaning of an idea lies in the response to which it leads—what he called "pragmatism"—inspired social behaviorism through William James and Charles Horton Cooley and, through John Dewey and George Herbert Mead, modern symbolic interactionism. Peirce saw the relationship between social process and cultural program as what he called *semiosis,* "an influence which is, or involves, a cooperation of three subjects, such as a sign, its object and its interpretant, this tri-relative influence not being in any way resolvable into actions between pairs" (Peirce 1931, p. 484).

At about the same time, the 1870s and early 1880s, the great Swiss linguist Ferdinand de Saussure (1857–1913) developed much the same view. He divided *signs* into *signifier* (perceived form) and *signified* (mental construct) and distinguished between what he called *langue,* the abstract and infinite potential of any natural language, and *parole,* its concrete and finite manifestations in particular speech (Saussure 1916). The latter distinction might be drawn, for example, between a particular algorithm for control, on the one hand, and its realization in a particular program or processor on the other. Saussure's aim to found what he called "a science that studies the life of signs within society" suggested that the programming of language and culture might be studied in its own right, despite the gap between it and Davis's "factual order embodying what is," a position identified with the "new structuralism" of the 1960s (Piaget 1970). According to the molecular biologist Gunther Stent,

> The emergence of structuralism represents the overthrow of "positivism" (and its psychological counterpart "behaviorism") that held sway since the late 19th century and marks a return to Immanuel Kant's late-18th-century critique of pure reason. Structuralism admits, as positivism does not, the existence of innate ideas, or of knowledge without learning. Furthermore, structuralism recognizes that information about the world enters the mind not as raw data but as highly abstract structures that are the result of a preconscious set of step-by-step transformations of the sensory input. Each transformation step involves the selective destruction of information, according to a program that preexists in the brain. Any set of primary sense data becomes meaningful only after a series of such operations performed on it has transformed the data set into a pattern that matches a preexisting mental structure [what we here call "preprocessing"]. These conclusions of structuralist philosophy were reached entirely from the study of human behavior without recourse to physiological observations. As experimental work shows, however, the manner in which sensory input into the retina is processed along the visual pathway corresponds exactly to the structuralist tenets.   (1972, pp. 24–25)

Most extensive application to cultural control of Saussure's "science of the life of signs within society" and Jakobson's binary phonological code has been made by the French structural anthropologist Claude Lévi-Strauss (1958, 1962). Saussure's *langue* has informed similar concepts throughout the behavioral sciences: cognitive psychology's internalized frameworks of rules and plans (Miller et al. 1960), the linguistic "competence" (as opposed to *parole* or "performance") of Chomsky (1968), and the heuristic strategies of artificial intelligence (Newell and Simon 1972). Both Saussure and Peirce are today considered cofounders of *semiotics* (Eco 1976, pp. 14–16), the study of the social production of meaning—and hence the possibility that sign systems might program social control—popularized in the 1960s by the French literary critic Roland Barthes (1957, 1968, 1977), himself a prominent member of the French structuralist school including Lévi-Strauss, psychoanalyst Jacques Lacan (1966), Marxist theorist Louis Althusser (1965, 1971), and essayist Michel Foucault (1966).

Attention to symbols rich in meaning shifts the essence of control from communication to the prior programming of its receiver. For the engineer, communication involves a quantifiable amount of information that "flows" from A to B; for the semiotician, A communicates by "pointing" (by whatever means) to information already stored at B. Clearly these are two different ways to view the same phenomenon: pointing must involve information flow (symbols must be sensed before they communicate), and the engineering model requires "decoding" of information by its receiver for communication to occur (Shannon 1948). Nevertheless, the possibility that a symbol might be simple in information content but rich in meaning or part of a highly complex system of meaning like a grammar gives human communication transcendent power. Although the engineering model better captures what occurs after one routinely asks an authority for the German name for Oswiecim, Poland, for example, the semiotic model probably better captures what will occur after you read the answer here: Auschwitz. Because of the prior programming that is meaning, some nine-letter words can "communicate" more than others of the same length, a fact not captured by any measure of information flow alone.

Before Lévi-Strauss began to concentrate on the explication of mythology, he devoted nearly twenty years to the simultaneous application of cybernetic and semiotic models to anthropology more generally (1958, pp. 55–97, 277–323), perhaps the most systematic effort to date to reconcile the engineering and symbolic approaches to social control. A related effort to combine symbolic and processing models began at the University of Chicago in the late 1920s and 1930s. Linguistic anthropologists Edward Sapir and Benjamin Whorf developed the idea that humans build up their sense of reality through the unconscious workings of their language to organize perception, a symbolic processual model that recalls the Humboldtian idea of language as *energeia* (Sapir 1949; Whorf 1956). Meanwhile, Peirce's irresolvable "tri-relative influence" inspired—through Dewey—Mead's *Mind, Self, and Society* (1934), which argues that shared symbols (Peirce's "signs") enable individuals to take the perspective or "role of the other" and thereby develop a "generalized other" (Peirce's "object") or internalized social conscience as well as self and mind (Peirce's "interpretants"). This view, similar to that of the structuralists in that it posits an innate "I" in contrast to the objective social self or "me," survives in modern symbolic interactionism (Blumer 1937, 1969; Swanson 1968).

Social Control: Programming or Process?

The distinction between programming and process, blurred among brain, mind, and language and between other symbolic cultural systems and the observable behavior they generate, raises a new question: what is the nature of order and control in societies more elaborate than those based largely on reciprocal altruism? Does control depend on prior programming—encoded in each individual either genetically or as learned cultural norms—that gives social life the appearance of being governed by consensual rules? Or are these apparent rules merely regularities in the interactions of self-seeking individuals, who might cooperate for a time through enlightened self-interest (as a stable strategy) but who would violate any rule the instant it no longer furthered their own ends?

Obviously this must be a fundamental question for social theory. It has echoes in the arguments of "formalists" versus "substantivists" in economic anthropology and in the split between "normative" and "interpretive" approaches to sociology. Some social scientists (Dahrendorf 1959; Giddens 1979) find similar divergence of explanation across the entire body of Western social theory. Talcott Parsons (1961) traced the split to differences between French Catholicism and British Protestantism beginning in the seventeenth century, the former being more "collectivist" and presumably more inclined toward explanations involving normative programming, the latter more "individualistic" and hence predisposed toward processual models. Parsons (1967) identifies the latter approach as one of "English utilitarianism," which he links with "economics, with biological thought and the beginnings of anthropology as a discipline." A British anthropologist, Malcolm Crick (1976), has developed implications of the divergence for the epistemology of social science more generally.

Social scientists sometimes dismiss the difference between programming and process as the taken-for-granted distinction between "ideal" rules and "real" behavior. The difference must be real at the level of material control, however, where the question remains: are more elaborate social systems controlled primarily by concrete programming, shared by most individuals either genetically or through socialization, or are such systems controlled primarily through the process of social interaction, each individual impinging on all others as an opportunity, constraint, or threat? Nor is this merely an academic question, void of political implications. At least since Hobbes's *Leviathan* (1651), those disposed toward the programming model, which suggests social conflict to be pathological, have favored a strong centralized authority. Those who find norms to be mere abstractions from the interactions of self-interest, by contrast, have inclined toward doctrines of *laissez-faire*, for example, in the utilitarian political theories of Bentham (1789) and Mill (1848, 1859).

Nowhere do the differences between programmed and processual control of societies appear more clearly than in the most consciously formalized institution of human social control: law. Anthropologist Laura Nader describes the difference as a discontinuity between "legal procedure" and "conflict resolution" (1965; Nader and Yngvesson 1973). Scholarly research in the tradition of law as programmed control began with Sir Henry Maine's *Ancient Law* (1861) and is most often associated with the ethnography of A. R. Radcliffe-Brown (1933, 1952). Research in the tradition of law as control through processing seems generally agreed to have begun with Bronislaw Malinowski's *Crime and Custom in Savage Society* (1926) and is most often associated with that anthropologist's subsequent writing (Malinowski 1934, 1945).

Perhaps the most extreme example of the view of law as programming is Isaac Schapera's classic *Handbook of Tswana Law and Custom* (1938), which presents for the cluster of chiefdoms centered on the Botswana–South Africa border a detailed inventory of rules organized and presented in categories similar to those of Western legal systems (Nader and Yngvesson 1973, p. 889). These rules, which the Tswana discuss explicitly and freely and invoke whenever disputes arise, Schapera finds mutually consistent and exhaustive, so that they might causally account for all legal procedures and outcomes, which he finds no need to explain further. Radcliffe-Brown (1952, pp. 195–199) generalizes such systems of rules into a separate and concrete normative programming of the "social physiology" analogous presumably to the genetic programming of the individual physiology.

By contrast, the processual approach to law owes little to theories of legal scholars like Maine and often ignores rules altogether. Like the ethologists and utilitarian economists, researchers in this tradition view behavior as constrained primarily by social relationships—reciprocity, obligation, and expectation—and by the more concrete aspects of social interaction. As described by Moore (1969, p. 258), "Malinowski's view made it difficult to separate out or define law as any special province of study. Law was not distinguished from social control in general."

As with the analogous distinctions among brain, mind, and language, however, and with the other symbolic cultural systems already discussed, law cannot be social control by either program or process alone. If collective rules do not control individual behavior, as the work of Maynard Smith, Axelrod, and the ethologists might also suggest, then the question remains why such rules exist at all, often in explicit and elaborate repertoires, even in relatively simple societies. If individuals are self-seeking and cooperate only as a form of enlightened self-interest, as a stable strategy, how might we understand the relationship between individual action and the sociocultural order or indeed explain any institutionalized social arrangement? On the other hand, if social control is programmed in a system of normative rules, either genetically or through socialization, why are such rules themselves the object of negotiation, or sometimes seen (Comaroff 1978) as a resource to be managed to one's own advantage? No researcher in the programming tradition has yet to elaborate the causal connections between prior law and the resulting legal procedures and outcomes, and at least two (Krige 1939, pp. 114–115; Gluckman 1967, p. 95) seem to imply that the connection involves more than formal logic: colonial administrators, they concede, probably could not adjudicate a complicated dispute among their subjects no matter how the native "law" might be codified.

Once normative and legal rules are seen to constitute *symbolic* control systems much like languages, the apparent contradictions between transactional and programmed or cybernetic control—as well as between human and nonhuman social order—can be resolved. Because

Homo sapiens alone can generate an infinite number of different outputs from a restricted number of principles, according to Chomsky (1968), and because we learn to speak correctly from a severely degenerate sample of examples, the rule structure of language must have two dimensions: one concerned with inputs and outputs, the other of a more abstract and possibly logical nature. H. L. A. Hart (1961) has found much the same structure in law: a surface structure of rules relating actual events to types of actions and a logically distinct generative or deep structure concerning precedents, interpretation, and changes in the surface rules. Because of this dual dimensionality, Crick (1976, p. 99) suggests, "law is a limited set of principles generating a vast range of particular cases," a fair description of a program in general.

Much the same view of law as a symbolic control system has been advanced by John Comaroff, an American anthropologist, and Simon Roberts, a British legal scholar. From their joint restudy of Schapera's rules of the Tswana, which are known collectively as *mekgwa le melao ya Setswana*, Comaroff and Roberts conclude:

> *Mekgwa le melao* represent a symbolic grammar in terms of which reality is continually constructed and managed in the course of everyday interaction and confrontation. Far from constituting an "ideal" order, as distinct from the "real" world, the culturally inscribed normative repertoire is constantly appropriated by Tswana in the contrivance of social activity, just as the latter provides the context in which the value of specific *mekgwa* may be realized or transformed. In short, notwithstanding the classical opposition drawn between them, norm and reality exist in a *necessary* dialectical relationship, a relationship that gives form to the manner in which Tswana experience and navigate their universe. As this implies, changes in the repertoire occur when transformations in the sociocultural system impinge on indigenous consciousness; and, when they do, they have consequences for future social processes. Viewed from this perspective, one major and troublesome problem of jurisprudence may be seen to be illusory. We refer to the "gap" problem—so called because apparent disparities between rules and behavior are allegedly incapable of elucidation—which ceases to exist once rules and behavior are seen to be generated from the same systemic source. (1981, pp. 247–248)

Our earlier example (Chapter 2) of a traffic control system not only can inform this conclusion but also suggests how both the programming and processual approaches to law can be misleading. As Hart (1961, pp. 55–57) points out, an ethnographer of traffic control could not fail to notice the correlation between the changing colors of traffic lights and the flow of motor vehicles. The obvious conclusion that the lights *causally* influence traffic patterns would not be strictly correct, however, because drivers control their own vehicles and are perfectly free to disobey the signals (subject, of course, to constraints imposed by other traffic). That drivers choose instead to obey the lights depends on the nonobservable fact that these have meaning—quite literally as signs—in a much more comprehensive symbolic system of control involving law, convention, and etiquette, among many other constituent systems. Once signal changes and traffic patterns "are seen to be generated from this same systemic source," to paraphrase the words of Comaroff and Roberts cited above, the relationship between signals and traffic no longer *requires* elucidation (it is causally spurious).

The more general point is that symbols do not effect control by causing but by meaning. Cause is effected by individuals who recognize the symbol *qua* symbol and hence its meaning. Meaning might be defined—following Peirce and the pragmatists—as the output that results from that meaning as input. Such definition involves a black box, but not one that contains a cause: relations to an action that are internal to an actor—like intention or motive—are not considered causes but rather part of the action (Ryle 1949, p. 113; Crick 1976, p. 92). Language, law, and other symbolic cultural systems do not control individuals through causation but through meaning. Although human beings do not differ from other species in that their worlds have meaning for them, in this narrow definition, we are unique in that our meaning systems are artifactual (cultural), not entirely learned, infinite in some capabilites, and otherwise reflective of the special generative structure of our brains that is language.

Parallels to language in other cultural control systems can be overdrawn, however. Although language must be integrally bound up in the foundation of human cultural programming, not all symbolic cultural systems can be expected to reflect even the most basic aspects of language (Lyons 1970, chap. 1). Even Chomsky (1968, pp. 64–65) has some doubts about the more general applicability of his generative models. Nevertheless, such models of the essential programming of language remain the most obvious place to seek the uniquely human basis for social control: complex systems of shared symbols and their culturally programmed meanings.

Cultural and Higher Levels of Control

With the emergence of language human societies acquired the programming basis for many overlapping layers of social control. Lost in prehistory, between three thousand and ten million years ago, are the origins of the major means of social programming, processing, communication, and control still operating today: ritual, exchange, generalized media, social networks, markets, and money, among many others.

Although ethologists since Julian Huxley (1923, p. 278) have used the word *ritual* for stereotyped displays of other species, these are mostly discrete signals that convey limited meaning (comparable to human nonverbal gestures) and at their most elaborate have only immediate signal value. A ritualized social interaction, by contrast, provides each participant with a repertoire of parts and roles from which to select, a formalized procedural framework or "program" (Scheflen 1967). When a particular framework is available for use by many or even all members of a society, the roles belong to the event and not to individuals and thereby serve for social control—in effect processing people through the interaction. In what Goffman (1971) calls a "remedial interchange," a ritual processes any two people through almost any inadvertent transgression in four simple steps: the role of transgressor apologizes, that of the violated accepts, the former thanks the latter, who replies, "You're welcome." As Smith (1977, p. 236) notes, "Nonhuman primates use formalized interactions that are analogous to at least the first two steps of this human procedure."

Durkheim (1915) stresses that human ritual not only labels, communicates, and controls individuals but also rejuvenates and reaffirms as well. Sacred rituals appear to be the most distinctly human, serving to replace the separate contents of individual minds with a shared program. "Ritual is strongest when it is most perfunctory and excites no thought," William Graham Sumner (1906, p. 61) observed. "Ritual is something to be done, not something to be thought or felt. Men can always perform the prescribed act, although they cannot always think or feel prescribed thoughts or emotions. The acts may bring up again, by association, states of the mind and sentiments which have been connected with them, especially in childhood, when the fantasy was easily affected by rites, music, singing, dramas."

Ritual can also preprocess thought more directly, however. Much as both language and music enable us to hear—as discrete—sounds that would otherwise be perceived as continuous (Mattingly et al. 1971; Cutting and Rosner 1974), and as culture imposes color categories on the continuous electromagnetic spectrum (Berlin and Kay 1969), so too does ritual render ambiguous social continua into discrete and "processable" categories. Rites of passage, for example, render the gradual physiological transformation of child into adult an unambiguous dichotomy (Van Gennep 1909), important preprocessing for any additional cultural programming that differentially processes children and adults. Such ritual preprocessing may accommodate a more fundamental linguistic programming that appears to depend on a uniquely human system of cerebral dominance (Marler 1975) that makes possible the continuous categorization and processing of essentially similar inputs.

In a critical review of the literature on sacred rituals, anthropologist Roy Rappaport (1971b) demonstrates that they can mobilize, label, and display societies to their biological, economic, and political advantage. Among the Maring of New Guinea, for example, Rappaport (1971a) demonstrates that ritual plays a crucial role in regulating energy flows as intensive agriculture substitutes for the complex natural plant community a more simplified system of only a few crops. Ritual also assures that the various products of a Maring pig are distributed among the society's members in a way that optimizes caloric and nutritional as well as more social benefits (Rappaport 1968). Because the Maring have no leaders who command allegiance during war, ritual dances enable individual men to indicate their willingness to lend military support by attending or not, so that the strength of each faction can be determined by head count. Bureaucracies first arose, in the nation-states of Mesopotamia and ancient Egypt, to fill many of the same information-processing, decision, and control functions that ritual provides in simpler societies like that of the Maring.

Although the causal sequence is probably lost forever in prehistory, economic exchange may have arisen from ritual exchange like the Kula or "circular exchange" that Malinowski (1922, p. 81) observed among the Trobriand Islanders: "The Kula is a form of exchange of extensive, inter-tribal character; it is carried on by communities inhabiting a wide ring of islands, which form a closed circuit [along which] articles of two kinds, and these two kinds only, are constantly travelling in opposite directions." Individual Trobrianders exchange the two articles, necklaces and armshells, with each transaction involving the giving of a single item of one kind and the receiving of a single item of the other, a relationship that, once begun, continued regularly for a lifetime, "The rule being 'once in the Kula, always in the Kula' " (Malinowski 1922, pp. 82–83).

Because "the Kula is not done under stress of any need, since its main aim is to exchange articles which are of no practical use" (Malinowski 1922, p. 86), the exchange is not an economic but a purely ritual one. Although any two partners exchange face-to-face, a relationship that Malinowski felt gave psychological satisfaction, necklaces and armshells move in opposite directions along the entire circuit of islands, thereby maintaining a bond of social control, the most incipient (binary) form of Durkheim's organic solidarity in an otherwise homogeneous or mechanically solidary society, that is, one with little division of labor or other aspects of individualism (Malinowski 1922, p. 510). This system-level communication and control function no Trobriander consciously planned or intends, however. "Not even the most intelligent native has any clear idea of the Kula as a big, organized social construction," Malinowski (1922, p. 83) concluded, "still less of its sociological function and implications."

At the same time, however, no Trobriander confuses the ritual exchange with an economic one. "The natives sharply distinguish it from barter, which they practice extensively," Malinowski (1922, p. 96) noted, "of which they have a clear idea, and for which they have a settled term." Unlike bartered items but like money, the necklaces and armshells constitute generalized symbolic media of exchange, something we will repeatedly find, as this discussion proceeds, to be crucial to social communication and control. As described for money by theorists (Mill 1848; Keynes 1930) who view it in terms of political and social control, a generalized symbolic medium has three essential characteristics: (1) it has no value in use but only in exchange; (2) it can translate value or meaning by making otherwise heterogeneous things comparable in terms of itself; and (3) it can serve as storage or memory to preserve value or meaning over time.

The necklaces and armshells of the Kula clearly meet these criteria for generalized symbolic media, although for social values only, not economic ones. The goods make disparate social relationships like kinship and power (sometimes superiority, sometimes subordination) comparable, for example (Malinowski 1922, pp. 91–98, 175), and in at least this limited way they serve to store meanings. Money similarly stores meaning that, for example, enables the separation of the selling decision from the buying one and thereby transcends the *coincidence of wants* that controls and often frustrates true barter. But the Kula exchange items "are neither used nor regarded as money or currency" and have value only when given to a partner (Malinowski 1922, pp. 91, 511).

Rather than serving as a proto-market in the development of an economic system, moreover, ritual exchange like the Kula probably provided the social integration and control that first made barter and money markets possible. As Lévi-Strauss (1949, pp. 138–139) argues for the classical example of ritual exchange, namely kinship, "the exchange relationship comes before the things exchanged, and is independent of them." In distinguishing such behavior from exchange for economic motives, Malinowski established ritual exchange as an infrastructure of communication and control in large-scale societies. Because necklaces and armshells have no economic value, they must be exchanged out of *social scarcity* (Lévi-Strauss 1949, pp. 32–35), which differs from economic scarcity in that it is purely *symbolic*, created by diminishing the value or otherwise altering the meaning of the individual's own outputs relative to the economically equivalent outputs of others. A birthday gift we buy for ourself, for example, simply *means* less than the same item given to us by a friend. Kinship is ritual and not economic exchange, Lévi-Strauss argued against Frazer (1918), because it is not economic but social scarcity—an artifact of rules concerning exogamy and incest—that governs the supply of possible spouses.

Ritual exchange also differs from economic exchange in that the costs of the purely social transactions are not charged against their immediate material beneficiaries. We attribute the costs of holiday gift-giving to that institution, for example, and not to those who receive our gifts. Similarly, we engage in ritual exchange not for economic but for social motives: to build and maintain *social networks*, the infrastructure by which we obtain *socially scarce* commodities. Judging from the research of social psychologist Stanley Milgram and his associates (Milgram 1967; Travers and Milgram 1969; Korte and Milgram 1970), social networks aggregate in what is indeed a "small world": an average chain of only four to seven intermediary contacts links any two individuals chosen arbitrarily from the entire U.S. population. This finding suggests that, via overlapping social networks, ritual exchange provides integrative control and communication for even the largest and most rationalized economic systems, a possibility explored formally by Pool and Kochen (1978).

Since Malinowski's pioneering work on the Kula, ethnographers have discovered the two types of exchange systems existing side-by-side in a wide range of societies. In Nigeria, for example, anthropologist Paul Bohannan finds:

> Distribution of goods among the Tiv falls into two spheres: a "market" on the one hand, and gifts, on the other. The several words best translated "gifts" apply . . . to exchange over a long period of time between persons or groups in a more or less permanent relationship. The gift may be a factor designed to strengthen the relationship, or even to create it . . . A "market" [by contrast] is a transaction which in itself calls up no long-term personal relationship, and which is therefore to be exploited to as great a degree as possible.   (1955, p. 60)

The ephemeral and potentially exploitative nature of purely economic exchange in even relatively small societies suggests that ritual exchange may be a further precondition—as an additional information-processing capability—for human reciprocity selection of the type discussed by Maynard Smith and Axelrod. Without the overlay of ritual exchange, moreover, economic transactions—because they would become more visible and salient—would threaten the stability of an entire exchange system whenever they became exploitative. Ritual exchange, by contrast, can be what Lévi-Strauss (1949, p. 220) called "univocal" or "generalized," that is, reciprocal only indirectly as, for example, in the Kula circular exchange. Nigerian sociologist Peter Ekeh, who calls Lévi-Strauss's distinction between restricted and generalized exchange "by far the most important development in social exchange theory" in an exhaustive critical survey of the field (Ekeh 1974), notes that "exploitation in univocal reciprocity and generalized exchange situations is much more subtle: the source of exploitation is less easily detectable and exploitation therein will contribute to social disruption less readily" (1974, pp. 208, 213).

In addition to circular systems, generalized exchange can take two other forms: individual to network and network to individual (Ekeh 1974, pp. 52–56). The latter prevails, for example, in the Nigerian *esusu*, where the entire group pays visits of solidarity to each of its members, in turn, over an extended period. Many American offices display network-to-individual ritual exchanges, for example, in taking up collections for individual workers on special occasions or during times of bereavement or hardship. The structural reverse, individual-to-network ritual exchanges, were observed by Herbert Spencer (1876–

1896, vol. 3, pp. 390–391) in "the drinking of men in a public-house." In what Spencer described as "usages curiously simulating primitive usages," each member of a social circle, in turn, buys a round of beer for all the others (the words "circle" and "round" take on their Kula meanings here). Spencer struggled to describe the distinction Malinowski would draw a quarter-century later between ritual and economic exchange: "We have here, indeed, a curious case, in which no material convenience is gained, but in which there is a reversion to a form of propriation from which the idea of exchange is nominally, but not actually, excluded."

Not only do men in British pubs consume drinks through ritual exchange, the principle of univocal reciprocity requires that they "drink level" throughout each round. Douglas and Isherwood (1979, pp. 124–125) cite a British periodical from the 1930s reporting that even with blind men included among the drinkers, "gulp for gulp they drink level to within a quarter of an inch throughout" (Mabey 1970, p. 47). Anthropologist Mary Douglas pioneered the study of economic demand and consumption as communication and control systems more generally:

> Man is a social being. We can never explain demand by looking only at the physical properties of goods. Man needs goods for communicating with others and for making sense of what is going on around him. The two needs are but one, for communication can only be formed in a structured system of meanings. His overriding objective as a consumer, put at its most general, is a concern for information about the changing cultural scene. That sounds innocent enough, but it cannot stop at a concern merely to get information; there has to be a concern to control it. If he is not in any position of control, other people can tamper with the switchboard, he will miss his cues, and meaning will be swamped by noise. (Douglas and Isherwood 1979, p. 95)

As network size increases among systems of generalized exchange other than circular ones, interpersonal communication begins to collapse into forms more suggestive of mass communication. Individual-to-network ritual exchange progressively deteriorates into publicity ploys (as in advertising's "free gift" offers) while network-to-individual exchange becomes increasingly rationalized in public awards (for society's creditors) and public charity (for society's debtors). Granovetter (1973) argues—from the fact that friends of friends tend to be themselves friends—for the importance to larger networks of what he calls "the strength of weak ties," the casual acquaintances that are more likely to express economic rather than affective relationships. The importance of these weak ties in linking more affective but "ingrown" networks and thereby facilitating social diffusion of controlling information and influence suggests that the multiplicity of casual acquaintances in modern society (Gurevich 1961) may at least partly substitute for wide-scale ritual exchange like that of the Kula.

As human social exchange has become both generalized and rationalized over the past ten million years, so too have its symbolic media of exchange. John Maynard Keynes, whose "genealogical tree" for money distinguishes twenty-two different functional forms (1930, pp. 9–11), concludes, "Money, like certain other essential elements in civilization, is a far more ancient institution than we were taught to believe some few years ago. Its origins are lost in the mists when the ice was melting, and may well stretch back into the paradisaic intervals in human history of the interglacial periods" (1930, p. 13). Keynes models the modern economy as a complex cybernetic system in which money in its various forms provides the medium of communication and feedback control. Today even economists do not agree on how to define money, much less measure it. The Federal Reserve provides three measures of money supply (M-1, M-2, and M-3) and one of "liquidity" (L). Talcott Parsons finds that, in addition to language and money, "social systems and other systems of action" are cybernetically controlled by "generalized symbolic media of interchange": political power, influence, value-commitments, and intelligence (1969, 1975). To this list George Homans (1961, p. 385) adds the generalized exchange medium of "social approval."

Table 3.2 illustrates the place of cultural and higher levels of control in living systems, including the type of social system possible at each level (a view of life itself from the information perspective can be found in the appendix to this chapter). As can be seen in Table 3.2, each level exists in a concrete sense, with matter and energy organized into both processor and programming: DNA and genetic programming, vertebrate brain and learned programming, formal organization and rules, machines and algorithms. As argued earlier in this chapter, all four levels of control involve three different temporal dimensions or address three distinct control problems: existence or being, experience or behaving, and evolution or becoming.

All three of these dimensions or problems involve a mechanism, its behavior, and the processes of its programming and reprogramming. Organization can be maintained, counter to entropy, by a hierarchy of

Table 3.2. Analytic dimensions and empirical properties of the four levels of control, including the type of social system possible at each level

Level of control	Organization of matter and energy		Analytic dimensions of the resulting control systems		
	Processors	Programming	Existence	Experience	Evolution
(Inorganic universe)					
Intermediate: Replicating molecules					
Level One: Life Genetically based sociality (many animal species)	Molecules of DNA	Genetically inherited programming	Replication of programming, protein synthesis	Response, adaptation of genetically controlled processes to environment	Organic evolution through natural selection
Level Two: Culture Culture-based social structures; human societies	Vertebrate brains	Learned behavioral programs	Behavior controlled by programming stored in memory	Learned responses, adaptation to environment through reprogramming	Cultural diffusion, change through purposive innovation, differential adoption
Level Three: Bureaucracy Bureaucratically controlled social systems	Formal organizations of individuals	Explicit rules and regulations	Rationalized processing, record keeping, hierarchical decision, formal control	Organizational response to environment, adaptation through reorganization, procedural changes	Diffusion of rationality with differential adoption as culture, selection pressure on organizations
Level Four: Technology Technobureaucratic Information Society	Mechanical and electronic information processors	Purposively designed functions and programs	Informational inputs, processing and storage, programmed decision and output	Interaction of processor and environment, adaptation through reweighting, reprogramming	Diffusion of processors and programs with differential adoption as culture, government, and market selection
Intermediate: Genetically engineered organic systems					
(Synthetic life)					

physical control: protein synthesis controlled by DNA, cultural behavior by learned programs, organizational processing by formal rules, machine processing by still more intentionally designed algorithms. Preprogrammed goals can be pursued, despite variation and change in external conditions, by a hierarchy of behavioral mechanisms: adaptation, learning, reorganization, reprogramming. Less successful goals can be modified or dropped, while still preserving or improving more successful ones, through a hierarchical process of system changes: organic evolution through natural selection, cultural diffusion through differential adoption at the individual and organizational levels, government and market selection at the technological level.

The currently precarious state of human culture and higher culture-based levels of human control can also be seen in Table 3.2. Today we stand between the inorganic universe, on the one hand, and the prospect that our technology might soon create new life—thereby replicating all four levels of control at the synthetic level—on the other hand. Intermediate forms now exist on both sides: self-replicating polymers on the inorganic side and genetically engineered systems on the organic one. In view of the continued development and proliferation of nuclear weapons technology, the next major stage in the evolution of life on earth might hinge on the question of which end-state of Table 3.2 will be achieved first: synthetic life, certainly one possible product of our evolving technobureaucratic control, or the return of the entire planet to the inorganic level, another quite possible outcome—ironically enough—of the same revolution in control technology.

Reductionism and Synthesis

Because we initially defined technology as the intentional extension of natural processes like respiration and metabolism, it would seem reasonable to discuss the Control Revolution using concepts that apply to all living systems. Similar reasoning is commonly used in environmental anthropology, economic geography, and ecology, among other academic fields. Unfortunately, when we seek to reduce any human phenomena to a set of simpler and more general concepts, we confront an objection that seems more doctrinaire than practical: many social scientists and humanists dismiss even the *possibility* that such commonalities might exist, the so-called *holist* position that opposes biological and other forms of *reductionism*.

Holists argue that unique phenomena "emerge" at successive levels of a hierarchy: animals, primates, our own species, formal organization, nation-states, the world order. At each of these levels and many in between stand specialists ready to defend an undeniable truth: the whole is often more than the sum of its parts. As biologist John Bonner assesses criticisms made by social scientists of human social biology,

> they are almost entirely related to the idea that a reductionist approach will not be useful in the social sciences . . . In their view, human societies are too complex, too special, too different from anything found in the animal world to be interpreted in any meaningful way by biological analysis . . . The reductionists tend to be contemptuous of all holists, for they feel they alone have the key to the universe. Holists know they have a broad perspective, a large insight, whereby they can see all the riches missed by the single-minded reductionist. In principle it would appear so easy to be both at once, but human nature is such that it enjoys taking positions on philosophical or political dichotomies. (1980, pp. 7-9)

As a result of this unwillingness among social scientists to concede that one subject matter might be reduced to some other, a multitude of overlapping concepts and propositions—differing even among journals and theoretical camps—has proliferated in increasingly fractionated subdisciplines and specialties. Thirty years ago, for example, the anthropologists Alfred Kroeber and Clyde Kluckhohn had already found 164 different definitions of "culture" in print, "probably close to three hundred"—in their judgment—if all variations had been taken into account (1952, p. 149). A more recent study (Dance and Larson 1976) identified 126 different definitions of "communication" without any special effort to compile an exhaustive list.

Apparently even as social scientists profess to be awaiting their Newton, a revolutionary figure—as defined by Thomas Kuhn (1957, 1962)—who will reduce their disparate efforts to a few basic concepts and principles, they continue to eschew such reductionism in the name of "emergence," phenomena found at one level of specialization and held to be irreducible to any others. As a result, more and more behavioral science journals have become what George Miller described for psychologists as "catalogs of spare parts for a machine they never build" (1956b, p. 252).

Today, continuing technological development—more than any charismatic thinker—suggests the integrative machinery we might build from the spare parts amassed by our various disciplines. The rise of the Information Society, more than the corresponding development of information theory, has exposed the centrality of information processing, communication, and control to human society. It is to these fundamental processes and not to any particular level in the hierarchy of living systems that we might hope to reduce our accumulating knowledge of human organization and society.

APPENDIX: What Is Life? An Information Perspective

What is life? Despite more than twenty centuries of sustained attention to the question since Aristotle's observation "Soul is better than body and the living, having soul, is thereby better than the lifeless which has none" (1912, p. 731b), biologists still do not agree on the criteria that distinguish the subject matter of their discipline. Most do not even attempt to define life. "Life does not exist in the abstract," writes biologist Helena Curtis. "There is, in fact, no 'life.' What exists and what can be studied are living systems, individual living organisms" (1975, p. 26).

But how do we recognize a living organism? By what criterion do we say, for example, that the blue-green scum on the side of a swimming pool has life and an intricate, growing crystalline structure does not? It is tempting to cite some physical constituent of organic matter, for example, *protoplasm*. Reputable scientists now discuss the possibility of life on other planets, however—life that may have a radically different chemical basis from that on earth. One molecular chemist (Cairns-Smith 1971) even speculates that our own earliest ancestors might not have been based on organic molecules at all but on tiny mineral crystals capable of self-replication. Whether or not such radically different life forms actually exist, we do not hesitate to use the word *life* to describe them. Clearly we mean something more abstract than any particular organic substance.

The problem becomes more than academic as we begin to explore into deep space, where something we would want to call *life* may be based on silicon rather than carbon or on ammonia rather than water—where organisms may be discovered, for example, that boil to death below the temperature at which oxygen boils. When we come to program a computer for an unmanned expedition to detect signs of life in these distant regions, for what do we instruct our machinery to look? All the laws of physics and chemistry by which we construct the equipment are assumed to hold throughout the universe—must we expect less of the principles by which we define the behavioral and life sciences?

In the absence of any generally accepted definition, many biologists simply list five to ten properties that they consider most basic to life. On the basis of several recent texts (Curtis 1975, pp. 27–32; Keeton and McFadden 1983, p. 1; McNally 1974, pp. 3–4; Nason and Dehaan 1973, pp. 3–4; Villee 1972, pp. 23–27; Weisz and Keogh 1982, pp. 19–27; Wilson et al. 1978, pp. 6–7), six properties can be singled out as among those most frequently cited: organization, metabolism and growth, responsiveness, adaptability, reproduction, and evolution. In the descriptions of these six properties, each considered in at least several textbooks as essential to life, we come perhaps as close as possible to biology's consensus definition of living systems in general.

1. *Organization.* Unlike the inorganic world, which consists of relatively random mixtures of fairly simple chemical compounds, living systems combine the same atoms and molecules in precise, organized, and complex ways. As we saw in Chapter 2, even the simplest microorganism can be distinguished by the relative complexity of its structure from, say, the rock in which it might be embedded. This explains why observation of structure remains the principal means biologists use to find microfossils. Essential to the organization of living systems is *homeostasis*, the maintenance of a relatively stable internal environment. We mammals normally maintain a fairly constant body temperature, for example, despite wide ranges in outside temperatures. In chemical composition, at least, all living things are homeostatic.

2. *Metabolism and growth* through destruction and reconstruction of organic material—that is, the breakdown of matter to produce energy (catabolism) and the synthesis of nonliving materials into living matter (anabolism). All living systems, as we have seen, are fueled by energy from the sun, either directly as in the photosynthesis of green plants or through the ingestion of other plants and animals; a few bacteria are capable of *chemosynthesis*, the manufacture of carbohydrates directly from inorganic compounds. In energy conversion and system growth, metabolism approximates what Aristotle saw as the essence of life: "The power of self-nutrition . . . is the originative power, the possession of which leads us to speak of things as living" (1931, pp. 413a–b).

3. *Responsiveness*, the capacity to respond to stimuli. Even unicellular organisms contract when touched, a property often called *irritability*. Much the same capacity can be seen at progressively higher

levels of complexity: bacteria move toward or away from certain chemicals, plants bend toward light *(phototropism)*, flies shoot out their probosci in the presence of sugar water, dogs salivate at the mere smell of food. Sometimes the response, although innate, automatic, and fixed, provides for *alternative* behavior as, for example, when a fly moves either toward or away from a light depending on its intensity. Honey bees exhibit a *continuous* range of responses to signals indicating quantitatively different directions and distances of a food source.

4. *Adaptability*, including the ability to learn. Unlike responsiveness, which involves genetically fixed responses to external stimuli, adaptability refers to environmentally determined, relatively enduring changes in an individual that occur in response to environmental factors rather than through maturation. Our skins darken upon prolonged exposure to the sun, for example; people who move to higher altitudes develop additional oxygen-carrying red blood cells. *Learning*, a particular type of adaptation that involves enduring changes in behavior, covers a range of responses that vary widely in complexity: *habituation*, a gradual decline in response on repeated exposure to insignificant and unrewarded stimuli; *imprinting*, a learned attachment to an environmental stimulus that can occur during only an extremely short period, usually early in an organism's life; *conditioning*, the association through reinforcement of a response with a new stimulus; and *trial-and-error learning*, an increase or decrease in the frequency of behavior according to the relative advantage or disadvantage of its outcome.

5. *Reproduction*. Living things reproduce themselves, in whole or in part, copy after copy and generation after generation, with relative copying fidelity. This holds true for a wide range of reproductive processes, including both sexual ones involving internal and external fertilization and several types of asexual processes: mitotic cell division, nonmitotic division, fragmentation of filaments, and budding. A similar capacity for self-reproduction in copies virtually identical in size, shape, and internal structure through numerous "generations" has not yet been achieved in any nonliving entity, even though von Neumann (1966) has proved that, above a certain finite number of parts, an automaton might be built that could reproduce others of complexity greater than its own, and these in turn others of still greater complexity. A. G. Cairns-Smith, a molecular chemist, has argued the same possibility for self-replicating polymers (1971, 1977).

6. *Evolution* through natural selection, the continual readjustment

to environmental variation and change over successive generations. Because organisms so often seem ingeniously suited to their particular ecological niches, it was long believed that each type must have been specially created. Darwin was first to attribute the result to differential reproduction of inheritable variation over generations of individuals. Although not a property of any particular living thing, such adaptation ranks among the properties most often cited by biologists as crucial to life itself. The possibility of life on other planets, for example, has prompted one zoologist to ask: "Will there be any general principle which is true of all life? Obviously I do not know but, if I had to bet, I would put my money on one fundamental principle. This is the law that all life evolves by the differential survival of replicating entities. The gene, the DNA molecule, happens to be the replicating entity which prevails on our own planet. There may be others. If there are, provided certain other conditions are met, they will almost inevitably tend to become the basis for a evolutionary process" (Dawkins 1976, p. 206).

These six properties, then, are most frequently cited by biologists as most essential to life. Does this simple list suggest any more general insight into the nature of concrete open systems, including our own society and culture, and into the problems of control in such systems?

Toward an Analytic Model

Although none of the biologists included in our survey attempts to integrate the manifest properties of life into a more general model of living systems, the six properties they identify most frequently do turn out to represent distinct dimensions of existence, experience, and evolution, the three functions of control that we noted in Chapter 3. Organization and metabolism can be seen to represent the static and dynamic aspects of existence, responsiveness and adaptability describe these two dimensions of experience, and reproduction and natural selection do the same for evolution. Relationships among these control functions and properties of living systems are shown schematically in Table 3.3, which lists each of the biologists' six properties as either static or dynamic aspects of the three control functions.

The distinction between static and dynamic aspects of a system recalls the work of Auguste Comte (1798–1857), student of Saint-Simon and generally considered the founder of modern sociology, a term he coined in 1838. Comte attempted to establish sociology in two branches:

Table 3.3. Analytic dimensions and empirical properties of living systems

Static	Dynamic
Organization	Metabolism

Existence (being)
in the absence of external change,
counter to entropy

Need: Maintenance of matter and energy processing
Control: Fixed programming distributed throughout system

Responsiveness	Adaptability

Experience (behaving)
during the life of one system,
in response to external variation and change

Need: Goal-directed response to external conditions
Control: External input and output with feedback, ability
to reweight or reprogram

Reproduction	Selection

Evolution (becoming)
across generations of programs,
through the differential selection of systems

Need: Preservation of programming with advantageous modification
Control: Ability to replicate or otherwise communicate programs to new
generations with high fidelity, some variation

"The one, the statical, will treat of the structural nature of Humanity, the chief of organisms; the other, the dynamical, will treat the laws of its actual development" (1852, p. 1). Unfortunately, sociologists continue to see the difference between static and dynamic, like the one they draw between structure and process, as a matter of *absolute* continuity or change, thereby reducing an analytically powerful distinction to one of mere description.

The folly of considering categories like structure, process, and even history as mere temporal distinctions can be seen in Table 3.3. All three functions of control have both static and dynamic dimensions that are obvious enough to rank among the properties most often cited by biologists as fundamental to living systems. Even the individual temporal dimensions can be seen to have both static and dynamic aspects: reproduction, for example, identified here as the static component of evolution, is probably a more dynamic process in any absolute sense than the dynamic aspect of existence, namely metabolism.

One can even imagine an alternative scheme in which reproduction is seen as the dynamic aspect of one analytic dimension and the static component of another. Certainly nothing about reproduction or any other empirical phenomenon makes it either static or dynamic per se but only relative to some analytic function. For this reason, we will consider *static* and *dynamic* to be defined only with respect to a particular dimension of analysis, thereby avoiding many classical controversies—for example, the one concerning the supposedly static nature of the term *social structure*, which at least two eminent anthropologists, Meyer Fortes and S. F. Nadel, have maintained involves "a sum of processes in time" (Nadel 1957, p. 128).

Table 3.3 also demonstrates that *programming* is central to control at all three levels. For existence, where the problem is to maintain organization by controlling matter and energy processing, the solution involves programmed control distributed throughout the system. For experience or behaving, where the problem is to pursue goals in interaction with the external environment, the solution involves programming with environmental inputs, outputs, and feedback, plus the capability to reweight contingencies within a program. For evolution, where the problem is to preserve programming across generations with goal-directed modifications, the solution involves replication of programming in new generations with relative copying fidelity but at least some variation that might be differentially selected.

Ernst Mayr (1974a) makes the useful distinction between *open* and *closed* programs: a closed program is one entirely encoded prior to the beginning of the behavior that it controls; an open program incorporates additional inputs as, for example, in conditioning or learning. Once filled with the requisite information, an open program becomes the functional equivalent of a closed one. In living systems, closed programs can be seen to control existence or being, open programs to control experience or behaving, and external selection of programs—whether purposive or nonpurposive—to control evolution or becoming. To generalize to all control systems, three different actions involving programming are essential: maintenance of closed programming throughout the system, modification of open programs in interaction with the environment, and evolution of all programming through differential selection.

As we saw in the example of rush-hour traffic in Chapter 2, living systems have thus far evolved four levels of programmable structures and programs: DNA molecules encoded with genetic programming, brains with cultural programming, organizations with formal written procedures, and mechanical processors with algorithms. Because the first level emerged at the origin of life and the second with the higher vertebrates (birds and mammals), these first two control revolutions are lost to prehistory. The third and fourth levels, however, are unique to *Homo sapiens*, with formal organization first appearing in Mesopotamia and ancient Egypt about 3000 B.C. and merging with mechanical control to form the technological basis of the nineteenth-century Control Revolution.

Because programming is essential to all four levels of control, it will be generally useful to consider its oldest and most basic form, at least that survives on earth: the macromolecules of DNA that we share with all living things. Not only does DNA provide a further test of our three-dimensional control model, it also enables us to reduce that model to the level of programming itself.

Three Functions of DNA

Even in living systems that have evolved brains, culture, and control technologies, most control remains programmed at the genetic level, a fact that we commemorate in expressions like "genetic programming" and "genetic code." All living systems maintain control through three genetic processes: replication, regulation, and reproduction. As we might expect, these processes correspond to the same three dimensions of control—existence, experience, and evolution—that we have already identified:

1. *Existence* (being) through *replication* of programming for distributed control. At least in multicelled organisms, existence depends on communication and coordination among individual cells. The human body may contain a quadrillion cells, including more than a billion qualitatively different ones; many organisms have even more. How do such staggering numbers of cells, each acting independently and many with highly specialized functions, come to achieve the coherence of a single organism?

Most biologists believe that at least in fungi and plants and possibly also in the animal kingdom, multicellularity first occurred through amalgamation. Several independent single cells came together, began to live as a group, evolved a division of labor, and eventually formed the integrated structure of a single multicelled organism. Even if this were not the case, coordination of functionally differentiated cells must have been as crucial to the evolution of animal species as coordination of functionally specialized individuals and roles has been to the development of human society.

Internal coordination of a multicelled organism depends on the exact replication of its genetic program or *genome*, the complete set of operating instructions for the *entire* organism, in each of its individual cells. Repeated replication of the genome is made possible by DNA's remarkable capability to break in half lengthwise, and by the capability of each of the strands thus separated to connect to freely floating molecular complements and thereby reconstruct two new copies of the original double helix. After these are complete, the entire cell divides into two new ones, each controlled by one of the two identical programs. What begins at fertilization as a single cell becomes two, these four, these eight, and so on into the quintillions (including the replacement of lost cells) if necessary, each cell containing exact copies of the original chromosomes.

With programming virtually identical—except for a rare copying error—in each of an organism's individual cells, the problem of their coordination is greatly reduced. Even in the vertebrates, which have the most centralized control systems in the spinal cord and brain, considerable autonomous control remains at the cellular level. If a vertebrate's brain is likened to the central processing unit (CPU) of a computer system and its spinal cord and peripheral nervous system to a network for communicating with various input and output devices like organs and muscles, then each cell can be seen as an intelligent microprocessor or minicomputer, effecting considerable autonomous control at the local level.

In this way each living thing constitutes an information-processing and communication system in which control of its internal structure and processes is both hierarchical and—at least at the genetic level—distributed as well. Such control accounts for the "soul" or "animus" (Latin root of "animal") that has been postulated by vitalists since Aristotle's *De Anima*. In this sense, at least, life has long been appreciated for the combination of hierarchical and distributed control that bureaucratic organizations have only recently come to appreciate in information-processing and communications systems. Organization and distributed control cannot account for adaptation, learning, and other forms of open-ended behavior, however; these characteristics derive from the second of DNA's three control functions.

2. *Experience* (behaving) through *regulation* of physical processes with feedback. Astounding though it may seem, adaptive control by DNA is effected through the manipulation of only twenty different amino acids. These compounds, either synthesized by the organism or provided in the diet, are recombined by DNA in chains of fifty or more to form proteins. Proteins, in turn, control the various chemical processes in a cell by switching them on and off at precise times in precise places. Particularly effective in control are the *enzymes*, proteins that often use negative feedback in the regulation of cellular processes. Branson (1953) concludes that the structures of proteins are highly efficient with respect to information processing: none of those he studied had less than 70 percent maximum information content and many approached their respective limits.

As Schrödinger had perceived by 1944 and Gamow stated explicitly a decade later, even as few as twenty essential substances can combine in many different ways, especially in protein chains ranging up to ten thousand links. With a conservative estimate of average length of 150 links, a count characteristic of hemoglobin, the number of different proteins possible—assuming that every combination of amino acids can serve—is 20^{150} or a hundred thousand quadrillion quadrillion . . . (the word *quadrillion* repeated a dozen times). This far exceeds the total number of elementary particles—protons, neutrons, electrons, and scores of others—estimated to exist in the entire known universe. Although many of the combinations could not exist, more than enough real possibilities remain to control even complex organisms at the cellular level.

In order to be adaptive, such control must involve information processing—the translation of genetic chains not into identical ones as in replication but rather into the amino acid sequence that they encode, which characterizes the protein required at a particular time and place in a life process. Here the role of information is straightforward: the four different genetic particles or *nucleotides*, usually represented by A, C, G, and T (abbreviations for their identifying bases), serve in triplets that encode one of sixty-four different "words," one to six of which represent each of the twenty protein-forming amino acids. The ordering of particles counts (AAG and GAA code different amino acids), with two of the sixty-four triplets functioning as "punctuation marks" to denote the end of the molecular instructions to build a particular protein.

In most cells the process of control is also simple: the double helix unwinds, generates a chain of 150 or more particles (aptly named *messenger RNA*) to instruct the protein-synthesizing machinery, and then rewinds, thereby preserving all of its many volumes of instructions for future use. This much-repeated activity, which requires only about a minute, constitutes adaptive behavior at life's most basic level, in cells so tiny that a hundred may be needed to cover the period at the end of this sentence.

Genetic control has one shortcoming: the genome is fixed forever at the instant of conception. Even though its programming routinely controls adaptation to the environment, it cannot itself be reprogrammed to accommodate gross changes. In this, at least, the genome resembles *firmware*, computer hardware's functional equivalent in programming, which cannot be changed or removed from a system but which must instead be used by all other software.

As in the design of new computers, natural selection had to decide which functions to build into hardware (programmable structure), which to program in as firmware (the genome), and which—in the case of the higher vertebrates—to leave to subsequent development in software (culture). In all species evolution seems to have favored programming—both firmware and software—at the expense of hardware, basing genetic control in the simplest imaginable structure: a single universal molecule that can be programmed with only four different symbols.

The advantage to this is one of *preprocessing*, identified in the previous chapter with a reduction in the total amount of information to be processed. Because each chemical base can bond with only one other, and the entire chain of nucleotides with only a complementary chain of messenger RNA, only minimal information needs to be processed in the latter's synthesis (called, appropriately enough, *transcription*). When, in turn, messenger RNA manufactures noninformational ribonucleic acids using adaptor molecules composed of twenty different types of *transfer RNA*, each of which can bond with only one of the twenty essential amino acids, information processing is similarly straightforward.

In other words, the advantages of preprocessing, realized much later in human technologies through rationalization, were first achieved in information processing at the genetic level. In sharp contrast to the simplicity of its hardware and processing, however, the information content of genetic programming can reach several million bits.

Like some recent computer firmware, genetic programming is not always closed but may be open as well. The young of many species, for example, can recognize their own kind at birth, obviously the result of a closed genetic program. The young of some species, however, learn to distinguish their own parents from other adults shortly after birth, once and irreversibly—the work of an open genetic program that, once completed, is indistinguishable in function from the closed program for species recognition. The result is *imprinting*, a term that itself suggests addition to an open program.

It may seem that similar programmed adaptability can be found in the various game-playing and related computer programs that progressively modify themselves through interaction with humans and other computers. Although such programs are commonly said to "learn," their development in fact more closely approximates our definition of evolution. So-called learning occurs as each branch of a decision structure is weighted up or down, depending on whether it has led to success or failure—a clear case of selection by the environment. Although such programs are not reproduced themselves, strictly speaking, their essential strategies are communicated from one game to the next with relative fidelity and purposive selection—the two features that distinguish evolution.

Most other computer programming does indeed resemble open genetic programs, at least in that it allows for a finite number of discrete inputs which, once determined, cause the program to operate in a way that is indistinguishable from its "hard wired" or closed equivalent. Whether closed or open, however, genetic programs cannot evolve unless they can be *reprogrammed*, the third of DNA's three control functions.

3. *Evolution* (becoming) through *reproduction* of programming in new entities with differential survival. Like replication and regulation, reproduction is essentially an informational task: to communicate genetic programming to each new generation. Unlike DNA's other two functions, however, sexual reproduction preserves only half of an organism's programming, not the entire genome. This enables new organisms to receive programming equally from each parent, thus making possible the continual reshuffling of genetic instructions into new patterns and their modification in new directions that may be more advantageous to the species. Each newly fertilized egg constitutes a kind

of trial balloon, in other words, to test whether its genome contains programs meriting more widespread adoption by the species as a whole.

This continual shuffling of genetic programming from one generation to the next is possible only because the chains of genes (named *chromosomes* in the late nineteenth century because they absorb colored laboratory stains) store information in structures that are both divided and paired. Each coherent division (gene) represents one *allele*, a member of a mutually exclusive and exhaustive set of rivals for control of the same trait (Mendel's term for alleles translates as "antagonistic factors"). For each such trait, at least two alleles—inherited from separate parents—occupy separate chromosomes. A combination of any of three possible alleles, for example, determines the 2^2 or four human blood groups: A, B, AB, and O. When alleles contain different instructions for the same trait, only one set (called *dominant*) can affect protein production; hence at least half of the programming contained in the genome is not activated for control but remains *recessive*.

In the formation of both ova and sperm for reproduction, chromosomes pair up—allele to allele—before splitting lengthwise. While lying side by side, strands originally from the father may exchange strings of nucleotides with programming originally from the mother, a process called *crossing over*. After replication of the separate strands, the resulting cells split again to form four sex cells, each containing only a half set of chromosomes, no one of which—owing to crossing over—may be like any others or ever duplicated again. In this way each new organism results from the chance combination of an astoundingly improbable egg and an equally improbable sperm and is unlikely to be remotely approximated in genetic programming even by its own siblings (unless, of course, they are identical twins).

Crossing over does not always occur and is not essential to evolution. Male fruit flies never do it, for example, and their species has a gene for suppressing it in females as well. Crossing over does leave open the possibility that any given chromosome might be a wholly new mosaic of genes—and even individual genetic instructions—taken from both maternal and paternal programming. In other words, even though crossing over does not increase the carrying capacity of the genetic hardware, it does increase the *degrees of freedom* available in the combination of messages and therefore the amount of information contained in any one message.

Despite this great freedom of choice in messages, the fidelity of their reproduction, by any standard of human engineering, is truly remarkable. Suppose that the seven hundred thousand manuscript pages of information coded in our own genes had been given to a battery of typists, who were asked to retype each page twenty-five times, about the average number of replications made by a single cell producing human ova or sperm. This amounts to about 17.5 million pages, in the typing of which we might reasonably expect at least one typographical error every twenty pages, probably more; let's say an even one million errors for the entire job. By comparison, the tiny cell in the human ovary or testis will make the equivalent of one typographical error—a mutation—only about ten times, roughly once in every 1.5 to 2 million manuscript pages. In other words, the copying of genetic programming in human sex cells is about one hundred thousand times as accurate as the typists it may eventually produce.

Despite the potential advantages of chance mutation, then, its probability has been kept very low. What randomness is necessary for evolution through natural selection has been largely relegated to higher levels—to the pairwise redundancy of alleles, to the shuffling of chromosomes, to crossing over, even to mate selection. Apparently it is much more important to copy faithfully and thus hope to preserve useful programming, however rarely it appears, than it is to generate a greater amount of new programming—only potentially useful—through more frequent errors in replication.

The result is that genetic programming can adapt only slowly to external conditions, a rate of change limited by the length of generations and hence much faster in fruit flies than in human beings—not to mention the giant sequoia. Only at the levels of brains, culture, and control technology has evolution enabled some species partially to short-circuit, in effect, the relatively slower process of reprogramming through natural selection.

That the process continues to operate, however, points up a more general conclusion: teleonomic *programming* need not imply a teleonomic *programmer*. As we saw in the case of game-playing computer programs, for example, a strategy for winning at chess can improve without anyone instructing it how to do so. All that is required for evolution to occur are the static and dynamic aspects of its essential control function: replication of programming and its differential selection relative to other programs. It is the emergence of precisely this capability, in the earliest ancestors of DNA that marked the origin of life on earth, that is the beginning of evolution through natural selection.

II

Industrialization, Processing Speed, and the Crisis of Control

4

From Tradition to Rationality: Distributing Control

Roll on, thou deep and dark blue Ocean, roll!
Ten thousand fleets sweep over thee in vain;
Man marks the earth with ruin, his control
Stops with the shore.

—Byron, *Childe Harold's Pilgrimage*

IN VIEW OF the cultural and market control achieved in even the simplest societies using kinship, religion, ritualized exchange, law, and social networks, as we saw in Chapter 3, it is certainly not obvious why the nineteenth century should bring a crisis of control to the industrializing countries of Europe and America or why the twentieth century dawned on a revolution in control technology. Commercial capitalism, born in the Mediterranean, had sustained an increasingly international exchange and processing of material goods since the thirteenth century. As a result of a succession of innovations in navigational and seafaring technology, this exchange embraced the entire world by 1519, when Fernando Magellan boldly set sail from Portugal to circumnavigate the planet. Colonization of Africa, Asia, and the New World under the control of nascent bureaucracies in the new nation-states of Europe brought these continents into a single world system by the eighteenth century—even as the colonizers competed and warred among themselves.

Control of the world system under commercial capitalism resided in stay-at-home merchants who conducted their business through similar merchants living in other trade centers and willing to execute transactions on commission (Gras 1939, chap. 3). These resident merchants introduced a series of innovations in the control of commerce, beginning in fifteenth-century Venice, including the systematic collection and processing of market information, the standardization of commission rates, reliance on private arbitration to settle disputes, and formation of joint ventures to spread risk and preserve mobility (Lane 1944, pp. 93–136). This last characteristic of the resident merchant, diversity of investment and function, constitutes a major difference between commercial capitalism and the industrial capitalism that would begin to supplant it in the nineteenth century. Medieval merchants who "put out" wool or cloth for spinning, weaving, or finishing notwithstanding, little specialization of investment occurred before the eighteenth century (Gras 1939, chap. 3).

Diversity of function resulted, in part, from the relatively low volume of trade under mercantilism, the dominant economic doctrine of the sixteenth and seventeenth centuries, under which each European power pursued its own most favorable balance of trade by limiting imports (Schmoller 1884, pp. 47–80). The primitive telecommunications of the same period, however, which made control at distances greater than a few days' ride by horse exceedingly tenuous, constitute a less often cited but equally important cause of the resident merchant's diversification of investment and resistance to specialization. As Byron observed in the passage that opens this chapter, even for a global power in the early nineteenth century "control stops with the shore." Lacking control at a distance, the merchant engaged in international commerce diversified his investments and economic activities for much the same reason that the modern investor maintains a well-balanced portfolio: to minimize risk.

Problems associated with attempts to maintain control at a distance can be seen in the largest centrally controlled system of the period, the Hapsburg Empire of the sixteenth and seventeenth centuries, which eventually stretched from California to Argentina and from the Hapsburg lands in Belgium to the Philippines and other Pacific islands. The Emperor in Spain often had difficulty maintaining control over the more remote territories. Even though the Laws of 1523 and 1526 strictly forbade enslavement or maltreatment of natives in the Indies (roughly the territory of Mexico, the Caribbean islands, Peru, and lands between), for example, and had the full support of Charles V, "the intervening distance made it impossible for the Emperor to enforce the laws," according to R. B. Merriman. *"Obedezcase pero no se cumpla*—let it be obeyed but not enforced—became the formula for the colonists' reception of unpopular decrees from Spain" (1925, p. 659).

Because directives from the Escorial often arrived in a colony long after local changes had rendered them obsolete if not counterproductive and even foolhardy, the formula did not always work to the Empire's disadvantage. Indeed, the so-called "colonists' expedient" proved useful enough to be officially adopted, two centuries later, in another situation where imperial control frequently broke down because of inadequate communication: on the battlefield. In 1757 Empress Maria Theresa, locked in the Seven Years War with Frederick the Great, instituted her Order of Maria Theresa for officers who turned the tide of battle by purposively disregarding orders (Boalt et al. 1971, p. 63). Because the Empress took pride in her centralizaton of military authority, disobedience without battlefield success invariably led to court martial. Her new order represented an effective means to distribute control without unduly disrupting the more centralized system.

The resident merchants who dominated world trade as early as 1300 for a period lasting some five centuries (Gras 1953, p. 73) confronted much the same control problem as did the Hapsburg emperors of the same period. As F. C. Lane describes the fifteenth-century merchants of Venice, their control over commerce amounted to little more than making "controlling decisions," after which they "had to leave much to the man on the spot" (1944, p. 99). In view of the primitive information-processing and communication capabilities that prevailed well into the nineteenth century, not to mention the highly speculative nature of venture trading, it is small wonder that, for the fifteenth-century merchant, handling far-flung commission agents "constituted a very large part of the problem of business management" (1944, p. 97).

A quantitative study indicates that this problem—essentially the inability to centralize decision making with sufficiently short turnaround times between command and feedback—continued to plague merchants three centuries later: "Shippers did not search for trade among overseas areas and ports with which they were unfamiliar, and there was a limit to the number of ports and areas that they knew well and in which they had reliable commercial contacts. Information costs and the risks associated with overseas trade were very high during the eighteenth century, and unfamiliarity raised these costs even higher, slowed the process of exchange and lengthened port times" (Shepherd and Walton 1972, p. 157).

Continuing diversity of investment and resistance to division of labor and specialization of function, although protecting the individual merchant against risk, only compounded the problem of controlling the world commercial system. That specialization increases control has been demonstrated by James Grier Miller, a behavioral and medical scientist who finds specialization to be one of a half-dozen common responses to the growth of information in systems ranging from individual cells to formal organizations (1978, chap. 5). Although specialization might have increased the resident merchant's control of preindustrial commerce even in the absence of adequate communication, the relatively low volume of both information and trade discouraged specialization. Thus the world commercial system foundered for centuries in a vicious cycle in which poor communications and the resulting lack of information prevented the increased specialization and control that would have made specialization itself less necessary.

The Solution of Distributed Control

Merchants and shippers sought to resolve this control problem much as did Charles V and as do managers today using microprocessors and microcomputers: they attempted to distribute control. Unable to be themselves everywhere that decisions had to be made or to be in communication with these places with turn-around times sufficiently short to effect adequate control, merchants enlisted in major commercial centers the preindustrial approximation of a portable, programmable information processor: the commission merchant. Replacement of profit sharing or *commenda* by the system of paying commission agents a fixed percentage of turnover distinguished the Commercial Revolution, after the fourteenth century, as concomitant with the replacement of traveling by resident merchants (de Roover 1953, p. 82).

That the commission agent served as a means to distribute control rather than merely to carry out the resident merchant's instructions can be seen in Lane's analysis of Andrea Barbarigo, a fifteenth-century merchant of Venice: "If he was to profit, his agents must give him the advantage of market situations if and when they developed and became known to the agent on the spot. A good agent was one who sold quickly and reinvested at good prices, obtaining high quality wares, and who found the cargo space in which to ship the investment. In regard to all these four desiderata, quick turnover, good prices, high quality, and adequate transportation, Andrea Barbarigo could give exhortation more easily than definite instructions" (1944, pp. 98–99).

Both resident merchants and their agents preferred commissions to profit sharing because of inadequacies in information processing: "As a practical matter," notes Lane (1944, p. 95), "it was very difficult for the stay-at-home merchant to obtain an adequate accounting for the funds or goods entrusted under profit sharing agreements." Profit sharing had worked as the information-processing system of the traveling merchant only because the merchant ordinarily went out with his wares, so that profits could be calculated among partners after goods had been at least partly sold and unsold goods thereby appraised at current market price (de Roover 1941, p. 92). Commissions afforded a much better information-processing system, although not without a half-dozen other informational innovations: invoices, bills of lading, ships' manifests, a postal service, systematic accounting methods, official clerks *(scrivani)*, and standardization of prices, shipping practices, and handling charges. Lane describes the relative advantages of commissions over the profit-sharing system:

> A resident merchant acting through agents residing abroad could keep track more accurately of what his agents were doing if their sales were in his name and on commission. The agents' accounts then consisted of records of sales (or purchases) at specified prices of specified amounts, and of relatively small and well standardized handling charges. Once trade connections were so firmly established that handling charges and shipping practices were generally standardized, and once prices, season by season or even day by day, were a matter of common knowledge among the mercantile community, the stay-at-home merchant could check fairly well on his commission agents' accounts . . . This advantage depended in turn on the contemporaneous development of a clear method of keeping agency accounts and of such commercial conveniences as the invoice, the bill of lading or the ship's manifest and reasonably dependable postal service . . . Invoices *(fatture)* were regularly sent with covering letters, and gave all the information necessary for checking with the records of the *scrivani* (the quasi-public officials who kept the ships' manifests and whose activities, especially on the galleys, were carefully regulated). (1944, pp. 95–96)

Even with the many innovations in information processing and communication made possible by the commission system, agents retained the one shortcoming common to all controllers (like bureaucracy in general) based on individual human brains: independence of purpose. A commission agent might profit at the expense of his principal, for example, by buying the consignment himself, before an expected price increase, and selling afterward, pocketing the difference. Although the general laws of agency protected the resident merchant, they remained a court of last resort in a relationship so dependent on the agent's own intelligence, autonomous judgment, and ability to make quick decisions. "Fear of a lawsuit might cause an agent to follow instructions," Lane (1944, pp. 98–99) concludes, "but the faithful performance of precisely defined tasks was not the characteristic most needed in the conduct of agents."

The problem of distributed control, in the absence of telecommunications sufficient to keep a constant check on one's agent (a technology that, because it would have made centralized control possible, would have rendered distributed control less important), was to find agents one could trust: "Even when he had no thought of a lawsuit," according to Lane (1944, p. 98), "an exporter and importer such as Andrea examined his agents' accounts and letters with great care before deciding whether to give him more business or to shift to another agent." An even better solution was to distribute the most dependable information processors available: an immediate family member, relative, or at least a legal partner, friend, or close acquaintance. Alfred Chandler writes as follows of the American colonial merchant:

> He tried, where possible, to have members of his own family act as his agents in London, the West Indies, and other North American colonies. If he could not consign his goods and arrange for purchase and sale of merchandise through a family member or through a thoroughly reliable associate, the merchant depended on a ship captain or supercargo (his authorized business agent aboard ship) to carry out the distant transactions. Even then, the latter was often a son or a nephew. The merchant knew . . . the shipbuilders, ropemakers, and local artisans who supplied his personal as well as his business needs. Finally, he was acquainted with the planters, the farmers, and country storekeepers, as well as the fishermen, lumbermen, and others from whom he purchased goods and to whom he provided supplies. (1977, p. 18)

Given the importance to a merchant of distributing his control globally, despite the inadequacy of other control technology, it is hardly surprising that family partnership became the essential form of business organization, not only in fourteenth-century Venice (Lane 1953, pp. 86–102) but throughout the world system for the next four hundred years (Heaton 1948, p. 358). As economic historian John Killick notes of a large group of merchants engaged in Anglo-American trading at the end of the eighteenth century, "The essential links that held such firms together despite the problems of communication across the Atlantic were ties of family and background. Indeed, the essential commercial unit before incorporation was frequently not the simple partnership but often a chain of mutually supporting partnerships reflecting a kinship network" (1974, p. 503).

Even when kinship or friendship did not exist, principals often attempted to evoke such feelings. As Lane (1944, p. 99) notes of Andrea Barbarigo, "There were in his business letters more protestations of personal affection than can be taken seriously . . . Conventional references to a loving concern for the agent's future honor and profit, references found even in brief instructions to agents hitherto unknown to Andrea, may be interpreted as promises to give more business if the present commission be well handled."

The apparent rationality of kinship- and friendship-based commercial networks like those described by Chandler, Killick, and Lane, when considered in terms of the need for distributed control in the absence of alternative information-processing, communication, and control technologies, provides new perspective on the role of traditionalism in the development of industrial capitalism. It affords us new insight, in particular, into that great historical transformation of traditional into rational society.

Control and the Rationalization of Commerce

As Max Weber stated the late nineteenth-century view, which is still generally accepted, "the traditionalistic attitude"—which he associated with dependence on family and personal loyalties—"had to be at least partly overcome in the Western World before the further development to the specifically modern type of rational capitalistic economy could take place" (1922, p. 71). Here Weber credited the pioneering work of the German sociologist Ferdinand Toennies (1855–1936), who saw the crucial precondition of the capitalist transformation as a shift in social relationships from those traditional and intimate *(Gemeinschaft)* to those impersonal and limited *(Gesellschaft)*, roughly the replacement of community by association (Toennies 1887).

Similar views of the great historical transformation appeared in the work of several other social theorists writing at the height of the Control Revolution—in Otto von Gierke's contrast of "communal" Germanic and "impersonal" Roman law (1880, pt. 2, chap. 6; 1881, 1913), for example, or in Durkheim's distinction between mechanical and organic solidarity (1893). Weber generalized Toennies's dichotomy to the extremes of a continuum running from *communal* social relationships "based on a subjective feeling of the parties, whether affectual or traditional" to *associative* relationships oriented toward "rationally motivated adjustment of interests" (1922, pp. 40–43).

Once traditional and personal loyalties are seen as rational means to distribute control in world trade following the Commercial Revolution, simple means-ends distinctions like that made by Toennies between the *Gemeinschaft*, in which he found family relationships to be ends in themselves, and the *Gesellschaft*, in which associates use each other as means to achieve economic ends, no longer appear tenable. As Weber himself argued, "Communal relationships may rest on various types of affectual, emotional, or traditional bases . . . but the great majority of social relationships has this characteristic to some degree, while being at the same time to some degree determined by associative factors" (1922, p. 41).

The point informs several controversies concerning the rise of industrial capitalism, in particular the one centered on Weber's own *Protestant Ethic and the Spirit of Capitalism* (1905) and the role of religion and ideology in societal transformation. In his analysis of Protestant sects Weber found that—contrary to the usual in-group cohesion, out-group exploitation—Puritans especially took pride in their trustworthiness among the "sinful children of the world." By demanding depersonalized family and community ties so as not to jeopardize the work of members' "callings" and by preaching the similar dangers of hatred, the Puritans fostered not only trustworthiness but reciprocal trust of nonbelievers, standards that Weber thought were significant for the development of modern commerce.

How, then, do we account for the obsession of the American colonial merchant with kin and personal relationships in his business or for the persistence of the family partnership as the dominant form of commercial organization well into the nineteenth century? The answer, as we shall see, lies in the need to control widely dispersed transactions without adequate telecommunications or effective legal sanctions. If lack of sufficient information-processing, communication, and control technology caused the retention of traditionalist values in commerce, it seems reasonable to expect the converse: that rationalization of these values will follow improvements in the same technology.

Killick (1974, p. 505) explains why the need to distribute control contributed to the retention of traditional relationships in commerce. In the late eighteenth and early nineteenth centuries "the effective conduct of the foreign exchange business required that the balance of the firms' liquid capital be shifted back and forth across the Atlantic as necessary," certainly a major control problem with the technology of the day. Well into the nineteenth century, Killick concludes, merchants like Bolton Ogden & Co. based their control on whatever communication they could manage, which usually meant hiring and distributing family members in key commercial centers, communication through kinship networks, and reliance on a traditional code of family and commercial "honor":

> Higher interest rates available in America meant that the firms kept proportionately more of their liquid capital there. Thomas Bolton therefore had to trust [his American partners'] control of his money without effective legal sanctions and without precise knowledge of current trends. Obviously information, intelligence, and experience would obviate the worse risks (information was provided by a huge and regular correspondence with firms all over the United States and Europe), but competition was bound to drive firms to the margin, and risks had to be taken to make profits or save a situation. Partners and agents overseas needed operational latitude to meet local circumstances, and although the law and mercantile ethics usually imposed some degree of trustworthiness even on strange agents, far greater reliance could be placed on confidence resting on friendship and kinship. The mechanics of business therefore explain the closeness of many transatlantic family relationships, why new partners were chosen with considerable care, and why "family" and commercial honor were such important precepts.   (1974, p. 505)

Just as the need to distribute control in the absence of any better control technology caused traditional values to persist in commerce, the rapid disappearance of at least the American general merchant and his family partnership in the early decades of the nineteenth century seems to be due less to ideological or religious changes than to the development of alternative and less centralized means of control. Although little specialization of investment occurred before the eighteenth century, by the beginning of the American Revolution certain "American houses" in London had become market if not commodity specialists. A French *Encyclopedie*, published in 1751, listed several advantages to the specialization of labor; Adam Smith included three— improved skills, time savings, and the increased likelihood of technological applications—in the first chapter of his *Wealth of Nations* (1776, pp. 9–14). By 1819 a Baltimore house had specialized sufficiently to be the obvious choice for a consignment of brandy (Bruchey 1956, p. 372). Killick (1974, p. 519) finds that specialization of commercial functions, well under way in the 1830s, increased "professional competence and institutional efficiency," at the expense of personal relationships, with the result that even a vastly increased commerce could be controlled "more cheaply, more safely, and *almost automatically*" (emphasis added).

In addition to specialization, American merchants relied on commercial control technologies developed earlier by the British and Dutch, including formal exchanges to conduct market transactions, concepts and usages of commercial law, and more sophisticated instruments of credit (de Roover 1963, pp. 49–58; North and Thomas 1973, pp. 53–156). The half-century following the American Revolution brought rapid development of at least the major underpinnings of a commercial control structure: commercial banks (1780s), a federal banking system (1791), state regulation of insurance (1799), federal bankruptcy law (1800), fire insurance joint-stock companies (1810), commercial newspapers (1810s), state marine insurance (1818) and insurance of bank deposits (1829), a commercial credit rating company (1841), and a credit protection group (1842).

Were these and many similar innovations in the national system for the control of commerce not sufficient to reassure merchants in their transactions with strangers, a parallel system of telecommunications developed in the early nineteenth century by which they might maintain command and feedback links with even a large number of far-flung associates. This system arose from the continuing coevolution of transportation and communications, each growing in response to a series of impressive innovations in the other: a federal postal service (1791), the first turnpike (1795), coastal steamboat travel (1809), mail delivery by steamboat (1813), regular packet service to England (late 1810s), steam railroads and Atlantic clipper ships (early 1830s), local postal delivery service (1836), regular transatlantic steamship service (1838), express delivery using railroad and steamboat (1839), a national turnpike to the West and the first interregional rail line (1840), commercial telegraph (1847), and regular steamboat service to California (1849).

The remainder of this chapter examines the transformation of American domestic and transatlantic commerce—most often away from tradition toward more rational modes of operation—in the half-century preceding America's Industrial Revolution. Even in this preindustrial period, as we shall see, rationalization of commercial structures paralleled the development of a national and international infrastructure of transportation, communication, and control of commerce. At least a half-dozen commercial institutions of the period, including the factorage, brokerage, auction, wholesale jobbing, and agricultural marketing systems, served to integrate the systems of technological and economic control also developing in the early nineteenth century. These same institutions constituted important transitional stages between traditional and modern control and thus served the transformation of American capitalism from a commercial system to an industrial one.

The Three Reciprocal Flows of Matter and Information

During the first decades of American independence, the personal networks of the colonial merchant gave way to increasingly specialized and impersonal market exchanges. Growing chains of such exchanges, involving both commercial transactions and material transshipments and opposite flows of matter and information, moved goods through increasingly denser networks of specialized middlemen. Considered at the most macro level, the new commercial infrastructure can be seen to have sustained three major exchanges of material goods for information—in the form of commercial paper—about their movements, or three predominant flows of matter and energy through the international processing system: First, cotton, tobacco, rice, and other agricultural commodities, grown mostly in the South, moved from plantations to river ports, then on to New York and other northern coastal cities, and eventually to Great Britain, where they fed the machines and workers of Britain's Industrial Revolution. Simultaneously with this flow, textiles and other machine-made goods passed from Britain back to American markets through similar chains of ocean and waterway shipments and commercial transactions. Finally, as the farming frontier moved across the Appalachians and into the Ohio and Mississippi valleys in the early nineteenth century, extensive interior networks developed to move agricultural commodities—provisions, horses, mules, and whiskey—south and especially eastward.

After completion of the Pennsylvania and Ohio canal systems in the 1830s, the interior movement included an increasing stream of Mississippi Valley wheat, flour, and other grains, most of it bound for domestic consumers. As Alfred Chandler describes the rapid development of this new continental commercial infrastructure and the world system into which it became increasingly incorporated, "Nearly every plantation, farm, and village in the interior came to have direct commercial access to the growing cities of the East as well as to the manufacturing centers of Europe. The output of millions of acres moved every fall over thousands of miles of water. Dry goods from Manchester, hardware from Birmingham, iron from Sweden, the teas of China, and the coffees of Brazil were regularly shipped to towns and villages in a vast region which only a few years before was still wilderness" (1977, p. 27).

The rapid growth of this continental commercial infrastructure had been paced by cotton. Not grown commercially in America until 1786, cotton began to restructure the nation's commerce after Eli Whitney introduced his gin in 1793. Two cotton mills had been built in the United States by 1795, four by 1804, only fifteen by 1808. The boom came in 1809 when an additional eighty-seven mills began operations, nearly quadrupling the nation's capacity—from eight to thirty-one thousand spindles—in a single year. Over the following two years, to the brink of war with England, capacity almost tripled to an estimated eighty thousand spindles (North 1966, p. 56).

The Factorage System of Distributed Control

Like the other two predominant movements of goods through the American economy, the flow of cotton from southern plantations to British factories quickly differentiated a half-dozen intermediaries to conduct some of the first specialized information work: factors, commission merchants, financiers, brokers, exporters, and manufacturing agents. First in this lengthening chain of middlemen leading to Liverpool stood the cotton factors, usually operating out of river ports like Memphis or Montgomery, who received shipments of raw cotton from the plantations and either sold them directly to agents of manufacturers or consigned them to other middlemen in coastal ports or in England (Bruchey 1967, pt. 4, chap. 3; Woodman 1968, pt. 1).

Because cotton prices could fluctuate wildly as a result of instability in worldwide supply and demand, even providing advances carried certain risks if prices fell during transit (Chandler 1977, p. 22). For this reason, cotton factors and other firms specializing in agricultural trading did not normally risk taking title to goods but worked instead for a flat commission (usually 2.5 percent). Early differentiation and predominance of a commission trade, therefore, might be directly attributed to the spread of commercial agriculture—especially cotton—in the South and West.

The factorage system for the movement of American cotton served much the same purpose as the commercial system that developed in the twelfth century when merchants began to conduct business through others in distant ports who executed orders on commission. By the fifteenth century such commission merchants had become institution-alized in the *factor*, from the French and Latin for "doer," a word that had entered English by 1485 (1491 with the commercial meaning); the word *factory*, for the office or position of the factor, had appeared by 1560 (Murray 1933, vol. 4, pp. 13–14). As one nineteenth-century Charlestonian recalled, "These businessmen of a high type . . . were indeed people who 'did' things for the planter as their designation imports" (Woodman 1968, p. 4). Indeed, it would be difficult to think of a more appropriate word for a programmable information processor and decision maker than "doer" in view of the implied purposiveness and the centrality of purpose to programming, decision, and control.

Drawing on several American legal works of the early nineteenth century, N. S. Buck defines the factor more formally as

> an agent empowered by an individual or individuals to transact business on his or their account. Usually he was not resident in the same place as his principal, but in a foreign country, or at a distance. The business which he transacted depended on the authority given him by his principal, and might be limited to a particular and specific transaction, or might be more extended, and comprise buying and selling, shipping, negotiating insurance, discounting bills, making payments, etc. He could, and in many cases did, do anything which the principal himself could do through an agent. A distinctive mark of the factor seems to have been that he was permitted to transact business in his own name and as if on his own account. (1925, pp. 6–7)

The American cotton factorage system evolved from the seven-teenth- and eighteenth-century system for marketing southern tobacco and West Indian sugar in Britain, a development that had four stages. Initially planters consigned their tobacco directly to British merchants who sold the crop, purchased and shipped manufactured goods in re-turn, and advanced credit when required. As tobacco production in-creased and the population moved inland, British firms located representatives in various colonial ports, who established themselves as intermediaries between American planters and merchants in Eng-land. When trade with that country was interrupted during the Rev-olutionary War years, many Americans replaced the British resident merchants or "factors," developing ties with the southern planting class and capital sufficient to finance the tobacco trade domestically (Gray

1941, chap. 18; Boorstin 1958, chap. 18). With the appearance of domestic cotton after the war, a parallel factorage system grew up for cotton, "not from a conscious desire to emulate the system of tobacco marketing," according to Harold Woodman, "but rather as a practical effort to meet certain immediate needs" (1968, p. 10).

Foremost among these were the need to gather market information, process it to make price predictions, and arrive at selling and buying decisions that maximized profits. For example,

> early in February 1800 a factor wrote to a planter that his cotton had been received but not sold. The shipment was being held, he said, because the market was uncertain but he predicted a price rise later. Unless given orders to the contrary, he would continue to await improved conditions before selling. Thus, the planter had not been forced to sell in an unfavorable market, nor had he been required to be on the scene awaiting the propitious moment to dispose of his crop. The factor had handled all this for him.   (Woodman 1968, p. 10)

That the antebellum southern planter intended his factor to be an all-purpose controller finds reflection in the memoirs of a South Carolinian, I. Jenkins Mikell, whose boyhood spanned the Civil War. Mikell refers to the position with the alliterative "factotum," from the Latin *Dominus factotum*, "one who controls everything" (Murray 1933, vol. 4, pp. 14–15), a word that southerners used for a head servant of a plantation who arranged his master's daily affairs. Mikell described the corresponding commercial functions, several of which might now be filled by personal computer software:

> The Factor was the factotum of our business life, our commission merchant, our banker, our bookkeeper, our advisor, our collector and disburser, who honored our checks and paid our bills. Many of the planters did not really always know what money they possessed. One year's accounts would overlap another's and sometimes years would pass before the accounts were balanced and settlement made. The planter did not worry, so long as he could draw what money he called for. It was a terrible jar when the old order of things gave place to the new, at the close of the War Between the States.   (Mikell 1923, pp. 200–201; Woodman 1968, p. 4)

Clearly the factor, as both longstanding family acquaintance and business agent, represented another intermediary phase in what Weber saw as the gradual transition from twelfth-century traditionalist ("internal") control to the rationalist ("external") control of the late nineteenth century. Internal to the southern aristocracy, the relationship between planter and factor was bound by a chivalrous "gentleman's code," what one historian of the period described as "that quick sensitiveness upon all questions of personal integrity and honor which they had inherited from their fathers," a sensitivity frequently reinforced by the duel (Osterweis 1949, p. 87). Externally, the international factorage system became bound up in what Justice Joseph Story called in an important 1840 Supreme Court decision the "ordinary relation of principal and agent," that is, the general laws of agency (Woodman 1968, chap. 6).

The legal concept of agency dates from the Roman Empire, although it never achieved effective form there, in part because the use of sons and slaves as instruments of contract made a law of agency less necessary for the Romans (Thomas 1968, pp. 13–14). Agency reflects precisely the compromise between autonomous and purely mechanical control over transactions that we might expect when even a single round of query, reply, command, and feedback required several months. A New York merchant described the problem in 1806 in a letter to his partner in Halifax:

> The order for cottons on a/c of Richard Roberts . . . I received . . . The commissions are very liberal but how to carry the order into effect I know not. People in giving such orders never reflect that Charlestown is near 900 miles distant from New York and that a month must elapse before you can get an answer to a letter . . . or that during that time prices are continually fluctuating and are not to be calculated upon with any degree of precision. You must be sensible Hindley [a Liverpool cotton buyer then in Charleston] must have every shilling remitted to him before he would undertake to buy cotton . . . and supposing that we had the means it can never possibly be worthwhile for us to have money lying idle in Charlestown, unless we certainly knew that the money would be invested.   (Killick 1974, p. 511)

Planters who attempted continual control of a factor from their plantations with telecommunications such as these "were at a distinct disadvantage," according to Woodman. "News about market conditions was often stale by the time it arrived in the mails. Information about a rise in prices might not reach the planter before a contrary tendency governed the market and the advantage lost" (1968, p. 22).

That the general laws of agency reflected the need for partially programmable, partially autonomous control can be seen in their codification by commercial nations and in the interpretations of these nations' courts. Although the factor conducted business in his own name, neither American nor British law allowed him to extend credit for the commodities that he sold. In transactions where a mercantile custom of credit prevailed, however, any loss through the failure of a factor's creditor fell to his planter (Buck 1925, pp. 7–8).

In short, the laws of agency made the planter himself responsible for anything his factor did in accord with tradition or upon his principal's direct command and for nothing else—laws that assured that the factor would serve as an intelligent information processor but not as an autonomous one. In this way the laws of agency rendered the factor a mechanical and hence programmable controller, much as the formal organization and rules of bureaucracy would do for large collections of individuals in the latter part of the nineteenth century. As the 1840 Supreme Court decision described this mechanical and programmable control, "the consignor had the privilege of setting conditions controlling the sale of his goods 'according to his own pleasure,' and the factor was bound to obey any instructions given him" (Woodman 1968, p. 60).

In the absence of any competing interests of his own, the factor served well as programmable controller, motivated by the fact that the higher the price he got for his planter's cotton, the greater his commission and commercial reputation. Here reputation and honor, enforced through interpersonal networks as the traditionalist control mechanism of the feudal South (Osterweis 1949, pp. 16–17), still prevailed over more rationalized control: "Given the importance of the factor-planter relationship," Harold Woodman concludes, "it is surprising how little litigation did take place, how general were the laws and court decisions, and how much of the relationship was governed by custom and tradition" (1968, p. 60).

The factor's purely mechanical processing broke down, however, as opportunities arose for him to profit at the possible expense of his planter. As the Alabama *Tuscumbian* warned in 1825, factors' "locations in seaport towns gave them an advantage of receiving advanced news of market quotations as well as contemplated changes"; hence they might buy cotton consigned to them before a price rise and sell it afterward, keeping the difference (Davis 1939, p. 145). Based on similar reasoning, *Farmer and Planter* advised its readers in 1858 to "scan with a suspicious eye" the operations of their factors (Gray 1941, p. 711). The mere fact that commercial newspapers and trade journals appeared in the South after 1815 signaled the beginning of the end of that region's largely traditional control of commerce.

As might be expected, the breakdown in trust between planter and factor brought increasing pressure for more rationalized control, in this case functional specialization of factors. In Louisiana in 1834 and North Carolina in 1835 courts ruled that the same firm could not serve as factor for both buyer and seller. They cited as their reason the incompatibility of programming: custom and law both bound a factor to sell at the highest attainable price and to buy at the lowest one possible. Some northeastern and British firms even announced that they would not engage as buyers any factors who served planters as sellers. By the late 1840s factors had begun to advertise that they did not *buy* cotton, and Woodman finds that most did not, even though buying remained—on separate transactions—legally permissible. A commercial magazine in 1853 clearly differentiated factors from buying agents and speculators from factors, who it reported "usually pledged not to speculate in produce" (Woodman 1968, pp. 69–71).

The growing communications infrastructure gave the planter more direct control of his factor by the early 1850s. Telegraph lines in operation by 1853 linked all of the major southern cities and much of the rural areas with northern ports (Pred 1980, p. 154); a transatlantic cable directly linked America and Europe after 1866. Meanwhile, commercial newspapers brought the planter estimates of the quantity of cotton produced and prices paid for it in southern, northern, and European markets. The *Price Current* of various market cities printed abridged editions, reduced in size and available in bulk orders, which planters expected their factors to send them regularly. Some *Price Current* editions even printed letter sheets, folded pages with reprinted prices on one side, the other side blank for the factor's letter; almost every collection of planter's papers contains such correspondence (Woodman 1968, p. 20).

Should more precise monitoring of the factor's sale of a consignment be desired by its owner, commercial journals reported the daily docking times of every ship and the arrival of every conveyance into the market so that planters might know exactly when their shipment had been received and could correlate this with the daily prices for cotton reported by the same publications (Woodman 1968, p. 71). Continued growth of such commercial media and telecommunications capabilities after the Civil War, along with various innovations in the control of distribution, brought the beginnings of the end of the factorage system by the early 1870s.

On the eve of industrialization, however, the southern planter still lacked the capability to process system-wide information that especially the cotton exchanges would provide the next generation. In the absence of such information, and because he also lacked sufficient speed of transportation and communication, the planter could not centralize control of his affairs but depended on the distributed control provided by his factor and a chain of other middlemen stretching through the river and coastal ports to Liverpool and Manchester. "Lacking the means of communication that would allow him to have precise and timely knowledge of market conditions," Woodman concludes, "the planter came to trust the judgment of his factors in most of his business dealings" (1968, p. 71).

Control Distributed to Commission Agents and Brokers

Continued growth of the international markets in cotton and other agricultural commodities during the 1820s and 1830s also encouraged commercial specialization in information work in the northeastern ports. Soon the commission merchant, a functional equivalent of the cotton factor, began to replace the general merchants of the eastern coastal ports. Although Buck (1925, pp. 6, 10) finds "no practical difference" between commission merchants and factors, the former did not ordinarily serve as factors exclusively but imported or exported goods on their own account as well.

Robert Albion, historian of early New York shipping, argues that the cotton trade helped to make that port America's most important by hastening the demise of its general merchants (1939, chap. 6)—those who, unlike the factor, traded on their own account, usually in many different commodities and to all parts of the world (Buck 1925, pp. 4–5). Because cotton and other agricultural commodities dominated trade through New York by 1840, and because of the risks associated with instability in their prices, nearly 70 percent of the port's thirteen hundred merchant establishments operated on commission in that year. In southern ports, more dependent on agricultural commodities, commission trading developed even faster: all but eight of the 383 New Orleans houses worked on commission in 1840. In Boston, by contrast, where agricultural trade never became as important, less than 40 percent of some 230 firms had established themselves on a commission basis in the same year (Pred 1973, p. 195).

Even as commission houses began to replace the more general merchants, the commission merchants themselves began to specialize in various ways that increased their control of a market. As early as 1795, for example, Everitt of London described his business as "principally the export of woolens to Ireland" (Buck 1925, p. 9). Although a New York City business directory of 1846 listed 317 "general" commission merchants, it also showed that ninety-one specialized in dry goods, eighty-six in flour and other produce, and eight in domestic hardware (Albion 1939, p. 275).

Some commission houses, in addition to dividing their labor according to categories of commodities, reduced their information-processing burdens still further through specialization along various other dimensions of trade, including regions (continents or countries) and, to a lesser extent, direction (imports or exports). William Higgins of London, for example, proclaimed in 1812 that he "dealt with Malta only, and with no other part of the Mediterranean." By 1826 Sir Claude Scott, a prominent British wheat merchant, imported exclusively and would not return any portion of a client's earnings in manufactured articles (Buck 1925, pp. 8–12).

Commission merchants in New York and other American coastal cities sold directly or on consignment to several different types of buyers: to speculators, many expecting to profit from shipping the cotton to other markets; to representatives of European cotton houses; and to agents of the Lancashire spinning firms (Buck 1925, p. 80). These last *manufacturers' agents*, who also worked on commission for their services, ended the chain of intermediaries between southern plantation and British factory.

Some speculators retained the option to sell either to New York commission merchants or in European markets by shipping cotton directly to Europe but sending samples from each bale to New York. This innovation, called selling "in transit," saved the extra expense of shipping cotton first to New York and then to Europe, at a time—the early 1850s—when three bales out of four eventually left the country anyway (Watkins 1908, pp. 29–31). The farther a port from New York, as might be expected, the more likely were its speculators to sell in transit: those in New Orleans, Mobile, and Savannah were the greatest users of the method. By 1858 in-transit sales in New York accounted for some two hundred thousand bales per year (Woodman 1968, p. 28).

In order to sell in transit or to be successful as a factor, commission merchant, or buyer, one had to be able to classify, grade, and sample cotton. Because its value varied markedly according to quality and grade, an error in classification would either prove costly to a seller or land him in litigation. This problem of preprocessing bales into their essential market information came to be resolved by yet another specialized middleman and information worker: the cotton broker. In larger ports where trade volume permitted, individual brokers served as intermediaries between buyers and sellers, helping them to find one another for a commission and to agree on a shipment's quality and grade; associations of brokers standardized market price quotations and sales reports and adopted—in the words of *Hunt's Merchants' Magazine* in 1860—"other rules amongst themselves to secure a more uniform and satisfactory method of receiving cotton" (Woodman 1968, p. 25).

Despite the important preprocessing functions of classifying and grading, the broker's primary informational tasks remained those of search and coordination. Brokers helped factors to dispose of odd lots, commission merchants to fill out shipments, and agents to complete manufacturers' orders, not by serving clients of long standing but by bringing buyers and sellers together for a fixed commission (one-half to one percent). When a merchant informed a broker that he wished to buy a certain quantity of cotton, for example, the broker would go to the market to learn the current price and where bales might be purchased. *Bankers' Magazine*, a New York trade journal, described in 1854 the information-processing tasks involved in binding buyer and seller legally:

> When the terms of purchase were approved by the merchant, and the purchase arranged, it was then the broker's duty to make a note of the purchase in a book of his own, stating the names of the buyer and seller, the quality and description of the article bought, the price for which it was bought, and any special or unusual conditions or terms of purchase. Having made this note, he signed it, and went to the selling broker to compare it with a similar note made by him. If the notes corresponded exactly, the buying broker sent an exact copy of his memorandum to the broker of the seller, whose duty it was to exchange it for an exact copy of his own memorandum. Each copy was, or should have been, signed by the broker who sent it. They were then called "bought and sold notes," the bought note being that which was sent to the seller's broker. If the same broker acted for both the buyer and seller, after making the entry in his book it was his duty—and it was generally the practice—to call on

both parties to the transaction, in order to ascertain that he had made his entry correctly, according to the views of both parties. He then sent a copy of the memorandum to buyer and seller. (Buck 1925, p. 21)

Because the various brokerage functions required detailed knowledge of a commodity and its market, most brokers specialized in ways that paralleled the specialization of the commission merchants. The cotton and wool trades had large numbers of specialists; sugar, spice, coffee, tea, and tallow brokers were not uncommon. At least two large brokerage houses specialized in West Indian produce by the 1820s; others concentrated solely on American, South American, or East Indian produce. An 1823 commercial directory listed only four brokers in Boston and none in New York; Boston's total had risen to twenty-three by 1850. Only in the wheat trade, which required little preprocessing of information because most grain was substitutable, did no brokers appear—factors sold directly to millers (Buck 1925, pp. 19–21, 28–29).

Credit Flows in the Control of Exports

In the cotton trade, unlike that of wheat, a considerable amount of capital could be tied up in the advances and discounted bills of exchange that kept the bales of raw cotton moving through the long chain of middlemen between plantation and factory (Buck 1925, pp. 23–27; Woodman 1968, chap. 10). Yet another set of intermediaries, the financiers, often the older resident merchants or establishments in Liverpool and London, provided the flow of capital to fuel the movement of cotton through possibly five transshipments: from plantation overland to the nearest river port, then down river to a coastal port like New Orleans, Mobile, or Charleston, by sea to New York or Boston, across the North Atlantic to Liverpool, and finally to the manufacturers themselves. These movements required at least three commercial exchanges: between planter and factor, between factor and manufacturer's agent, and between agent and manufacturer. With each separate movement of cotton toward England, an advance against its sale moved in the opposite direction, so that the total voyage generated at least four commercial transactions and four different bills of exchange.

After shipping his cotton, a planter could draw from his factor, usually two-thirds to three-fourths of the current value, thereby creating a domestic bill for discount. The factor reimbursed himself for the advance, after consigning the cotton to a ship bound for England, by drawing on his Liverpool branch, thus creating a second bill for discount. After a British manufacturer purchased the consignment, the agent sent a sterling bill for the proceeds—minus the advance drawn against them—back to the American office. The factor sold this sterling bill to a local bank for dollars and then authorized the planter to draw another bill to cover the remainder of his payment.

Except for their opposite directions of movement, cotton and money, representing material processing and the information used for its control, played equivalent roles in the system. Anyone with a few bales could easily get an immediate cash advance from the agent to whom he consigned it, whether in the interior, in the southern ports, in the North, or in England (Woodman 1968, pp. 34–35). Conversely, it was virtually impossible to obtain goods on consignment without giving advances (Buck 1925, p. 13). Thus the commercial infrastructure also served as a credit network, with advances based on the value of cotton moving in the opposite direction of its shipment, so that the entire cost of its transportation, processing, and distribution could be financed almost entirely on credit.

Rapid growth of this infrastructure might seem a direct result of Britain's Industrial Revolution, with the mounting demand of its new textile machinery for raw cotton. Most historians of the period concur, however, that this development of the quaternary or control sector had begun well before the cotton trade and would in any case have come by the time of America's industrialization of the 1840s. "Even without the boom in cotton and textiles," argues Chandler (1977, p. 19), "specialization in commercial business enterprises certainly would have come to the United States in the fifty years after 1790." By that year, other historians report, structural differentiation had already begun in the commerce of larger towns like Philadelphia (Warner 1968, chap. 1) and New York (Harrington 1935, chap. 2), where shopkeepers who bought from local wholesalers and sold only at retail could be distinguished from suppliers who bought goods directly from abroad— merchants who were themselves beginning to specialize in particular lines.

Distributed Control of Imports

Retail sales by the newly specialized shopkeepers soon constituted only the final transaction in a steadily expanding distributional system that brought an increasing variety of foreign goods to American markets. Like the opposite movement of cotton and other agricultural commodities for export, the system for the import, processing, and distribution of especially European goods quickly differentiated a half-dozen intermediaries engaged in specialized information work: British manufacturer's agents and exporters, American importers, auctioneers, jobbers, and interior merchants, as well as the newly specialized shopkeepers. Rapid increases in imports, especially of British textiles following the War of 1812, provided the essential impetus for this structural differentiation.

Despite the war, virtually total control of the importation and distribution of foreign goods in America remained—after 1815—in British hands (Cohen 1971, p. 510). British exporters purchased the bulk of domestic textiles from manufacturers' agents located in London, Manchester, and other commercial centers; some had begun business as early as the 1740s and sold exclusively to exporters. These merchants usually consigned goods to America at their own risk, either to their own agents, to branch houses, or—in the case of less important exporters—to an American house for sale on commission. Only a few American importers, residing in Great Britain, purchased on their own account from British merchants or from manufacturers directly (Buck 1925, pp. 103–108).

Before the war American imports from Europe had been moved by the same commission houses that handled southern cotton—import and export functions had not yet begun to differentiate. With the increasing volume of foreign trade after 1815, however, the shipment and distribution of imports to the United States rapidly specialized in two distinct ways. First, the older firms that continued to export cotton on commission progressively narrowed the range of their imports. Second, new merchants increasingly specialized in the purchase of goods in Europe and their import and sale to American manufacturers and wholesalers (Killick 1974; Perkins 1975, chaps. 2–4). Because prices for these goods fluctuated less than did prices for agricultural commodities in the international market, these more specialized importers—unlike exporters—often took title to shipments rather than selling on consignment or commission (Tooker 1955, pt. 2; Porter and Livesay 1971, chaps. 2–4).

In his study of Bolton Ogden and other large groups of merchants during the critical period from 1790 to 1850, John Killick concludes that specialization of the American import trade did hasten the replacement of traditional family relationships by more rationalized commercial associations:

> The relationships amongst the younger partners was not as warm as amongst the older generation . . . These estrangements . . . reflected the declining value of the chain of mercantile relationships, for which it no longer seemed worthwhile to struggle. A number of reasons account for this decline, which, repeated in many other firms, was probably apparent in transatlantic commerce as a whole. The shift of American interest westwards, the rise of manufacturing, the smaller proportionate importance of British commercial credit, and speedier transatlantic communication were all important . . . Increases in the volume and regularity of commerce were leading to specialization. Trade increasingly relied on professional competence and institutional efficiency rather than personal relationships. Many of the functions undertaken in 1800 . . . were by 1840 being split up amongst shipowners, manufacturers, bankers, credit rating agencies, speculators, and stockbrokers. Consequently, vast increases in commerce came to be handled more cheaply, more safely, and almost automatically. One important result was the decline of many traditional firms . . . as well as of the close-knit skein of mercantile relationships across the Atlantic and the economic and cultural unity that this had implied.   (Killick 1974, p. 519)

Thus did traditional relationships give way—in the course of a single generation—to what Weber would call "rationally motivated adjustment of interests." Central to the change was the auction, a dominant system for processing and communicating market information in the 1820s and 1830s. During this brief period the auction system served not only to control the allocation of commercial goods but to smooth the transition from traditional face-to-face exchange to more rationalized forms of distribution.

The Auction System in Control of Distribution

Auctions and similar bidding systems originated in the trading practices of the ancient civilizations and rank among the oldest forms of control over the allocation of resources (Amihud 1976, p. xi). Herodotus describes a Babylonian auction, about 500 B.C., in which women of marriageable age were sold—in descending order from the most beautiful—on condition that they be wed by their purchaser; money paid for the more comely subsidized those less attractive as bidding became negative (1910, pp. 100–101). The Romans adapted auctions to commercial trade and differentiated two of the few information specialists in ancient times: the *argentarius*, who organized, financed, and regulated the sale, and the *praeco*, who advertised and promoted the auction and served as auctioneer (Thomas 1957, p. 43).

The word *auction*—derived from the Latin *auctus*, "an increase"—entered English by 1595 and became commonplace in the late seventeenth century (Murray 1933, vol. 1, p. 559). Auctions had spread to New York City by 1676, when Matthias Nicoll was appointed vendee master with security proof of two thousand pounds sterling, an indication of the auctioneer's responsibility at the time (Westerfield 1920, p. 164). Of the better-known surviving British auction houses, Sotheby's began as Samuel Baker's, bookseller, in 1744, Christie's in 1766. The earliest English legal action to develop from an auction, *Daniel vs. Adams*, was settled in 1764 (Cassady 1967, p. 30).

Despite the ancient origins and widespread application of auctions, they did not become important means of allocation in the international commercial system until the Nonimportation Act of 1806, the Embargo of 1807, and the War of 1812–1814 disrupted the flow of goods from Britain to the United States. Except for a few months respite in 1809, the eight years brought increasing shortages of British goods in America. What goods did get through sold at auction, an especially attractive means of distribution in a seller's market—that is, when supply is low and demand high (Cohen 1971, p. 488). "The large sales and high prices attracted the attention of both buyers and sellers and gave them the habit of attending the public sales," according to Jones (1937, p. 33). "English merchants observed all this, and in preparation for the cessation of hostilities established extensive depots of goods at Halifax, Bermuda, and other British possessions."

Upon cessation of hostilities in the spring of 1815, the imperial warehouses dumped their contents on the American market. "All witnesses agreed that large profits were made on these shipments," according to Cohen (1971, p. 489). The size of these profits helped to establish the auction system as a legitimate commercial institution. "The auction offered the most convenient medium of sale and was used quite generally . . . The success which attended these first ventures encouraged the merchants and other speculators to send fresh orders to Great Britain, and so a stimulus was given to the British manufacturer to increase his output" (Buck 1925, pp. 137–138).

American importers, still few in number, their business devastated by the eight-year disruption of trade with Britain, now found that they lacked the processing speed needed to compete with the auctions as a distribution system on two counts. First, because of the slowness of transatlantic communications, importers had to order at prices already weeks old and had difficulty responding to the frequent oscillations in auction prices (Albion 1939, p. 276). Second, whenever an American importer sent an order to Britain for goods, the manufacturer or merchant could send a duplicate order on his own account back on the same ship, knowing that he could sell them at auction almost before the hapless importer could have his own goods unloaded (Buck 1925, p. 142).

The superior responsiveness of the auction as a processor of market information, in short, virtually drove the American importer out of business. Robert Albion, historian of New York port, described the problem for that city's importers:

> Merchants, having to order at prices determined months in advance, could not compete with the unpredictable fluctuations of the auction rooms; their countinghouses were deserted by the New York jobbers who, like many retailers from other parts of the country, haunted the auction sales in search of bargains. Country newspapers advertised the stocks of local storekeepers, "recently purchased cheap at auction." Articles of every sort—houses, land, tea, crockery, hardware, and much else were put up for sale, but British textiles overshadowed all the rest at the Pearl Street "vendues." (1939, pp. 276-278)

As a result Americans owned only a quarter of the dry goods imported to New York by 1819 and roughly the same fraction as late as 1825. A commercial newspaper in Baltimore claimed in 1824 that the city had hardly one large textile importing house because of foreign competition (Buck 1925, p. 140). The four major commercial centers of Boston, New York, Philadelphia, and Baltimore had 160 importing houses in the wool trade before the War of 1812–1814; in 1830 only about twenty of the same class remained, the reduction attributed in congressional testimony to the auction system (Jones 1937, p. 35). As an anonymous writer summarized American trade in 1820, "the manufacturers of Birmingham and Manchester, in England, have become the traders of New York" (Buck 1925, p. 140).

Although American importers successfully lobbied for restrictions on auctions in several states and localities, including Philadelphia and Boston, similar efforts failed in New York City, where in every year between 1821 and 1833 at least 30 percent of all imports sold at auction

(Cohen 1971, p. 496). During the peak sales period for New York auctions, 1825–1836, sales averaged $21.6 million per year, 38 percent of the value of all imports entering the port (Albion 1939, p. 410). By 1827, 26 percent of all imports to the United States sold at auction, up from 10 percent only a decade earlier; the fraction remained above one-fifth through 1830 (Cohen 1971, p. 496).

Auctions as an Information System

The auction has three features that distinguish it from other means of allocation among competing buyers: (1) goods auctioned are ordinarily indivisible; (2) their seller remains relatively passive; and (3) there are no provisions for *recontracting*, that is, canceling a sale if a better price is available from another seller or group of sellers. Recontracting will ordinarily achieve Pareto Optimality, the equilibrium state in which no improvement for any one actor could leave all others at least as well off, for a simple reason: whenever suboptimal allocations do occur, they can be recontracted until optimality is achieved. That auctions are *not* Pareto Optimal can be seen in tobacco auctions, where so-called "pinhookers" buy lots in the early rounds and resell them—right on the floor of the auction house—in later rounds. "If Pareto Optimal allocations are achieved in this manner," according to economist Andrew Schotter, "we certainly could not attribute them to the auction itself." As Schotter generalizes: "Not only do mistakes occur in auctions, but all mistakes are final. Therefore, it is not surprising to find sub-optimal allocations occurring in auctions" (1976, pp. 4–5).

Despite being a suboptimal means of economic allocation, auctions have several advantages as information-processing, communication, and control systems. First, auctions serve as a general means to process goods that for a variety of reasons contain more information than the larger distribution system has been designed to handle. As Buck (1925, p. 57) notes of the eighteenth-century British cotton auction, for example, "In many cases it is resorted to as a means of disposing of odd lots, damaged goods, or goods *difficult to classify*" (emphasis added). Goods "difficult to classify" represent a breakdown of preprocessing and a challenge to information processing for control of distribution, which the auction helps to meet even today in police auctions, liquidation sales, and auctions of antiques and art, rare books and coins, unclaimed baggage, and unimproved land (Cassady 1967, pp. 17–19). In the words of business economist Henry B. Arthur, "Auctions do have a place, especially where unique products, perishables, or unpredictable supplies have to be moved" (1976, p. 200).

Second, an auction system serves as a particularly effective means to control the distribution of material whose flows are sporadic or irregular. During the War of 1812–1814, for example, what few British goods entered America came mostly as prizes captured by armed privateers or from smugglers; the auction first established itself in American commerce as a means to dispose of these occasional items (Buck 1925, p. 136). A. H. Cole, historian of the American wool manufacture, suggests that an auction system may be necessary to distribute the sputtering output of any new industry: "The auction method, unorganized as it was in this period, was distinctly the sign of an immature market for wool fabrics. It suggested a lack of regularity in the demand for such goods and the failure of woolen factories to provide an even and orderly flow of products to the consuming areas" (1926, p. 218). Even in more mature industries like cotton textiles, where factory output proved steadier, "large producers discovered that they could, in slack season . . . dump their surplus through auctions" (Cohen 1971, p. 499). For this reason, argues Ware (1931, p. 176), the auction's "decline was coincident with the growth of selling houses, which were better equipped to gauge market conditions, and with the opening of foreign markets, which carried off surplus goods."

Third, an auction system serves as a direct substitute for information about new or unfamiliar markets, or indeed for a distributional infrastructure itself, an important function filled today by general retail chains, trade representatives, and marketing agents and consultants. "Foreigners selling in the United States had very poor knowledge of the size and nature of the American market, and they found auctions a convenient means of disposing of goods and realizing a quick return on their ventures," according to Ira Cohen, historian of the New York auction system. "Domestic manufacturers and interior retailers often found the auctions a useful distribution mechanism because the early domestic wholesaling system was rudimentary and imperfect" (1971, pp. 488–489).

Fourth, an auction system serves as a good information processor and controller because it can be readily programmed or reprogrammed to meet particular conditions. When Samuel Baker began the book auctions that would become Sotheby's in 1744, he published a list of five rules containing, among others, a price-escalator clause: "No Person advances less than Sixpence each bidding, and after the Book arises to One Pound, no less than One Shilling" (Brough 1963, pp. 25–27). A New York law passed in 1817 attempted to regulate the conditions for conduct of auctions by requiring that they be published in a newspaper of the port at least two days in advance (Cohen 1971, p. 494), and auctioneers routinely repeated the rules before the start of bidding (Westerfield 1920, p. 175). Even the particular form of bidding was open to reprogramming: although most nineteenth-century auctions used the ascending system common today, some adopted, for example, a descending system (called *mineing*) in which the auctioneer began at a high price and continually reduced it until a buyer called "Mine!" Auction houses also used combinations of descending and ascending bidding systems (Cassady 1967, p. 32).

Fifth, auctions served as a regular means by which manufacturers and distributors could communicate face-to-face, an important informational function now institutionalized in the annual convention. "The auction system for the American manufacturer was not only a means of distribution but also a method of educating retailers and jobbers as to what was available," writes Ira Cohen about the New York auction. "At the same time, the manufacturers were able to learn more about what the public would buy" (1971, p. 499). Geographer Alan Pred, in a study of the circulation of information in early America and the effect on urbanization, draws much the same conclusion about Boston's semi-annual auctions of the 1820s and 1830s: "Most important, in terms of the amount of information gathered and disseminated as a result of auction activities, retailers sometimes attended these sales 'for many days, and even for weeks together' " (1973, p. 151; quote from Flint 1822, p. 59).

Finally, auctions served to control distribution by providing working models of the larger distribution system upon which new products and styles could be tested, an important informational function filled today by attitudinal surveys, test marketing, and similar techniques of modern market research. For example, the textile auction in the early nineteenth century "offered a means of testing a market which, because of the slowness of communication and the lack of well organized selling methods, was otherwise difficult to gauge," according to Caroline Ware, historian of New England cotton manufacturing. "It was a good way in which to introduce new styles of goods and its prevalence around 1830 may have been partly due to the extension of calico printing at that time and the need of introducing a variety of prints to a market which was accustomed to handling principally standard gray goods"

(1931, p. 175). Cohen (1971, p. 498) draws the same conclusion for New York auctions: "To American manufacturers the auction room also offered the chance to gauge the readiness of the market for the introduction of new products and styles. The greatest period of experimentation in styles was in the years 1826 to 1832, which coincided with the maximum use of auctions by domestic producers."

Shortcomings of Auctions as an Information System

Despite these several advantages of auctions as information-processing, communication, and control systems, they also constitute—in a separate sense—a barrier to information. Because the identities of both the manufacturer and the seller of a good are easily concealed in an auction, the system undermined the communal relationships of traditional commerce. As a New York citizens' committee expressed this discomfort, auctions subverted "the mutual confidence and courtesy that subsisted, in our better days, between the responsible importer and his customers" (Westerfield 1920, p. 194). Whether this view was justified or not, American colonists considered the auction a disreputable way to sell goods, which led owners to develop various devious means to conceal their identities (Westerfield 1920, p. 164), so that suspicion and reality reinforced one another.

Because the auction concentrated trade in a few places and facilitated face-to-face exchanges, however, it may well have served as a necessary if temporary transition between traditional commerce and more rationalized forms of distribution. "It was a period when, the world over, people began to wear cheaper clothes introduced and made possible by the Industrial Revolution," Westerfield (1920, p. 196) notes. "Auctions probably did facilitate this change of custom in costume by breaking the rigid trade channels and giving the manufacturer a competing outlet for his new products."

The growing number of lesser quality, machine-made goods only fueled suspicions about the new form of business relationships. Importing merchants, hardly disinterested observers in that auctions directly threatened their commercial function, reported to Congress in 1819 that goods sold at auction "are of less than the usual length, deficient in breadth, of a flimsy texture; in short, inferior in every respect to the goods they are intended to represent; yet, so well dressed, and in other respects, so highly finished to the eye, that they generally escape detection till they reach the consumer, who too late discovers their inferiority" (Buck 1925, p. 149). To aggravate such misrepresentations, one newspaper charged in 1828, "At nine-tenths of auction sales, one minute, or even less, and scarcely ever so much as two minutes, is all the time usually allowed to a large company of perhaps two hundred buyers, to examine, in the twilight of an auction store, amidst noise and confusion, goods which they never saw before. The worse the goods—the shorter will probably be the time given" (Westerfield 1920, p. 197).

Several other complaints against the auction system involved abuses of information processing or communication: auctions obliterated the "distinctive character of goods," so that "they could no longer be bought simply by name and brand and number" (Westerfield 1920, p. 197); "the names and marks of well known manufacturers were counterfeited" (Buck 1925, p. 149); auctions involved publication of false news, fictitious bidding, and false reports of sales (Westerfield 1920, p. 198).

Because complaints about auctions centered on their shortcomings as an information-processing and communication system, it is hardly surprising that regulatory legislation centered on these same functions. New York's Auction Law of 1817, for example, required a minimum of two days prior notice of an auction in at least one newspaper of the port, during which time goods would be available for inspection; buyers had three days after a sale in which to return goods that differed from the terms of sale or to receive appropriate price adjustments. Auctions had to be held between sunrise and sunset "to prevent poor lighting from misleading buyers" (Cohen 1971, p. 494). Federal legislation proposed in 1829 by the House Ways and Means Committee would have required auctioneers to publish, at least two days before a sale, "schedules of the goods, name of the importer or consignee, detailed description of the goods and their marks, name of vessel by which they were imported and the time of their importation," with the same information to be posted at the time and place of sale (Westerfield 1920, pp. 207–208).

Meanwhile, the auctioneers, feeling the heat of public dissatisfaction and regulatory legislation recent and pending, had in 1821 presented Congress with their own view of how auctions might be improved—as an informational system—through their own self-regulation. Worth noting are the means by which goods might be preprocessed—a half-century before standardized quantities, grading, packaging, and trademarks had become commonplace—to facilitate their processing by potential buyers:

When the sale is of magnitude, it is generally advertised in the principal commercial cities, with an enumeration of the articles to be sold. Printed catalogues are prepared, specifying the term of credit, with other conditions of sale, and detailing the contents of each package, the number of pieces, the varieties of quality, by number or otherwise, and the lengths; all of which is guaranteed to the purchaser . . .

The packages are arranged in lots corresponding with their numbers on the catalogue, and are exhibited sometimes two entire days before the sale, sometimes but one . . . When the goods are prepared for inspection, the purchasers are invited by public notice in the papers to examine them. Where it is necessary for an advantageous examination, whole packages are displayed; where it can be made with more convenience from samples, one or more pieces of each quality are exhibited; and where there are many packages exactly corresponding one only is shown.

Pattern cards are exhibited displaying the assortment of colors, etc. The purchaser receives every information and facility that can contribute to his convenience and protect him from mistake. The goods are arranged with so much attention to the accommodation of the purchasers, that three or four hundred packages may be examined with care and accuracy in one day.

On the day of sale the purchasers assemble, each prepared with a catalogue marked with his estimate of the value of the articles wanted; a practice that not only guards the buyer against any disadvantageous excitement which competition naturally produces, and refers him to the deliberate opinion formed upon careful examination before the sale, but also promotes a general knowledge of merchandise in every variety, and creates a useful register of the fluctuations of the market, as these catalogues are generally preserved, with notes in the margin of the prices at which every article has been sold.

At the commencement of the sale the conditions are recapitulated by the auctioneer, among which is a provision that no allowance will be made for damage or deficiency after the goods have left the city . . . This being, however, a declared condition at all times, the publicity of the rule insures the prompt examination of the goods.　(Westerfield 1920, pp. 174–176)

Commercial newspapers served to inform potential buyers of three events: when auctions would be held, when catalogs would be available, and when goods could be examined. Auctioneers facilitated the processing of available goods by their customers, so that packages might be examined at a sustained rate of one every two minutes, with a half-dozen major information-processing devices: lists and catalogs, sequential numbered keying of items, samples and pattern cards, and published conditions of sale. Buyers were also encouraged in more rationalized information processing, including prior written estimates of value, on-the-spot inspection of purchases, and records of market trends. That the auctioneers intended their system to remedy problems of face-to-face commerce, what they called the "disadvantageous excitement which competition naturally produces," again suggests that the auction served as an intermediate stage between traditional trade and more rationalized forms of distribution.

Further Rationalization of Distributional Control

Before self-regulation could be widely established by the auctioneers and before federal regulatory legislation gained widespread support, the auction system faded from American commerce almost as quickly as it had appeared. During the mid-1830s jobbers began to purchase directly from domestic and then foreign manufacturing agents, with a resulting decline in auctions. As early as 1834 auctions accounted for only 7 percent of U.S. imports (Cohen 1971, pp. 495–496).

Causes for the decline in auctions can be found in technological innovation and increased control of both foreign and domestic trade in the 1830s. Clipper ships, streamlined vessels designed for speed, appeared in the early 1830s; regular transatlantic steamship service began in 1838. "Steam navigation brought together the agents of foreign commission houses and the jobbers of this country," notes Westerfield (1920, p. 208), "and the inducements for a speculative and uncertain market were lessened."

Because of the easier contacts between foreign merchants and American jobbers, many jobbers became importers, a commercial specialty that by the 1830s, according to Cohen (1971, p. 505), was "playing the major, if not decisive, role in handling America's international trade." A British parliamentary investigation in 1833 asked whether—because of improved transatlantic transportation—"a practice had not grown up of late years which had not existed before, of the Americans coming over themselves to see the goods and make their purchases" (Buck 1925, p. 153). By going directly to the source of supply in this way, a practice followed in the China trade, American importers managed to circumvent the auction system altogether (Cohen 1971, pp. 505–506).

Because the new American importers afforded a more regularized and predictable means of disposing of goods, European manufacturers turned to them in increasing numbers. Sales in the port of New York, which received more than half of all U.S. imports after 1830, indicate that auctions increasingly shifted over the 1830s and early 1840s from British dry goods to Oriental teas, silks, and chinaware and to hardware, groceries, and liquors. European dry goods accounted for more than half of New York auction sales until 1829, only a third in 1834; total sales of these goods dropped from a high of $14.8 million in 1827 to $7.2 million in 1840.

In 1846 a New York City business directory listed 40 percent as many importers as jobbers in dry goods: 89 versus 224 (Albion 1939, p. 421). The port had nearly 60 percent as many importers as jobbers in china, glass, and earthenware (53 vs. 93), 80 percent as many in drugs (37 vs. 46), about an equal number in silks and fancy dry goods (42 vs. 40), two-and-a-half times as many in hardware and cutlery (86 vs. 35), and four times as many in wines and liquors (89 vs. 22). Control of the importation of foreign goods to America had clearly shifted— between the War of 1812 and the advent of industrialization in the 1840s—from British to American hands.

The auction system, which had served to control distribution during this transitional period, did not disappear entirely after the 1840s. New York City's auction sales in the half-year ending June 30, 1851, for example, totalled $15.8 million, of which $3.5 million came from Erie Railroad bonds and other securities (Jones 1937, p. 40). Even today, according to a list compiled by Cassady (1967, pp. 17–19), 150 major types of commodities—from lumber to licenses—are still sold at auction. Each week the federal government auctions off more than a billion dollars worth of Treasury bills, not to mention military and highway construction contracts, oil exploration rights, surplus equipment, and a wide range of other commodities. "So many of the most important goods and services in our economy are allocated through a bidding or auctionary procedure," according to economist Oskar Morgenstern (1976). "Yet no theory of auctions exists. Moreover, hardly any of the leading textbooks on microeconomic theory even mention this important form of trading."

If the auction had little impact on economic theory, it left a lasting mark on the American distributional system in the form of a new controller: the wholesale jobber. That jobbers existed in 1800 cannot be directly documented, according to Jones (1937, p. 10), but a review of trade and commerce in 1820 reported that "during 1815 and 1816 they had extended credit from one end of the Union to the other." The rise of the auction system in the same years had, ironically enough, threatened to eliminate the jobbers by enabling retailers to purchase goods directly at auction. As Americans began to move westward in the 1820s and 1830s, however, the wholesale jobber assumed an increasingly important role in the lengthening chain of transactions extending from British factories across the Atlantic, over the Appalachians, and into the small country stores springing up in the newly admitted states of the American heartland: Kentucky (1792), Tennessee (1796), Ohio (1803), Indiana (1816), Illinois (1818), and Missouri (1821). By 1840 the wholesale jobbers of New York's Pearl Street had become the principal customers at the city's auction houses (Albion 1939, p. 280), a position they would enjoy until the Civil War (Taylor 1951, pp. 397–398; Cohen 1971, p. 507).

The Jobbing System in Control of Distribution

The jobber managed to survive along with auctions by spanning a gap that they left in commercial distribution with a new system of information processing and communication linking European and northeastern manufacturers, on the one hand, with rural and western storekeepers on the other. In effect, the jobber helped both manufacturer and retailer with informational tasks that the auction system made difficult for them to perform on their own.

Especially the foreign manufacturer, who resorted to auction sales in lieu of establishing information and commercial contacts in America, lacked the information needed to extend credit to distant storekeepers, many of whom had been in business only a brief time. "Whenever possible the auctioneer shifted the responsibility for approving credit to his principal, either the agent or manufacturer," Cohen (1971, p. 508) notes, and "the manufacturer frequently carried the burden of sometimes dubious debts." Even New England textile manufacturers, according to Ware (1931, p. 176), "had to carry the financial burden of possible bad debts and long and irregular credit."

The interior storekeepers, on the other hand, lacked the telecommunications necessary to follow fluctuations in northeastern auctions, could not afford to visit them more than once or twice a year, and could not always stay long enough to obtain a wide variety of goods at the most advantageous prices. A New York retailer, more fortunate than his rural and western counterparts in that he could visit the auction houses daily, lamented the need by 1828 to be continuously alert to market information:

The system is miserably changed: now I am obliged . . . to *live* in an auction store the best part of my time; I am there all the business part of the day, say from 8 to 2 and 3 o'clock, and all my business is done at home by my clerks . . . I have known an article sold in the morning at 65 cents per yard, and the same article precisely, from the same package, sold in the afternoon of the same day, at 43 cents. I have known prints sell at 28 cents and go down to 21 cents in less than five minutes. I have known an article that was not very plenty advance in a few days from 35 to 65 cents, by the competition in the auction room.   (Jones 1937, pp. 51–52)

Into this gap in distribution between manufacturer and retailer, an informational niche created by the auction and its breakdown of the commission merchant system and by the lengthening lines of communication created by westward expansion, stepped the wholesale jobber. Glenn Porter and Harold Livesay, historians of early American marketing, report the situation for dry goods:

Out-of-town retailers obviously could not loiter in New York waiting for appropriate goods to go on the block, and auctioneers could not extend credit to country merchants. The jobber did both, and his performance of these services made him the most important link in the web of agencies which distributed textile products throughout America. In the words of one contemporary observer, the jobber "ruled the trade." The jobber achieved his position of dominance because his ability to connect the mills to their ultimate consumers was "essential to the orderly marketing of fabrics."   (1971, p. 27; quotations from Cole 1926, p. 291)

This conclusion applies equally to the distribution of groceries, drugs, hardware, clothing, liquor, and a variety of other manufactured and imported items. It reveals the jobber to be not only a new information worker but also a material extension of the purely informational system of the auction. He helped to distribute in time and space what the auctions could merely allocate to the person willing to pay the highest price. By determining and paying that price at auction, the jobber also served the commercial communication system of money and credit by processing information from many scattered retail markets into single purchasing decisions.

Jobbers performed another informational function that stemmed directly from the sparseness of rural retail markets. Where the volume of business permitted, retailers ordinarily took advantage of the economies of division of labor and specialized in a particular line of goods. Rural and western storekeepers, however, supplied markets too small to permit much specialization (Jones 1937, p. 64); hence the name "general store" for their establishments. This situation reflects what economist George Stigler has more recently elaborated in an article that takes Adam Smith's original theorem (1776, chap. 3) as its title: "The Division of Labor is Limited by the Extent of the Market" (Stigler 1951). Nor could country merchants afford to stock much of any one item in view of the difficulty and expense of transportation before the railroad; the early storekeeper "kept all of his goods in a chest or box, which was opened whenever a purchaser would appear" (Jones 1937, p. 44).

This rapidly growing but widely dispersed retailing that could not specialize constituted another commercial niche that came to be filled by the wholesale jobber. Unlike country storekeepers, jobbers could process market information continuously, wait for favorable prices, and purchase in large quantities, either wholesale or at auction. They unpacked and reorganized these goods into many smaller and more varied lots, which they sold either to local retailers or to cotton factors and country storekeepers on semiannual visits to the Northeast (Jones 1937, chap. 2; Porter and Livesay 1971, chaps. 2, 3). By specializing in the processing of bulk goods into the four major store lines—dry goods, groceries, hardware, and earthenware—that dominated early nineteenth-century retailing, the jobber profited from, in effect, increasing the information contained in the commodities, much as, say, a file clerk gets paid for sorting stacks of papers into the appropriate folders. The information contained in the jobber's store lines would have cost most retailers much more than the commission—often as high as 20 percent—were they forced to gather it themselves.

This does not mean that many storekeepers did not try, on their visits to the Northeast, to buy goods directly at auction and thereby avoid the jobber's commission. To forestall this possibility, jobbers maintained elaborate intelligence-gathering and information-sharing networks that not only kept them current on each storekeeper's credit rating but also alerted them to his arrival in their city. As one British traveler described the system, "A Southern or Western merchant no sooner reaches New York, Boston, or other intended place of purchase, than his arrival is thus and immediately known, chronicled through the city, when, if by reference he is found to pass muster, that is, is considered of sufficient solvency, he is immediately waited upon by a numerous bevy of agents, who crowd these hotels in wait for the latest comer, and who are in nowise restrained within any considerate or reasonable bounds, in their importunities and mode of seeking orders" (Jones 1937, pp. 16–17).

"Competition for a visiting retailer's business was fierce," according to Cohen (1971, p. 507), "for the only time the supplier could sell his goods to the retailer was during the latter's semi-annual visit to the city." Not only did the jobber have to worry that an out-of-town customer might talk to other jobbers or frequent auctions—even contact with other visitors could result in lost business. As one early nineteenth-century retailer reported of a visit to New York, "I occasionally made the acquaintance of several grocers who wanted small lots of the goods offered for sale, and we frequently clubbed together and bid off a lot which, being divided between us, gave each about the quantity he desired, and at a reduced price from what we should have been compelled to pay if the goods had passed into other hands [the jobber's] and thus been taxed with another profit" (Jones 1937, p. 52).

In response to such diverse competition for his customers' purchases, the jobber introduced the practice of diverting their attention throughout a visit. This task fell to the junior partners and confidential clerks of the jobbing houses, a company of young men that in New York City came to be called the "Prime Ministers." According to Cohen (1971, p. 507), "Their function was to so monopolize the visit of the out-of-town buyer that he would be happy, kept from competitors, and inclined to buy from the right jobber."

As we might expect, wider power soon accrued to the function of controlling new customers, whose continued business often required that the jobber be, in the words of one Pearl Street merchant, "very liberal of his money in paying for wine, oyster-suppers, theatre-tickets, and such other means of conciliating the favor of the country merchant as are usually resorted to" (Greene 1834, p. 60). As Westerfield (1920, p. 186) writes of the Prime Ministers, "they had entire control of all country buyers who visited the city for the first time; they were men of education and polished manners, superior to the merchants who employed them; and they soon acquired a powerful influence in the mercantile world."

Rationalization of the Jobbing System

By the 1830s, and especially after the decline of the auction system, jobbers began to abandon the haphazard practice of selling at whatever price the market would bear (Cohen 1971, p. 507). Reuben Vose, a

New York jobber who specialized in hats and shoes, introduced the "one price system" and eventually published a catalog describing some hundred different items, with fixed prices included, which he mailed to his customers. As a result, according to Westerfield (1920, p. 185), Vose "gained and kept the ascendency over all other New York jobbers in sales to Western and Southern merchants; his business was conducted on a strictly cash basis and he won from the credit houses their cash business."

The informational innovations of fixed prices and price catalogs— two decades before major stores adopted fixed prices and thirty years before the first mail-order catalogs—advanced the control of distribution for its three most important actors. For the jobber, the price fluctuations of the auction system had made planning and hence further rationalization of commerce difficult at best; fixed prices offered the possibility of much longer-range planning and hence market control. For the retailer, the jobber's catalog meant that he could make a good estimate of his purchases before leaving for the Northeast; it thus eliminated the uncertainties involved in negotiating with jobbers or buying at auction. For the manufacturer, greater rationalization of commerce meant better planning of production and hence further stability of distribution; "the danger of a downturn in the market before one's goods reached the auction block was very real," according to Cohen (1971, p. 508), so that "domestic manufacturers, like their foreign counterparts, naturally favored a more regularized and predictable means of disposing of their goods." Better information, in short, meant better control of the distributional system by all actors involved.

Although the jobber gained control through his continuing specialization in distribution, he could be totally eliminated by still greater control effected through integration at higher levels of the system. In textiles, for example, jobbers supplied stores that sold retail to consumers who, in turn, either made their own clothing or took the cloth to tailors. By the 1840s, however, the demand for ready-made clothing had been well enough established for manufacturers to integrate the importation of textiles and the production and retailing of apparel in a single firm. By integrating their specialized functions under one roof, they eliminated the importer, wholesaler, jobber, storekeeper, and tailor. In 1845 an American firm advertised that, in eliminating the importer and jobber, it had cut the price of clothing 24 percent (Jones 1937, p. 13). Clearly control of processing could be enhanced not only through division of labor and specialization but also through vertical integration of the separate functions—what Alfred Chandler (1977) would call control by the "visible hand."

In most industries, however, the specialized function of the jobber did not begin to be eliminated until the Control Revolution of the 1870s and 1880s, when the modern mass retailers—department stores, chain stores, and mail-order houses—started to bypass the wholesalers altogether. During the 1840s and 1850s, with the rapid development of a commercial infrastructure based on railroad and telegraph and the advent of American industrialization, the wholesale jobber completely replaced the traditional mercantile firms that had dominated the world's trade for half a millennium. As jobbers fought the vertical integration of manufacturers by bypassing importers to seek out their own foreign sources of supply, importers responded by attempting to sell directly to retailers so that, in the words of Jones (1937, p. 14), "importers became jobbers and jobbers became importers."

In this way the importing and jobbing functions became integrated in wholesale houses that bought directly from manufacturers and sold to specialized retailers, thereby eliminating many intermediate links in the chain of middlemen. "The new jobber created large buying networks through which he purchased directly from manufacturers at home and abroad," according to Chandler (1977, p. 216), "and he built extensive marketing organizations to sell to general stores in rural areas and specialized retailers in the cities." So great an advantage did this prove to be that by the 1860s Chicago alone had fifty-nine jobbers with annual sales of more than $1 million each (Twyman 1954, p. 31). A decade later, as we shall see in Chapter 6, the great wholesale houses like Chicago's Marshall Field were among the largest and most differentiated structures of bureaucratic control that would appear before the twentieth century.

During the 1840s and 1850s the role of the jobber changed in several other ways that increased his control of the nation's distributional system. With the penetration of the railroads into the interior in the late 1840s and early 1850s, jobbers themselves began to move westward from the coastal ports, so that the eastern seaboard lost its control over the distribution of American manufactures. No longer did southern factors and western storekeepers have to make their treks to the Northeast; the jobbers had come to them—or at least as close as the new regional wholesale centers like Cincinnati, Louisville, Chicago, and St. Louis. Even short trips to these cities became unnecessary for retailers after the Civil War, when the wholesale houses began to send

out traveling salesmen—by the railroads and then by horse and buggy—
to pay regular visits to even the smallest and most distant country
stores (Clark 1944, chap. 6).

Distributed Control of Agricultural Commodities

With the establishment of separate commercial systems for the movement and distribution of both exports and imports by the 1830s, a third
infrastructure began to develop to move agricultural commodities from
the American interior south and east, mostly to domestic markets.
Like the more developed systems for moving American cotton and
European manufactured goods, the interior networks eventually differentiated—as intermediaries separating western farmers and the
distant consumers of their products—a half-dozen middlemen with
specialized informational functions: country storekeepers, warehousers, commission merchants, freight forwarders, wholesalers, and exporters.

Just as the cotton factor provided the first step in the lengthening
chains of distribution stretching from southern plantations, country
storekeepers played a similar role for small farmers in the South and
West (Atherton 1949; Woodman 1968, chap. 7; Danhof 1969, chap. 2).
Economic descendants of the scattered storekeepers in the interior of
colonial America, western storekeepers made half of their income by
obtaining local crops and produce, consigning them to commission merchants in ports downstream, and arranging for their delivery via flat-
boat. Commission houses, in turn, forwarded the shipments to other
merchants in coastal ports or in the North, providing advances against
the sale much as in the cotton trade (Clark 1970, pp. 301–304).

The other half of a storekeeper's income came from retail sales of
manufactured goods and staples from the East. As we have already
seen, country storekeepers normally visited eastern cities twice a year
to replenish their stocks and arrange for transshipment overland to
Pittsburgh and then downstream to their local port. Soon eastern
wholesalers provided western storekeepers more credit than did the
houses to which they consigned crops and produce (Chandler 1977, pp.
23–24). Because auctions so dominated New York markets in the 1820s,
that city quickly became the wholesale center for the country trade,
mecca for rural shopkeepers on their semiannual pilgrimages, and focus
of a flourishing credit network to support the distributional networks
of the West. Only coffee, sugar, and molasses did not normally sell at

New York auctions, moving instead directly to the West through New Orleans (Albion 1939, pp. 178–187; 410).

As with the factorage, brokerage, auction, and jobbing systems already examined, the general country merchant system of distribution constituted an important intermediate stage in the transition from traditional to modern control of commerce. As agriculture moved westward in the 1820s, the first farms on the frontier were necessarily small, nonspecialized, widely separated, and often semisubsistent—they produced little beyond their own needs, quantities tellingly referred to by western farmers as "surpluses" (Danhof 1969, p. 17). Even where farmers might have desired to produce for the market, money remained scarce; what little existed was generally suspected to be counterfeit (Jones 1937, p. 45). Into this traditional society, little changed from that of the Middle Ages, came the country merchant, usually in the earliest wave of western settlement (Danhof 1969, p. 29). By accepting farm goods in barter for supplies from the East, the frontier storekeepers provided the sole link not only between the agricultural and market sectors but also between traditional and rational commerce.

As long as the horse powered transportation, according to Clarence Danhof, historian of nineteenth-century American agriculture, "an overnight trip of perhaps 12 to 15 miles represented roughly the distance within which frequent contacts with markets could be maintained" (1969, p. 27). Farmers who could reach a public market within a day enjoyed the advantage of maintaining personal networks of buyers and consumers. Virtually all cities and towns maintained their own public markets, patterned after those in Europe and administered by a clerk or market master. These rank among the first information workers in the American West, even though they performed many noninformational functions in addition to enforcing market ordinances, testing weights and scales, collecting rents, and settling disputes (Jones 1937, p. 59). For the majority of American farmers before 1820, however, and for most western farmers until the 1860s, the nearest storekeeper remained the route to market. Many stores even maintained wagon yards—occasionally providing firewood, shelter, and meals—where back-country farmers could stay when they arrived to dispose of a crop (Atherton 1949, pp. 48, 95).

Because agricultural communities are notoriously short of ready cash, a problem aggravated by the undeveloped state of banking in the early nineteenth century, western storekeepers had little choice but to accept farm products in exchange for merchandise from farmers who had no other means by which to pay. As one British traveler on the Ohio and Mississippi rivers wrote in 1806, "the entire business of these waters is conducted without the use of money" (Jones 1937, p. 45, quoting Ashe 1808, p. 51). Although some so-called cash articles— mostly imports like tea and coffee and processed goods like iron, gunpowder, and leather—could not be purchased except with money, storekeepers accorded the status of money to many farm commodities that met the same informational criteria: standardized, light-weight, durable, readily transportable, and generally accepted for cash by wholesale dealers downstream. These frontier equivalents of money included linen, cloth, feathers, beeswax, deerskin, and furs (Jones 1937, p. 45).

We have seen that the jobbers who supplied the storekeepers reorganized bulk purchases into smaller and more varied lots for sale in the general stores. The storekeepers themselves performed the reverse function for material flows in the opposite direction: they collected a wide variety of commodities from small, nonspecialized, and widely scattered farmers and stored these goods until they had sufficient quantity of a like kind to consign to commerce. Butter, for example, "typically moved to market through the country store," according to Danhof (1969, p. 40), "where it was collected in small quantities and indiscriminately mixed, the accumulations being sent to market a few times a year as quantities and opportunities permitted."

At first the new middlemen kept commercial products of the Ohio and Mississippi valleys—corn, pork, tobacco, hemp, whiskey, and lead— flowing downstream to New Orleans and the East, and supplies of manufactured goods and imported staples moving back through the Appalachians to Pittsburgh and then downstream to western storekeepers. After completion of the Pennsylvania and Ohio canal systems in the 1830s, western merchants not only had a shorter route to the East but also a cooler one for perishable products. Wheat and flour, which often rotted or soured in transit through the South, now passed eastward with less loss.

As with cotton, the distributional network for grain was also a credit system, financed by advances and the discount of notes on shipments in transit—grain moving eastward, credit westward. Although eastern exporters sometimes shipped grain or flour on consignment to a foreign house, most of the trade—and hence its financing—remained in the United States (Clark 1966). In structural differentiation of the commercial infrastructure, however, the pattern of "specialization in the grain trade followed that of the provisions and cotton trades," according to Chandler (1977, p. 24), "yet because of its smaller volume before 1840, it was less systematized and specialized than that of cotton," further evidence that division of labor and specialization depend mostly on the *volume* of flows through the system.

Lack of specialization in the movement of agricultural commodities before 1840 is reflected in the number of rural and western stores. In 1839 nearly fifty-eight thousand retail establishments of all kinds operated in the United States, about one store for each three hundred of the population (the ratio today is about 1 per 120). Although many of the stores were in eastern cities, fully 20 percent could be found in the six states of the western heartland: Ohio (8.0 percent), Indiana (3.1), Kentucky (2.9), Illinois (2.3), Missouri (1.9), and Tennessee (1.8). The importance of general stores in less settled areas can be seen in the fact that all three U.S. territories in 1839 had more stores per thousand population than the national average of 3.4: Wisconsin had 5.8, Florida 4.4, and Iowa 3.6 (Jones 1937, pp. 56, 71). Although the direct impact of the rural stores on the commercialization of agriculture is difficult to measure, the proportion of farm products not consumed within rural communities but offered for sale in urban markets increased from about 20 percent in 1820 to approximately 40 percent by 1870 (Danhof 1969, p. 2).

Certainly storekeepers played an important role in the development of at least one informational innovation for control of agricultural distribution: the preprocessing of farm products into standardized grades and types. The advantages of such grading in processing commodities for export had been recognized in the colonial period and had brought government inspection and grading requirements in seaport cities for major products like tobacco and flour. Traditional farming communities of the West began to feel pressure—through their storekeepers—for similar rationalization of their production methods. "It made no difference to a farm family if its potatoes were an assortment of white and red varieties," according to Danhof (1969, p. 43), "but such mixtures suffered significant discounts when offered for sale."

Quickly farmers became aware that, as a result of grading and typing, the premium market for high-quality products constituted a separate distribution channel through which they might secure greater returns for their output. As the more settled areas began to specialize, products of consistently high quality came to be marketed under the names of their region of origin: Genesee flour, Goshen butter, Herkimer cheese (all regions of western New York state). Because wholesalers accepted such names as designations of grade, they came to be applied to products of similarly high quality as agriculture moved westward: in the 1820s, for example, Michigan flour sold under the Genesee label; by the 1850s the Goshen name applied to the better butter of Illinois and other western states. Grading of wool remained a source of conflict between producers and merchants from the late 1830s through the early 1860s; grain grading caused controversy throughout the 1860s (Danhof 1969, pp. 43–44).

Although the country merchant continued to be an important agent for the handling of small lots from many farmers throughout this period, regional specialization in commodities and large volume production brought replacement of the storekeeper's marketing function by more specialized information-processing institutions and practices. As the grain traffic increased, for example, so too did the number of specialties in the string of middlemen linking western wheat farmers to northeastern markets. Country warehousers and millers joined local storekeepers in consigning grain and flour to commission houses. Freight forwarders more specialized in grain appeared in the lake ports, especially in Buffalo. With the development of a rail network in the wheat-growing regions in the 1840s and 1850s, a farmer could load his crop onto a grain car at the nearest station, consign it to a warehouse or elevator, and either sell it outright or take a receipt against later sale.

In Chicago and other distributional centers, warehouse receipts for stored grain became a new communications medium for its purchase and sale and thereby facilitated control of both its processing and financing in increasing volume. Agricultural historian G. A. Lee described this new system for the collection, storage, and marketing of grain:

> For the first time anywhere, it forced into the grain trade huge quantities of uniform warehouse receipts based on a definite quantity and grade rather than an individual lot in store. The use of these receipts to represent a particular article is hundreds of years old, but their use based on a standard known to a large market group, developed quite independently as a direct product of the Chicago elevator system. Grades were first indicated on the receipts at Chicago in 1857. The practice spread to other markets and became the keystone of a spectacular, worldwide, market structure.   (1937, pp. 22–23)

By 1859 total U.S. wheat production had reached 173 million bushels (U.S. Bureau of the Census 1975, p. 512), of which 46 percent came from the five Great Lakes states including Wisconsin (admitted to the Union in 1848). This region had more than double the per capita wheat production of the United States as a whole (Clark 1966, pp. 197–198). Small wonder that the U.S. census of 1860 declared the grain trade "the leading agency in the opening up of seven eighths of our settled territory" (Lee 1937, p. 16).

Thus the increasing speed and volume of movement of agricultural commodities, which resulted from the canals and railroads and was controlled by the western storekeepers and a growing chain of middlemen, provided the material and commercial metabolism for an entire continent—much as the cotton and import trades had done for the international system. Little different from that of the traditional societies of the Middle Ages as late as 1830, America's interior marketing system by the 1860s, "though still immature," according to Danhof (1969, p. 44), "comprised the basic elements that have characterized it since."

The rapidity of this transformation was in large part due to the country storekeeper, intermediate between traditional and modern institutions, who developed and diffused the transitional control system: acceptance of barter for credit, preprocessing of commodites into standardized grades and types, and consignment at commission. Danhof (1969, pp. 47–48) summarizes the degree of control thus effected in only thirty years from the viewpoint of the western farmer: "The typical farmer wished to sell his produce with as little effort and attention as possible. His ideal was to sell at his barn door, and certainly at a distance no greater than the nearest county seat. By the sixties, the great majority of farmers enjoyed such a situation."

From Tradition to Rationality: Weber Reconsidered

Thus did the agricultural marketing system of the American interior, like the systems for control of the cotton and import trades, bridge the transition from traditional to rational control of commerce in the half-century preceding America's Industrial Revolution. Contrary to Max Weber's interpretation, the great transformation of traditional into modern society, one based not on family and communal relationships but on what he called "rationally motivated adjustment of interests" (1922, p. 41), seems to correspond less to ideological or religious changes than to the development of alternative means of control. Merchants at the turn of the nineteenth century kept the traditional means of control—family partnerships, the southern "gentleman's code," the community relations that bound farmer and storekeeper—only so long as technologies of information processing, transportation, and communication remained underdeveloped.

Traditional control proved particularly useful, in the absence of adequate telecommunications, for *distributing* control, that is, for controlling through the use of other people—commission agents, factors, auctioneers—as portable and programmable information processors. But distribution of control became progressively less important over the first half-century of American independence, as we have seen, following a rapid series of innovations in transportation and communication and a parallel informational system of commercial control, which increased the merchant's ability to maintain distant and hence less distributed control. The transatlantic mails enabled foreign exchange partners to monitor one another; commercial newspapers gave plantation owners the opportunity to check up on their factors; auctions allowed textile manufacturers to test the market for new products and styles. Indeed, Clarence Danhof attributes the growing dissatisfaction of western farmers with the agricultural marketing system after the Civil War to "the fact that, as functions of distributing products over an ever-widening market were taken over by others, farmers became acutely aware of the wide margin between the prices they received for their products and the prices prevailing upon the central markets or paid by the ultimate consumer" (1969, p. 45), information that the growing transportation and communication infrastructure increasingly brought to the American interior.

Even in the preindustrial period, as we have seen, the development of society as a concrete open processing system paralleled the continuing rationalization of material distribution. At least a half-dozen commercial institutions of the early nineteenth century—including the commission agent, factorage, brokerage, auction, jobbing, and country merchant systems—served to integrate the developing systems of technological and market control. These same systems, it would seem, and not changes in ideology or religious belief, constituted the most important vehicles of transition from traditional to modern commerce and from commercial to industrial capitalism.

Even Weber, who never abandoned his argument that the Protestant ethic had contributed to the capitalist transformation, conceded in 1910 that he had not yet assessed the reciprocal relationship: the impact of social and economic arrangements on religion. Once he did, he added, his critics would probably accuse him of materialism, as they then accused him of idealism (Bendix 1960, p. 90). In his 1919–1920 lectures, published posthumously as *General Economic History* (1923), Weber suggested that traditional ties might serve as means to economic ends—as in the family partnerships that dominated the twelfth through early nineteenth centuries—in the transition to industrial capitalism:

> Originally . . . there is absolutely unrestricted play of the gain spirit in economic relations, every foreigner being . . . an enemy in relation to whom no ethical restrictions apply; that is, the ethics of internal and external relations are categorically distinct. The course of development involves on the one hand the bringing in of calculation into the traditional brotherhood, displacing the old religious relationship. As soon as accountability is established within the family community, and economic relations are no longer strictly communistic, there is an end of the naive piety and its repression of the economic impulse. This side of the development is especially characteristic in the West. At the same time there is a tempering of the unrestricted quest of gain with the adoption of the economic principle into the internal economy. The result is a regulated economic life with the economic impulse functioning within bounds.   (Weber 1923, p. 356)

Thus Weber clearly saw that internal (family) controls could be abandoned only as external (economic) controls became stronger. Internal controls, as we have seen in this chapter, can provide the programming for one's agents in distributed control, a transitional role for tradition in the rationalization of exchange. Industrialization of the external control technology of information processing, transportation, and communication, as we shall see in the remainder of this book, comes at the expense of that transitional control.

5

Toward Industrialization: Controlling Energy and Speed

Here is an enormous, an incalculable force practically let loose suddenly upon mankind . . . precipitating upon us novel problems which demand immediate solution . . . Not many of those . . . who fondly believe they control it, ever stop to think of it as . . . the most tremendous and far-reaching engine of social change which has ever either blessed or cursed mankind.

—Charles Francis Adams, Jr., "The Railroad System"

DESPITE the modest technological and economic innovations in control under commercial capitalism, the world distributional system, even in the early nineteenth century, simply did not require anything approaching the degree of control that would become necessary under rapid industrialization. Because capital remained so mobile under the centuries-old commercial order, it served as the major medium—in the form of money and commercial paper—for communication and control of the world system. Merchants in each major port could devote their capital and enterprise to ships and shipments, to a wide variety of wares, and to the provision of storage, loans, and insurance for merchants resident elsewhere. They thereby maintained both the liquidity and flexibility to help control, in small part, the larger system. Not until commercial capital came to be tied down in the British power-driven industries did profit begin to depend on the ability to manage not the totality of one's investments but the processing of relatively much smaller investments in raw materials. The faster one moved these throughputs past one's fixed capital, the greater the returns on one's investment.

If profit provided the incentive to process matter faster under industrial capitalism, steam power provided the means. The difference, in a word, was *speed*. As long as the extraction, processing, and movement of matter depended on traditional sources of energy (human, draft animal, wind, and water power), the material economy did not differ markedly in the speed of its throughputs from that of the Middle Ages.

Even a major technological innovation in transportation, the introduction of the Dutch flyboat or flute about 1595, had no significant impact on the speed of material processing in the seventeenth and eighteenth centuries. Lightly built without armament and with a flat bottom, elongated shape, and simplified rig, the flyboat not only moved faster than other contemporary ships but had peacetime operating costs one-third to one-half lower (Barbour 1930, p. 285). Nevertheless, imitations of the flyboat did not diffuse to many routes, including those in the Mediterranean, western Atlantic, Caribbean, and East Indies, because pirates and privateers frequented these waters; the armament of trading ships was necessary well into the eighteenth century (Shepherd and Walton 1972, pp. 81–82). After reviewing the dates vessels cleared and entered the ports of Boston, New York, Barbados, and Jamaica between 1686 and 1765, Shepherd and Walton found no significant increase in speed—an average of slightly under two knots— over the eighty-year period (1972, pp. 77–78).

Despite limits on the speed of shipping under sail, by the early eighteenth century innovations in material processing had already begun to reduce the time that ships had to remain in port. In the Chesapeake, for example, the arrival after 1707 of Scottish factors, who introduced the inventorying of tobacco in warehouses so that it could be more quickly loaded with the docking of a ship, helped to cut average port time from one hundred to forty-seven days over the eighty years; so, too, did the use of certificates of tobacco deposits as a medium of exchange, an innovation that led to inspection systems and, in effect, quality control of tobacco. Similarly, in Barbados average port time fell from forty-nine to twenty-three days with the growth of a more systematic market economy, including centralization of markets and standardization of quality, prices, credit, and rates of currency exchange (Shepherd and Walton 1972, pp. 78–80, 159–160).

Speed and Control

Because processing rates remained so woefully slow between the twelfth and nineteenth centuries, the mercantile system did not require much in addition to the traditional means of control: individual managers and the simplest of commercial and governmental organization. As one historian observed of Baltimore merchants in the 1790s,

> The nature of their business, and the way in which they conducted it, fall quite within the mainstream of commercial capitalism. Standing at the very end of that 500 year long period in economic history it is remarkable how similar they were in every important respect to the merchants who stood at or near its beginning. Their form of organization, and their methods of managing men, records, and investments would have been almost immediately understood by the 15th century merchant of Venice . . . They were a family partnership, the basic form of business organization not only at Venice but in every land. The central accounts in their double-entry bookkeeping records were Adventure and Merchandise Accounts . . . both well fit the pattern of medieval foreign trade.   (Bruchey 1956, pp. 370–371)

What meager trade statistics exist for the preindustrial period suggest a similar conclusion. Venice, at the height of its glory in the fifteenth century, boasted some three thousand ships with a carrying capacity of perhaps 150 thousand tons. In 1760, on the eve of the Industrial Revolution, the newly united kingdom of England, Scotland, and Wales had a combined merchant fleet only three times as large: 7,081 vessels with a reported capacity of nearly a half million tons. "In the century or so preceding the Industrial Revolution," according to Deese (1957, pp. 348–349), "the development of international trade was normal in degree, depending mainly on the increase of population and any cheapening in transport caused by improvements in the arts of navigation and shipbuilding."

That world trade grew relatively little between the Middle Ages and the eighteenth century reflects the limited dependence of preindustrial societies on long-distance processing and distribution of material goods. Even at the end of the eighteenth century most nations remained almost wholly self-dependent for the livelihood of their people; international commerce consisted mainly of traffic in luxuries for the few. Reinforcing the status quo was the mercantile system, dominant economic doctrine of the sixteenth and seventeenth centuries, under which each European power pursued its own most favorable balance of trade by limiting imports (Schmoller 1884, pp. 47–80).

Even the slow expansion of trade, however, had ramifications for system-wide control. Because of the growing number of ships moving throughout the system and the increasing regularity in their movement, Shepherd and Walton (1972, p. 55) conclude that "the speed of communications probably increased somewhat" despite little change in ship speeds. Certainly telecommunications in the eighteenth century left substantial margin for improvement: letters between America and London typically took two months in each direction, which meant that any response to a change in market conditions (already two months in the past) could only be effected a minimum of four months after the fact.

Economics of Market Information

To the extent that market conditions change faster than buyers and sellers are able to respond, using the infrastructure of information processing and communication available to them, or that the costs of information about prices remain high, markets will deviate from the classical economic model, which assumes free and perfect information and instantaneous transactions. Indeed, some economists (for example, Taylor 1969, p. 191) virtually equate the extent to which a market has achieved "perfection" with the quality of its communications, even though other factors are also involved. Needless to say, in view of the poor state of telecommunications in the eighteenth century, American colonial markets functioned imperfectly in the special economic sense that prices varied among regions in excess of shipping costs and others associated with distribution: insurance, inventorying, commissions, and other transaction costs.

These persistent regional variations in prices, a direct result of inadequate telecommunications, presented colonial merchants with the chance for both profit and ruin. As might be expected, the correspondence and accounts that survive reveal "continuous exchange of information about market conditions, prices, and expectations" (Shepherd and Walton 1972, p. 54), a struggle to overcome the conditions responsible for their opportunity but also much of their risk. By increasing the number of merchants in major trade centers and thereby increasing the density of their commercial information networks, the greater volume of shipping in the eighteenth century probably helped to reduce market imperfections as well as risks by fostering more and better market information.

Increases in both the number of merchants and the density of their interactions encouraged specialization in the information function. Such specialization depended on the extent of the market to absorb the increased output, as Adam Smith (1776, chap. 3) saw; it also depended on what Durkheim (1893, bk. 2, chap. 2) called the "moral density" of a community. As typically happens when inadequate communications create an imperfect market (Seldon and Pennance 1976, p. 186), the uncertainty of resident merchants generated an information "industry," including some of the first specialists in information processing and communication: brokers, agents, advertisers, wholesalers, and other commercial intermediaries.

Specialization of the trading function itself, a development that we have seen began in the mid-eighteenth century, also depended on the extent of the market, in turn limited by *distribution costs*, the difference between the price at which a producer sells an item and the price paid for it by a distant consumer. Even when production costs can be lowered through specialization, these gains may be overshadowed by the high costs of distribution. Because this differential includes the costs associated with risks in an imperfect market, as represented, for example, by the cost of insurance, poor communications must be considered an important cause of high distribution costs, limited markets, and underdeveloped division of labor. Indeed, insurance might be seen as a special type of *information cost*, more typically defined as the time, money, and effort required to obtain information (Seldon and Pennance 1976, p. 186). To the extent that insurers trade off higher rates with the acquisition of more information, however, or merchants trade off greater coverage with the maintenance of tighter control, we can say that insurance—like many other distribution costs—represents the cost of *not* having information.

From this perspective, the growth of trade can be seen to beget still more trade, at least at the low levels of the eighteenth century. Even slight increases in trade helped to expand markets by improving information—and thereby reducing distribution costs—in several ways: through faster mail, greater density of merchant exchange networks, and increased specialization in information processing and communication. Expanded markets and denser commercial networks, in turn, enabled increased specialization of the trading function; such specialization improved system control. At the same time, reduced distribution costs made trade even more attractive to merchants, so that its volume increased still further in a continuing spiral fed by positive feedback.

With the four-month lag in transatlantic communication, however, high information and hence distribution costs remained, in effect, a tariff on colonial trade, which reinforced the protectionism of the mercantilist system. For many goods and services, the high costs of distribution combined with high production costs exceeded any return they could bring in a distant market; hence they did not exchange in the larger system. By discouraging the movement of goods, high distribution costs encourage economic self-sufficiency (Shepherd and Walton 1972, pp. 53–55). This latter dynamic, combined with the mercantilist requirement that manufactures and not precious metals be exchanged for foreign goods, provided impetus for industrialization (Schmoller 1884).

Industrialization and Control

By the time Britain decided to enforce the mercantilist Navigation Acts in its American colonies in the 1760s, it was clear that many of the former luxury goods from the New World had already become for the growing populations of Europe and America today's everyday staples: sugar, tobacco, tea, molasses. Because such goods could be obtained only in exchange for others, the so-called nation of shopkeepers found itself transformed into what would come to be known in the nineteenth century as the "workshop of the world."

Factories with water-powered machinery, tended by hundreds of specialized workers, appeared in Britain after 1700, especially in the metalware and textile industries; by 1750 such centralized production gained at the expense of the "putting out" system of employment in individual homes. Cotton textiles led the revolution, spurred on by a rapid succession of innovations: John Kay's fly shuttle (1733), Richard Arkwright's water frame (1769), James Hargreave's spinning jenny (1770), Samuel Crompton's combined jenny and frame or "mule" (1779), Edmund Cartwright's power loom (1783), and finally—from America—the cotton gin (1793). In a Birmingham factory with two waterwheels and some six hundred workers, in 1775 Matthew Boulton and James Watt began the manufacture of patented parts for Watt's new engine, which transformed the power of moving water into the potential energy and speed of steam and helped to launch the material economy on a process of industrialization that would inevitably lead to a crisis in its control.

The need for sharply increased control that resulted from the industrialization of material processes through application of inanimate sources of energy probably accounts for the rapid development of automatic feedback technology in the early industrial period (1740–1830).

Although feedback control dates at least from the waterclock of Kte-sibios of Alexandria, in the third century B.C., no new feedback system appears to have been invented between antiquity and the thermostatic furnace devised by Cornelis Drebbel of Holland about 1620—a hiatus of some four centuries (Mayr 1970). Moreover, "we know of only three relevant inventions in Europe prior to 1745," writes Otto Mayr, his-torian of feedback technology, "in contrast to no less than 14 between 1745 and 1800," the advent of centralization and industrialization of cotton textile production. "In the middle of the 18th century," Mayr concludes, "the application of feedback in technology achieves a veri-table breakthrough" (1970, p. 127).

Table 5.1 lists nineteen major innovations in the development of automatic feedback control in the early industrial period. Although some applications were to preindustrial technologies like windmills, float regulators, and watches, two-thirds of the inventions came in new industrial processes—eight in steam technology directly (Fig. 5.1). Almost all of the innovations appeared in the industrializing coun-tries—almost two-thirds in England. Almost three-fourths controlled temperature, pressure, or speed, the crucial parameters of steam-powered production. The parameter that this book finds to be most central to the Control Revolution, namely speed, drew the most in-novations in its automatic control (six)—all after the mid-1780s. In short, Table 5.1, insofar as it establishes at least a temporal correlation between early industrialization and what Mayr calls a "veritable break-through" in feedback technology, bolsters a central argument of this book: the Industrial Revolution and the harnessing of inanimate sources of energy to material processes more generally led inevitably to an increased need for control.

Even with enhanced feedback control, industry could not have de-veloped without the enhanced means to process matter and energy, not only as inputs of the raw materials of production (including the sustenance of an urbanized workforce) but also as outputs distributed to final consumption. Commercial capitalism, although it did not mark-edly change the speed and volume of material processing, did establish the essential infrastructure for the movement of matter and energy on a worldwide scale. This included not only material technology like port facilities and ships but also nonmaterial infrastructures like inter-personal channels of information gathering, processing, and exchange, programming in the form of commercial law, and innovations in infor-mation technology like credit instruments and inventory techniques.

Table 5.1. Major innovations in the development of automatic feedback control in the early industrial period, 1740–1830

Year	Innovation	Application	Parameter controlled	Innovator or adopter	Country
1745	Fan-tail	Windmill	Direction	Edmund Lee	England
1746	Float valve	House water reservoir	Level (of water)	William Salmon	England
1751	Temperature regulator	Chicken incubator	Temperature	Prince de Conti	France
1758	Float valve	Steam engine boiler	Level (of water)	James Brindley	England
1765	Float valve	Steam engine boiler	Level (of water)	Ivan I. Polzunov	Russia
1771	Sentinel Register	Chemical furnace	Temperature	William Henry	America
1777	Temperature regulator	Chicken incubator	Temperature	Bonnemain	France
1784	Float valve	Steam engine boiler	Level (of water)	S. T. Wood	England
1787	Centrifugal pendulum	Windmill	Speed	Thomas Mead	England
1788	Centrifugal governor	Steam engine	Speed	James Watt	England
1789	Roller reef-ing sails	Windmill	Speed	Stephen Hooper	England
1790	Float valve	Steam engine	Speed	Brothers Perier	France
1793	Pendule sympathique	Watches	Speed (syn-chronized)	A.-L. Breguet	France
1799	Pressure regulator	Steam engine boiler	Pressure (of steam)	Robert Delap	Ireland
	Pressure regulator	Steam engine boiler	Pressure (of steam)	Matthew Murray	England
1803	Pressure regulator	Steam engine boiler	Pressure (of steam)	M. Boulton and J. Watt	England
1804	Centrifugal pendulum	Windmill	Speed	John Bywater	England

Year	Innovation	Application	Parameter controlled	Innovator or adopter	Country
1820	Pressure regulator	Furnace system	Pressure (of steam)	William Brunton	England
1830	Thermostat	General	Temperature	Andrew Ure	England

Source: Adapted from Mayr (1970).

By developing this infrastructure, essential for the processing and distribution of matter and energy, the Commercial Revolution helped to establish necessary preconditions for the Industrial Revolution, in effect the application of inanimate energy to the material processing system.

This explains why industrialization quickly undid most of the constraints on trade perpetuated by the mercantile system. By 1800, even after severe losses in the war with France, British shipping capacity exceeded one-and-a-half million tons, a threefold increase in just forty years; actual tonnage moved between British and foreign ports increased fourfold during the same period (Deese 1957, p. 349). Between 1820 and 1860 Britain conscientiously dismantled its protectionist system in favor of free importation and opened the nation's ports to raw materials and foodstuffs in exchange for manufactures. By mid-century, as a result of proliferating railroads and low ocean freights, wheat grown in the Mississippi Valley sold in Britain at prices that drove local farmers to remove arable land from cultivation (Rees 1920, pp. 119–121).

The corresponding transformation of the American economy, the more limited focus of this chapter, reflects similar stages in the development of a social system for the processing of matter and energy: innovation in the movement and exchange of material goods in the early nineteenth century preceded widespread application of inanimate energy; industrialization, in turn, brought control of the system into question in the 1840s through 1880s and began a revolution in information processing and communication that continues to this day. Much as the Commercial Revolution established preconditions for the Industrial Revolution, the latter provided both the necessary and sufficient conditions—the nineteenth-century crisis of control—to produce the Control Revolution.

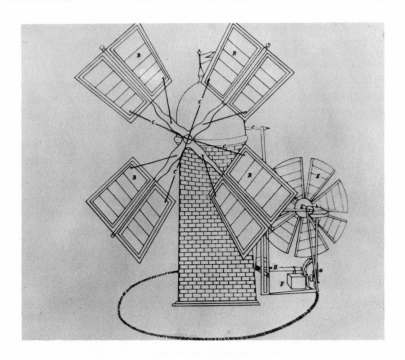

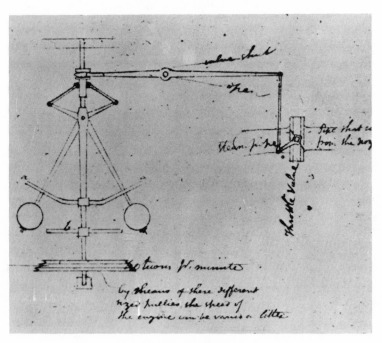

The Material Economy as a Processing System

To understand the sequence of developments leading from the industrialization of the American economy to the crisis in its control, the purpose of this chapter, it will be useful to reconsider the model of society as a processing system. The essence of a human society, as we saw in Chapter 2, is the continuous processing of matter and energy, which sustains individual members and their social organization, for a time, against the inevitable degrading of energy and eventual heat death. No living system can counter entropy except through the extraction of matter and energy from its environment and through their processing, distribution, and final consumption. Economic science, as represented by a tradition extending from Quesnay (1758) and Walras (1874) to Clark (1940) and Leontief (1941), has for centuries been devoted to the study of material flows through social systems.

As we might expect, an economy's major sectors, as delineated by Clark (1940), Hatt and Foote (1953), and Bell (1973), correspond to major stages in the essential life process. The primary sector—agriculture, fishing, lumber, mining, oil and gas—represents the extraction of matter from the environment to produce energy, including the calories to sustain individual organisms. The secondary sector—processing primary goods, as in construction and manufacturing—represents the synthesis of matter and energy into more organized forms (negentropy). The tertiary sector, including transportation and utilities, represents the infrastructure for distributing matter and energy about the system, while the quaternary sector—trade, finance, insurance, and real estate—constitutes a parallel infrastructure for the collection, processing, and distribution of information that is necessary in all living systems for the control of material flows. Finally, the "highest" of all sectors in its remove from the physical environment is the quinary sector, including government, law, and education, representing the societal programming—socialization, education, law making—and collective or representative decision making to effect control.

Figure 5.1. Early innovations in automatic feedback control. Top: windmill with fan-tail and self-regulating sails, from Edmund Lee's patent of 1745 (English Patents of Inventions, old series, no. 615, 1745). Bottom: James Watt's centrifugal governor (1788), from a design sketch, Boulton & Watt, 1798 (Dickinson and Jenkins 1927, table 81; courtesy of Oxford University Press).

Because America remained all but untouched by the several millennia of Western economic development when the London Company first settled at Jamestown in 1607, the American colonial period represents an unparalleled opportunity to study the conscious reconstruction of a material economy. Although the details are obviously beyond the scope of this book and have already been accorded lengthy treatment by many others (Clark 1916, chaps. 1–10; Williamson 1951, pt. 2; Boorstin 1958; Kranzberg and Pursell 1967, vol. 1, pts. 2, 3; Shepherd and Walton 1972), Table 5.2 presents a selective chronology of innovations in the American colonies, the first steps in the development of a concrete system for the processing of matter and energy and the information for their control.

As Table 5.2 suggests, the processing of matter and energy and of information develop simultaneously and in parallel—neither can be said to precede the other in any general sense. Innovations before 1650 in the harnessing of energy (windmill, power canal) and material processing (iron works, brick kiln, grain and cloth mills, brass and iron foundries) appeared to stimulate over the next century several business controls: machine patents, incorporation and capital stock sales, business manuals, fire insurance. Tertiary sector developments in transportation (highways, bridges, shipping) and utilities (municipal water supplies) correspond to parallel informational innovations in the control of transportation: a postal system, published road maps, lighthouses, insurance of ships and their cargo.

Colonial control of the material economy through information processing, programming, and communication also included developments in the quaternary subsectors of trade (chamber of commerce, trade registers), finance (banks, paper money), and insurance (marine, fire, and life). Also present before the War for Independence was control by the quinary sector: colonial governments (postal services, patent and copyright laws), the legal profession (law books, formal instruction), and public education (libraries, textbooks). Even mass publishing, usually associated with the nineteenth and twentieth centuries, had appeared before 1800, first as unfolded newspapers or "broadsides" (after 1689), then as periodicals (1741) and daily papers (1784); even newspaper advertising had begun by 1704 and could run to the half-page by 1743.

This five-sector model of America's material economy, on the eve of industrialization, will be useful in understanding the major developments leading over the next half-century to the Industrial Revolution and the crisis of control. Continuing innovation in the tertiary sector,

Table 5.2.   Selected events in the development of a national system for processing matter, energy, and information for control, seventeenth and eighteenth centuries

Year	Matter and energy	Information
1619	Iron works built near Richmond, operates only briefly	
1629	Brick kiln established (Mass.)	
1632	Windmill erected for grinding grain (Mass.) Va. establishes highway program	
1638	Cloth mill established (Mass.)	
1639		Post office (Boston) established for foreign mail Printing press imported from England (Mass.)
1640	Canal completed to create water power to run mill (Mass.)	
1643	Iron works established (Mass.); produces eight tons per week	
1645	Brass and iron foundry opened (Mass.); makes tools, utensils	
1646		Machine patent granted (Mass.)
1648		Compilation of colonial laws published (Mass.)
1650	American iron exported (Mass.)	
1652	Municipal (Boston) water supply system built of wooden pipes	
1672		Copyright law passed (Mass. General Court)
1673		Postal route established between New York and Boston; monthly service inaugurated
1675		Commercial corporation (N.Y. Fishing Company) chartered; shares of capital stock sold
1679	Commercial ship launched on Great Lakes (Erie, Mich.)	

Table 5.2 (*cont.*)

Year	Matter and energy	Information
1681		Shorthand used in court (Md.)
1689		Newspapers appear, first as unfolded "broadsides"
1690		Paper mill built (Pa.); produces 250 pounds per day Bank established, paper money issued (Mass.)
1692		Parliamentary act establishes post office in America
1698	Stone bridge completed (Pa.)	Road map listing distances from Boston published in an almanac Library—with commissioners and trustees—opened in Charleston
1703		Business manual published; contains examples of bills, bonds, deeds, letters, warrants, etc.
1704		Successful newspaper (*Boston News-Letter*) published, runs first newspaper advertising
1710	Iron slitting mill established to slit nailrods (Mass.)	
1714	Schooner built (Mass.)	
1716		Lighthouse built by Mass. at entrance to Boston Harbor
1721		Fire insurance offered on ships, cargo, merchandise (Pa.)
1732		Circulating library established by Benjamin Franklin (Pa.)
1735		Fire insurance company organized (S.C.)
1739	Iron casting introduced (Pa.)	
1741		Periodicals issued (Pa.)
1743		Half-page newspaper advertising appears (N.Y.)
1746	Iron rolling mill with three stacks established (Pa.)	

Year	Matter and energy	Information
1753	Steam engine imported, used to pump water from mine (N.J.)	
1755	Pumping plant built for municipal (Bethlehem, Pa.) water system	Law instruction offered in a college (King's College, N.Y.)
1756	Stagecoach service links Philadelphia and New York	
1759		Life insurance firm incorporated
1768		Chamber of commerce formed by twenty merchants (N.Y.)
1771		Permanent printing type foundry opened (Pa.)
1773		Trade register published (Pa.)
1774	Steel shovels produced (Mass.)	
1776		Engineering book published
1779		Law school (William and Mary) established (Va.)
1780		Ratio of paper money to silver in America becomes 40 to 1
1781		Congress charters first bank (Bank of North America) with capital of $400,000
1783		Copyright laws passed by states (Conn., Mass.)
1784	Trade ship (*Empress of China*) visits China	Newspaper begins daily publication (Pa.)
1785		Decimal system of money adopted by Continental Congress City directories published (Pa.)
1786	Spinning, carding, roping machines built (Mass.); spinning jenny put in operation (R.I.)	
1788	Water power used at commercial worsted wool mill (Conn.)	Cotton goods trademarked by seal of corporation (Mass.)

particularly in the movement and exchange of material goods, preceded widespread application of inanimate energy. Subsequent innovation in the primary sector in the 1820s and 1830s produced a new source of energy—anthracite—and a new type of power: steam. Continuing developments in the secondary sector produced new materials—especially iron and steel—essential for harnessing the power of steam. Steam power applied first to the tertiary sector (steamboats, railroads) and then to the secondary (factory production) so speeded up the entire processing system that its continued control would come into question during the half-century of the 1840s through 1880s, as we shall see in the next chapter. This crisis of control, in turn, brought rapid development and innovation in the informational quaternary and quinary sectors, a revolution in information processing, programming, and communication that has continued unchecked to this day. The Information Society is one result.

The Clark-Bell Sequence Revised

The sequence of development just outlined, although not incompatible with the descriptions of other social scientists, will enable us to distinguish several new phases in the development of a national economy. Daniel Bell, for example, citing the work by Colin Clark (1940) just discussed, associates industrialization with a shift in the labor force from the primary (extractive) to the secondary (goods producing) sectors of the economy. A major dimension of the so-called postindustrial society, according to Bell, is the subsequent shift to the tertiary (service) and higher-order sectors, beginning with transportation and public utilities, distribution (wholesale and retail), and finance, real estate, and insurance (1973, pp. 127–128). "A large service sector exists in every society," Bell concedes, but "in a preindustrial society this is mainly a household and domestic class" (1976, p. xvi).

Although an accurate summary of aggregate statistical trends (domestics did constitute the single largest class in Britain until about 1870), Bell's account obscures an important precondition of industrialization. Because industrialization involves the utilization of inanimate sources of energy by society as a concrete processing system, it cannot develop unless an adequate infrastructure for the movement and processing of matter, energy, and information already exists. How else can we explain the appearance of formal organization and bureaucracy even before the Commercial Revolution and their continuing reappearance in undeveloped forms throughout the world down to the nineteenth century? Initial development of the tertiary (transportation and utilities) and quaternary (trade, finance, and insurance) sectors may just as well be viewed as a *precondition* of industrialization than as a harbinger of the postindustrial society. Indeed, processing structures, energy utilization, and control capabilities necessarily coevolve in all living systems.

What made the Commercial Revolution truly revolutionary was that, for the first time, distributional and control systems including information processing, programming, and telecommunications could be sustained indefinitely on a global basis. Industrialization became revolutionary when the energy harnessed vastly exceeded that of any naturally occurring or animate source; the resulting throughput and processing speeds greatly exceeded the capability of unaided humans to control. What made the Control Revolution in fact revolutionary was the development of technologies far beyond the capability of any individual, whether in the form of the massive bureaucracies of the late nineteenth century or of the microprocessors of the late twentieth century. In all cases it was not the novelty of the commodities processed (whether matter, energy, or information) that proved decisive, contrary to Bell, but rather the transcendence of the information-processing capabilities of the individual organism by a much greater technological system.

The first half-century of the new American republic illustrates this qualification of the Clark-Bell sequence. Although the Industrial Revolution cannot be said to have begun much before the 1840s, the decade that brought coal-powered factory production, the railroad, and the telegraph, America had already undergone extensive transformation of its transportation, commercial, and financial systems by that time. Key to the change was rapidly increasing division of labor in occupations and enterprises, specialization that lowered information and transaction costs but increased reliance on the market as a control system.

As we have already seen, the colonial merchant, a generalist who embraced all types of products and embodied all basic commercial functions, differed little from his counterpart in fifteenth-century Venice. Within two generations, however, these general merchants had been largely replaced by more specialized workers: shipowners, financiers, jobbers, transporters, insurers, brokers, auctioneers, retailers—a growing network of middlemen to process and move material goods. What merchants remained came increasingly to specialize in only one or two lines of goods, and to concentrate on a single commercial function: importing, wholesaling, retailing, or exporting. At the onset of the Industrial Revolution America's institutional system had already been reconstituted in the specialized types of service enterprises— shipping lines, freight forwarders, insurance companies, and banks— that characterize the tertiary and quaternary sectors of the economy.

Pre-Steam Innovation in the Tertiary Sector

In the tertiary sector two wholly new and modern types of transporation enterprises—common carriers and companies to build and maintain overland rights-of-way—helped to establish the distributional infrastructure necessary for industrialization. Because there were relatively few roads in colonial America, nearly all freight moved by water. This meant that common carriers—enterprises specialized in the transport of goods and passengers, with services available to all who could pay—first appeared in transportation along water routes, especially in transatlantic shipping. Because the eventual construction and maintenance of overland rights-of-way, in the form of turnpikes and canals, required large amounts of capital, these projects also led to new institutional forms: to incorporated joint-stock companies, forerunners of the modern corporation, and to state-controlled commercial enterprises.

Except for a few ferries, no common carriers traveled water routes in the eighteenth century. Most transatlantic shipping operated between broad regions, as between New England and the Mediterranean, with ships moving from port to port within each area as trading opportunities arose. Ships did not ply between prespecified terminuses, however, nor keep to any particular schedules. Although shipowners sometimes rented out space in their holds, they were under no obligation to carry cargo for other merchants and did so only when they had unused capacity (Shepherd and Walton 1972, chap. 4).

With continued expansion of transatlantic trade into the nineteenth century, the volume of traffic enabled certain shippers to become "regular traders," crossing back and forth between a major American port and a European one; New York had more than thirty such vessels making two round trips to Liverpool annually by 1815 (Albion 1938, p. 16). The trading function then differentiated into that of the general merchant, who could now charter ships rather than owning them, and the shipping agent or "husband," who might own and operate several vessels but conduct no trade of his own. Husbands specialized in the management of the shipping operation, including negotiating with merchants, scheduling departures, selecting and instructing ships' officers, deciding repairs, receiving and loading cargoes, arranging payment of customs and port duties, and keeping the accounts (Albion 1939, p. 268; 1941, p. 7). This division of labor proved so efficient in moving American cotton to England, for example, that by 1820 less than one-third of Liverpool's thirty leading importers owned any ships of their own (Williams 1969, p. 199).

Once the advantages of plying prespecified routes between major ports had been established by the regular traders, standardization of transatlantic shipping was quickly extended to departure schedules, even though crossings might take from three weeks to three months, depending on the vagaries of wind and weather (Chandler 1977, p. 35). In January 1818 owners of four regular traders in cotton and textiles founded the first scheduled packet service, the Black Ball line, which ran between New York and Liverpool with regular departure dates and times. Four years later a packet line linked Philadelphia and Liverpool; within a decade similar services plied the coastal lanes from New England south to Charleston and Savannah and even connected these Atlantic ports with Mobile and New Orleans (Albion 1938, pp. 33–35, 47, 60). Merchants who established successful lines either came to specialize in shipping or sold their interests to more specialized agents who might manage several vessels, each regarded as a separate business (Albion 1941). Typical of the new specialists was Charles Morgan, a Connecticut Yankee who between 1819 to 1846—although nominally a "grocer"—held equity in eighteen packets serving ten different lines from New York to New Orleans, Kingston, and Liverpool and in at least fifteen unscheduled tramps trading to the West Indies and Europe (Baughman 1968, pp. 9, 248–249).

Although shipping along water routes constituted the crucial area of growth for common carriers, the new type of enterprise did spread on overland routes as well. During the colonial period only a few stage-coach and wagon lines had operated as common carriers. Stagecoaches regularly carried passengers and mail but little freight and ran on mostly informal schedules. Charles Paullin describes transportation out of New York City at the turn of the nineteenth century:

Travel by stagecoach was limited to the area bounded by Boston, Bennington, and Albany on the north, Richmond on the south, and the Allegheny Mountains to the west . . . Buffalo was a nine-day journey via the Hudson-Mohawk route, and the trip over the mountains to Pittsburgh required ten days. It required 22–25 days to reach Louisville by way of the Ohio and 28–30 days to reach Cairo. In Tennessee and Kentucky travel on horseback was more rapid than water transportation, because of the circuitous courses of the rivers, and several well-defined overland routes such as the Natchez Trail had already developed.   (1932, p.133)

Even less scheduled than the stagecoaches were the wagon lines, usually operated out of rural towns, which picked up shipments from local merchants and hauled them to the nearest port, where the teamsters waited until a city merchant had a return shipment for their local areas. This system prevailed until the 1830s, Alfred Chandler reports, "even in Philadelphia, a city whose large hinterland was served by the best turnpike system in the nation" (1977, p. 32). The first turnpikes, in New England and the Middle Atlantic states, were built and maintained by private corporations, as were the few canals built before 1820.

After completion in 1825 of the 340-mile Erie Canal, a nine-year, $140-million public works project which the state of New York served both to organize and bankroll, merchants in other Atlantic ports clamored for their own connections with the West—while western businessmen sought similar links to the Ohio and Mississippi. Much too expensive to be built with private capital, even if pooled through incorporation, the new canal systems of Maryland, Virginia, Pennsylvania, and Ohio were largely financed by these states and their port cities and managed by representatives of their political bodies (Taylor 1951, pp. 48–52; Goodrich 1960, chap. 2).

Even though American businessmen increasingly relied on local, state, and national governments to fund canals and turnpikes after the 1820s, they did not suggest that the governments operate the common carriers that used the new routes and indeed government never assumed this role (Taylor 1951, chaps. 2, 3). Even the private corporations that had built and maintained the first turnpikes and canals rarely managed any of the transportation lines that used them. The stagecoach and wagon lines running on the new turnpikes as well as the barge lines shipping on the new canals operated much as did common carriers of colonial times. As on the transatlantic routes, however, overland transport increasingly differentiated separate commercial and shipping functions. Although the first canal boat lines were organized by merchants who needed to move their goods, these men quickly specialized in shipping or else sold their interests to a freight forwarder who might own and operate large fleets on the canals. By the

1840s these new transportation specialists offered regular through-freight arrangements with far-flung lines (Scheiber 1969).

Contrary to what we might expect from the Clark-Bell sequence of industrial development, therefore, industrialization of America did not begin before extensive innovation in the tertiary sector: common carriers, incorporated joint-stock companies, and state-controlled enterprises to build and maintain rights-of-way. Ship's husbands, regular traders, scheduled packet lines, turnpike and canal companies, and the freight forwarders all appeared in the first half-century of the American republic and flourished well before the onset of the Industrial Revolution in the 1840s. This squares with the systems view that tertiary services constitute a necessary precondition of industrialization—and not only because they facilitate the movement and processing of matter and energy by the system. Tertiary services also enable businessmen to specialize in only a few lines of goods and even to concentrate on a single commercial function (Pred 1973, chaps. 3–5); by reducing information and transaction costs they increase the efficiency of advancing division of labor.

Despite increases in complexity and specialization, the infrastructure of distribution remained unchanged on the key dimension: speed. Until the 1830s, when steam first began to power overland transportation on the railroads, goods still moved at the speed of riding horses, draft animals, and water and wind power. Although Robert Fulton and his financial backer, Robert Livingston, established the first steamboat line in 1807, persistent problems long confined the new technology to only the stillest waters. Along coastal routes the steamboat reduced travel times as much as one-half: between 1800 and 1830 the trip from New York to Charleston was reduced from ten to six days, to New Orleans from 26–28 days to two weeks (Paullin 1932, p. 133). Steamships did not ply the high seas regularly before the 1840s, however; Tyler (1939, pp. 18–28) reports fewer than fifteen transatlantic crossings using even auxillary steam power before 1838. Although a steam-powered boat, the American-built *Savannah*, first crossed the Atlantic in 1819, it used its collapsible paddle wheels only about eighty hours of the twenty-seven-day trip and soon returned to the coastal trade as a sailing ship without engine (Taylor 1951, pp. 113–114).

Table 5.3 summarizes selected events before 1850 in the development of that part of America's transportation infrastructure and its controlling services based on animate and wind power. The tertiary infrastructure developed rapidly on both land and water over this half-

Table 5.3. Selected events in the development of a national transportation infrastructure using only animate and wind power, 1789–1849

Year	Development
1789	Survey of U.S. roads—containing 86 plates—published
1794	Two-mile canal around falls of the Connecticut River at South Hadley Falls, Mass., opens to traffic
1795	Lancaster Turnpike, 62-mile road of macadam (pounded stone), links Philadelphia and Lancaster, Pa.
1796	Suspension bridge erected Coastal survey book (Furlong 1796) published; includes directions for entering, leaving major ports and harbors north of Virginia
1806	Congress appropriates $30,000 for Great National Pike (Cumberland Road) from Maryland to Ohio
1807	Congress appropriates $50,000 to survey U.S. coasts
1809	Railroad—0.75 mile, with cars pulled by horses—built at stone quarry
Early 1810s	Atlantic shipping volume enables traders to become common carriers between major American and European ports
1816	Coastal survey superintendent appointed by Secretary of Treasury
1818	Owners of four regular traders in cotton and textiles found packet service (Black Ball) between New York and Liverpool
Late 1810s	Packet ships—with regular service on a particular run—become predominant in Atlantic shipping
1821	Packet line links Philadelphia and Liverpool Canal tunnel—450 feet long—opened to traffic near Auburn, Pa.
1825	New York State's 340-mile Erie Canal—a nine-year, $140 million public works project—begins operations
1826	Railroad using horses to draw wagonloads of ore along cast-iron rails opens between Quincy, Mass., and Neponset River
1828	Baltimore and Ohio Railroad begins service with horse-drawn cars
Late 1820s	Canals extended to eastern Pennsylvania by owners of anthracite fields in the region
1831	Regular coastal shipping service links New York and New Orleans

Year	Development
c. 1833	Clipper ship—streamlined vessel designed for speed—constructed
1830s	Wagon lines carrying freight between rural towns and ports begin to operate on regular schedules
1840	Great National Pike (Cumberland Road) completed between Maryland and Vandalia, Ill.
1842	Overland wagon road (Oregon Trail) through Rocky Mountains to Pacific surveyed by federal government
1840s	Freight forwarders operate large fleets on canals, offer regular through-freight arrangements with other lines
Late 1840s	Clipper ships become predominant, especially on routes to California, China

century. On land, roads extended from the sixty-two mile Lancaster Turnpike (1795) to the Great National Pike connecting Maryland and Illinois (1840) and the Oregon wagon trail through the Rockies to the Pacific (1842); railroads developed from a three-quarter mile line powered by horses (1809) to steam power twenty years later. On water, transatlantic shipping evolved from the first common carriers (early 1810s) and packet service (1818) to clipper ships in the late 1830s and 1840s; canals appeared, flourished, and began to be abandoned over the same period. Boosting both land and water transportation, particularly in the new nation's first quarter-century, was the quinary service of the survey: of roads (1789), ports and harbors (1796), and the U.S. coast (1807); as early as 1816 the country had a permanent superintendent of the coastal survey.

Despite the fact that the tertiary infrastructure developed steadily throughout this preindustrial period, few major innovations must be left off the list—before the late 1830s—by excluding steam power (compare Table 5.6, later in this chapter, which supplements and extends the chronology in Table 5.3 for steam). Without steam power the infrastructure improved, as did commercial distribution, but neither got dramatically faster. Teams of horses and mules continued to power canal boat lines, on which sustained speeds as great as four miles an hour proved rare (Chandler 1977, p. 47). The trip from New York to Cleveland took ten days in 1830, to Detroit required thirteen;

from Detroit to Chicago took six days overland, about two weeks on the Great Lakes (Paullin 1932, p. 133). In 1840, after a decade of railroad development, more than 90 percent of Post Office routes still depended on the horse (Pred 1973, p. 85).

Such dependence on natural and animate sources of energy and on undeveloped road surfaces and waterways often put transportation at the mercy of weather and wind. This meant that, before steam power, transportation lacked two other prerequisites for the application of control technology: regularity and predictability. As we have seen, a transatlantic crossing under sail could be expected to take seven weeks, plus or minus a month. Outside of the South, winter freezes usually halted all shipments via canals and other inland waterways for several months each year; flooding and drought disrupted transportation during the other seasons. Heavy rains could keep interior towns mudbound for weeks; snows routinely isolated even the largest cities for days on end.

Not until the advent of railroads and ocean-going steamships in the 1840s, therefore, did transportation systems begin to achieve the speed, regularity, and predictability that would eventually challenge existing communication and control technologies—and indeed call continued control of the systems into question. Even early in the nineteenth century, owing to the increasing specialization of the tertiary sector, the movement, processing, and distribution of goods began to generate an increasing number of transactions and transshipments. As long as these exchanges and flows remained slow and irregular, however, they could be controlled by market mechanisms. Well before applications of steam power speeded up the system, in other words, market control had already improved through structural differentiation and specialization in the white-collar or quaternary sector of the American economy—changes that involved major innovations in commerce, finance, and insurance.

Innovation in the Quaternary Sector

Contrary to what we might expect from the Clark-Bell sequence of development, the industrialization of America did not begin before considerable differentiation and systemization of the commerical infrastructure took place, especially in the cotton trade but also in imports and grain. Even in the absence of Britain's Industrial Revolution and the resulting boom in cotton, as we saw in the previous chapter, specialization of American commerce would have come by the mid-nineteenth century. As it happened, a complex network of specialized middlemen—storekeepers and factors, commission merchants and exporters, wholesalers and manufacturing agents—arose well before the onset of the American Industrial Revolution in the 1840s. This growing commercial infrastructure, a complement of the new transportation infrastructure discussed in the previous section, stood ready to move, process, and distribute the matter and energy necessary to sustain the first phase of industrialization. Development of both these tertiary and quaternary infrastructures constitute necessary preconditions for industrial society.

Even more striking, however, is the mechanism for *control* of the commercial infrastructure. At mid-nineteenth century virtually every movement through the system—every transaction or exchange between middlemen—was still governed by the simple market mechanisms described by Adam Smith in his *Wealth of Nations* (1776). As Alfred Chandler describes these mechanisms, "Merchants who carried out the commercial transactions and made the arrangements to move the crops out and finished goods in did so in order to make a profit on each transaction or sale. The American economy of the 1840s provides a believable illustration of the working of the untrammeled market economy" (1977, pp. 27–28).

Despite structural differentiation and the specialization of merchants, the commercial infrastructure remained wholly unchanged since colonial times on what we have already found to be the key dimension for control: speed. Not until the application of steam power to processing and distribution in the decades 1840–1870 did the speed, regularity, and predictability of throughputs to the commercial system call into question its control by market mechanisms. Owing to the increasing systematization of American commerce beginning early in the nineteenth century, the volume of throughputs and the number of transactions and transshipments did increase steadily. This resulted in improvements and refinements of market control, however, not in the mounting crisis of control that would close out the century.

Differentiation of Finance and Insurance

Structural differentiation of the quaternary sector also occurred, well before American industrialization, in the ancillary services that support commerce, namely finance and insurance. Throughout the colonial period such services had remained parts of an undifferentiated commercial function. The colonial merchant not only served as shipowner, importer, exporter, wholesaler, and retailer for a wide range of goods, as we have seen, but also financed and insured their movement through the economy. With the rapid expansion of trade after independence, however, merchants began to find advantage in pooling their capital in a new institution, the joint-stock company, and to provide the infrastructure and support services useful to all commercial activities—not only canals and turnpikes but banks and insurance companies as well.

Because established merchants, who already financed the cotton export trade and, to a lesser extent, the return traffic in manufactured goods, naturally specialized in international and interregional banking, the first incorporated banks served mostly local businessmen who had less capital and therefore greater need to pool resources. Unlike the merchant banks, which were virtually all British, the incorporated American banks did not issue short-term bills of exchange for property in process and transition but rather provided much longer-term loans based on more fixed sources: mortgages, securities, and personal promissory notes. Incorporated insurance companies, begun in port cities by local importers, exporters, and shipowners to reduce the costs of insuring vessels and cargoes, also provided long-term loans, primarily on mortgages.

Both banks and insurance companies filled two other vital functions. First, their stocks could be held as investments during a period when few other opportunities aside from land and nonliquid assets had yet developed. Second, both institutions reduced information and transaction costs to the economy by entrusting most day-to-day operations to some of the first salaried information specialists: secretaries, clerks, cashiers, appraisers, and inspectors (Redlich 1951, chap. 2; Albion 1939, pp. 270–274).

Through the issuance of bank notes, state-chartered institutions also provided the generalized medium of exchange necessary for market control of a preindustrial economy. Unlike Europe, where merchant bank financing generated bills of exchange for market control, the United States relied largely on notes issued by various state-chartered banks for its standard circulating medium. This remained true until the American Civil War, when the federal government began to issue its own paper money in quantity.

America's first quaternary-sector enterprise to be incorporated was the Bank of North America in Philadelphia, chartered in 1781. By the time Congress approved Alexander Hamilton's proposal for a federally chartered institution in 1791, a half-dozen more banks had already opened in major ports: Philadelphia, New York, Boston, Baltimore, and Charleston. The first incorporated insurance company began operations the following year; by 1807 forty American firms had specialized in marine insurance. With the opening of the second federally chartered Bank of the United States in 1816, the country had nearly 250 banking enterprises; there were 307 by 1820 (U.S. Bureau of the Census 1975, p. 1018). Well before the advent of industrialization in the 1840s the quaternary sector of the U.S. economy had differentiated savings banks (1819), trust companies (1822), investment trusts (1823), and building and loan associations (1831) (Kroos and Blyn 1971, pp. 57–63); the insurance business included a few companies that specialized in both fire and life insurance by 1840 (Chandler 1977, p. 32).

Centralized control of national and international finance also appeared for the first time in the 1830s. Although many British merchant bankers maintained interlocking partnerships, these rarely encompassed more than three cities; most banks conducted business in distant ports through correspondents, independent merchants who worked on commission. By 1830, however, the Second Bank of the United States operated twenty-two branches in all regions of the country from its headquarters in Philadelphia. This banking system, probably the most competitive on earth for financing interregional and international commerce by the early 1830s, provided the administrative infrastructure for centralized control of credit and funds transfers among widely separated branch offices, which nevertheless maintained considerable local autonomy—one of the few distributed control systems established without benefit of electrical communication (the telegraph did not appear until after the bank ended its operations in 1836).

In essence, the Second Bank controlled a financial network that paralleled the commercial networks described in the previous section (Redlich 1951, chap. 6, sect. 3; Temin 1969, chap. 11). This new financial control system might be summarized as a circular exchange among branch offices in three cities: a western river port, a southern coastal port, and a northeastern commercial center. When planters and farmers turned over their outputs to the commercial system, they received bills on the proceeds which they sold for bank notes at the river port branch; this office remitted the bills to its branch in the coastal port to which the goods had been shipped. Merchants in that city exchanged the shipment for bills on the proceeds from other merchants in the northeastern commercial center where the goods were to be sold; these bills also entered the coastal port branch. After the planters and farmers spent their newly earned currency on manufactured and imported goods from the northeastern port, these bank notes moved back through the same commercial network, many eventually arriving at the city's branch bank. When this office called on its western branch to redeem the notes, the latter paid the bills remitted to the coastal port branch, which in turn paid with bills drawn on merchants in the northeastern city.

In other words, both western produce and bank notes moved eastward through the combined commercial and financial infrastructure; both finished goods and bills of exchange moved westward through the same system. As the bills and notes—in effect the media of communication and market control—reached the opposite ends of the exchange system, branch banks along the route (like the coastal port branch in the example here) effectively canceled out the countervailing paper, which greatly reduced information and transaction costs while maintaining the system of control by market mechanisms.

The efficiency of this distributed system of branch networks gave the Second Bank unprecedented control over a national economy and inevitably generated political controversy. In 1832, the year President Andrew Jackson vetoed recharter of the bank, it handled $22.6 million per month in domestic transactions, up from $1.9 million at the beginning of 1823; profits on these same exchanges exceeded $740,000 per month (3.3 percent) in 1832, up from less than $50,000 (2.6 percent) over the same period (Catterall 1903, p. 502). After expiration of the Bank's charter in 1836, merchant bankers resumed major financing of long-distance trade while local and domestic commerce continued to be served by state-incorporated banks, which by 1840 numbered more than nine-hundred—nearly a threefold increase in just twenty years (U.S. Bureau of the Census 1975, p. 1021).

Despite the headstart in capital accumulation that the Industrial Revolution afforded Britain's merchant bankers, local and regional financing of U.S. commerce came to be controlled by incorporated state banks, and the federally chartered Second Bank more than held its own in interregional and even international banking. These new financial networks, a complement to the growing transportation and commercial infrastructures discussed in the previous two sections, facilitated market control of the movement, processing, and distribution of the matter and energy necessary to sustain the first phase of industrialization. Growth of control by these higher levels of the quaternary or service sector, like the parallel development of the transportation and commercial infrastructures, constituted a necessary precondition of industrial society.

Table 5.4 summarizes selected events before 1860 in the development of America's quaternary sector, including trade and the controlling subsectors of finance and insurance. Again contrary to what we might expect from the Clark-Bell sequence of industrial development, even the higher-order control levels of the quaternary sector had developed and differentiated, with growing systematization, well before the advent of American industrialization in the 1840s. As shown in Table 5.4, not only trade but also finance and insurance developed throughout the preindustrial and early industrial periods as a result of a continuing series of innovations in information-processing and communication technologies in all three subsectors.

In trade, innovations included all of the major means to increase control: differentiation and specialization, programming, standardization and preprocessing, enhanced information processing and communication. Structural differentiation of commerce included that of city merchants into retail shopkeepers and their wholesale suppliers (late 1780s), of general merchants into shipowners, managers, and traders (early 1810s), of interior jobbers into more specialized middlemen (1830s). Functional specialization included that of older commission firms in particular imports (late 1810s), of the new importers in goods from Europe (late 1810s), and of the city of New York, first as a hub for the rural trade and center of credit (1820s), then as the major port for the export of cotton and other agricultural commodities (late 1830s).

Innovations in the programming of trade included a federal bankruptcy act (1800), codification and teaching of the principles of commercial law (c. 1802), and court decisions establishing the corporate form (1819), trust powers (1822), and regulation of interstate commerce (1824). Standardization and preprocessing innovations included the adoption of fixed prices by specialized jobbers (1830s), the appearance of companies to establish commercial credit ratings (1841), and the standardization of methods for sorting, grading, weighing, and inspecting agricultural commodities (early 1850s). Information-processing innovations included two major institutions, the auction (after 1815) and the commodity exchange (early 1850s), as well as the adoption of the

Table 5.4.   Selected events in the structural differentiation of commerce
and controlling financial and insurance sectors, 1789–1859

Year	Development
1780s	Half-dozen banks open in major ports
Late 1780s	Commerce in Philadelphia and New York begins to differentiate into retail shopkeepers and their wholesale suppliers
1791	Bank of United States chartered in Philadelphia, used by Secretary of Treasury Hamilton as fiscal agent
1792	Federal mint authorized; director appointed; building constructed in Philadelphia Price of gold fixed by Congress Stock exchange organized in New York by protective league of twenty-four brokers who fix commission rates on stocks and bonds Life insurance offered by a general insurance company Incorporated insurance company begins operations
1795	Business journal (*New York Prices Current*) begins publication
1799	Insurance companies regulated by state (Massachusetts)
1800	Federal Bankruptcy Act establishes a uniform system of bankruptcy for merchants, bankers, brokers, factors, underwriters, insurers
1802	Textbook on commercial law published
1804	Insurance agency—representing London firm—opened in New York
1810	Fire insurance joint-stock company organized
1811	Charter of the Bank of the United States expires, not renewed by Congress
Early 1810s	Function of general merchants differentiates into those of shipowners, managers, and traders
c. 1815	Newspapers proliferate, devote increasing space to commerce
1816	Second federally chartered Bank of the United States authorized Savings banks begun in Philadelphia, Boston
1817	New York Stock Exchange organized
1818	Ten percent of U.S. imports sell through the new institution of the auction Marine insurance law enacted by state (Massachusetts)

Year	Development
1819	Immortality of corporations set by *Dartmouth College vs. Woodward*
Late 1810s	Older commission firms narrow their range of imports; new merchants increasingly specialize in goods from Europe
1820	New York insurance firms organize board of underwriters, set rates for ships, cargoes, freight earnings Mercantile, mechanics' libraries opened in New York, Philadelphia for clerks, tradesmen, mechanics, and apprentices
1822	New York company (Farmers' Fire Insurance and Loan) becomes first corporation in world to be granted trust powers
1823	Federal assay office opens in Philadelphia Mint
1824	Interstate commerce case, *Gibbons vs. Ogden*, decided by Supreme Court
1826	More than half of imports through New York port sell at auction
1829	Legislation by state (New York) insures bank depositors; state bank commissioners, examiners appointed
1820s	New York becomes wholesale hub for rural trade, mecca for country shopkeepers on semiannual trips, center of credit network supporting distribution in West
1831	Trade journal (*Rail-road Advocate*) published Building and loan association formed to safeguard savings, secure mortgages at reasonable interest rates
Early 1830s	Control of national, international finance becomes centralized in the Philadelphia headquarters of the Second Bank of the United States, which operates 22 branches in all regions Interior commercial system develops to move western agricultural commodities to markets in South and East, differentiating a half-dozen specialized middlemen
1835	Mutual life insurance company chartered Mutual fire insurance company for factories incorporated
1836	With expiration of the Second Bank's charter, merchant bankers resume major financing of long-distance trade
1830s	Private semaphore systems built by shipping and mercantile firms to relay messages from incoming ships to counting houses Jobbers begin to purchase directly from domestic and foreign manufacturing agents; auctions decline

Table 5.4 (*cont.*)

Year	Development
1830s	Specialized New York City jobbers adopt fixed prices, publish catalogs, mail to customers
Late 1830s	Cotton, other agricultural commodities come to dominate trade through New York, hastening the demise of general merchants, helping to make port most important in North America
1840	Auctions fall to only one-eighth of U.S. import sales
1841	Commercial credit rating company (Mercantile Agency) begun Advertising agency (V. B. Palmer) established
1842	Credit protection group formed by importers, commission houses
1844	Credit report book published, distributed by subscription
1847	Health insurance company incorporated
1850	New York City directory lists more jobbers than importers in a half-dozen speciality goods
1851	Insurance board established by state (New Hampshire)
1852	Branch of federal mint established on west coast Insurance magazine (*Tuckett's Monthly*) published
1853	Federal assay office building authorized, erected on Wall St. Through bill of lading introduced Bank clearing house organized, exchange opened on Wall St. Trade association (American Brass Association) organized to regulate prices
1854	Firm (United States Trust) organized to act as executor, trustee
Early 1850s	Commodity exchanges opened; standardized methods adopted for sorting, grading, weighing, inspecting
1859	Insurance department established by state (New York)
1850s	Specialized jobbing begins to spread from large coastal ports Warehouse receipts with uniform grades serve as exchange media

through bill of lading (1853) and the use of warehouse receipts with uniform grades as a medium of exchange (1850s). Commerce also adopted a series of innovations in communication, including journals of prices

(1795), commercial newspapers (c. 1815), mercantile libraries (1820), trade journals (1831), ship-to-shore semaphore systems (1830s), agencies for advertising (1841), and credit report books distributed by subscription (1844).

In addition to this extensive innovation in commerce itself, increasing control of the processing and distribution of material goods also was seen throughout the preindustrial and early industrial periods in the financial and insurance subsectors of the quaternary sector. Innovations in finance included a half-dozen major institutions: general commercial banks (1780s), the stock exchange (1792), savings banks (1816), state boards of bank examiners (1829), building and loans (1831), and bank clearing houses (1853). Innovations in insurance included a half-dozen new types: general life (1792), joint-stock fire (1810), state marine (1818) and bank deposit (1829), mutual life and fire (1835), and health (1847).

Even the quinary sector, including the federal and state governments and the legal profession, played an active role in increasing control of commerce during this period. Well before the advent of American industrialization in the 1840s, the federal government had established two national banks (1791, 1816), a mint (1792), and an assay office (1823); it had also attempted to control commerce by fixing the price of gold (1792) and establishing a uniform system of bankruptcy (1800). Meanwhile, various state governments had also been active throughout the preindustrial and early industrial period in adopting innovations for the additional control of commerce: regulating insurance companies (1799) and marine insurance (1818), granting trust powers (1822), insuring bank depositors (1829), and establishing insurance boards (1851) and departments (1859).

Well into the 1840s, however, despite intervention by the federal and state governments, the mechanism for control of the American economy remained the market. Differentiation and specialization of commerce and the various quaternary and quinary innovations in control notwithstanding, market control remained unchanged on the key dimension: speed. Even in the distributed control structure of the Second Bank, with its extensive network of branches, bank notes and bills of exchange moved between offices at the speed of the riding horse, stage coach, flat boat, and sailing ship.

In 1840, as we have seen, virtually all Post Office routes still depended on horses. Not until the spread of railroads and ocean-going steamships in the 1840s and 1850s did the systems of transportation and the commerce dependent on them attain the speed, regularity, and predictability that called their continued control by market mechanisms and supporting quaternary services into serious question. Until such application of steam power to the material economy, the entire operations of the Second Bank, with twenty-two branch offices and profits fifty times those of the largest mercantile house in the country, could be run by just three people: Nicholas Biddle and two assistants (Redlich 1951, pp. 113–124). Clearly Biddle, who died in 1844, knew nothing of the crisis of control that would begin to challenge his fellow bankers of the succeeding generation.

Lagging Development of the Primary and Secondary Sectors

Although to give a precise date for the American Industrial Revolution or the resulting crisis of control would be specious, several technological and economic developments converged in the late 1830s and 1840s to give the process a relatively distinct beginning. Each of these developments involved one or both of two essential changes: increasing levels of energy utilization or *power* and the progressive translation of this power into the increasing *speed* with which matter, energy, and information moved through the system. All else being equal, increases in power will always result in increases in speed, which in turn increases the need for *control* and hence for communication, information processing, programming, and decision.

To power the extraction, processing, and distribution of material throughputs to industrial production, the system turned for its energy from sources animate (human and draft animal) to sources inanimate (steam-powered machinery) and from natural kinetic sources (wind and water) to sources chemical (coal and electric batteries). For this reason anthracite and iron mining gained in importance in the primary sector relative to agriculture, while textile and metals production—the most effective early applications of steam power—came to dominate the secondary sector. Steam-powered transportation, especially railroads and steamship lines, speeded processing and distribution; even faster electrical communication via a national telegraph grid helped to control the new systems of transportation and commerce. Although not all of these changes characterized the earlier Industrial Revolution in Britain, they had converged by the mid-nineteenth century in factors that began to push American industrialization toward its crisis of control.

As in Britain, where the proximity of large coal and iron deposits speeded industrialization, the Industrial Revolution did not take hold in the United States before domestic fields of the two key ores had been sufficiently developed. Although iron and coal were the only two materials mined in colonial America, the new continent did not yield either ore in quantity until the 1830s. Before that decade the former colonial mines of Virginia on the James River remained the only domestic source of coal available to New England manufacturers; for many years the James River mines measured total output not in tons but in bushels (Eavenson 1942, chaps. 6, 7). This lack of coal hindered the spread of steam power. Although operating a steam engine cost two-and-a-half times as much in the northeastern United States as it did in England in 1828, in Pittsburgh—where coal abounded—it cost only three-fourths the English rate (Chandler 1977, p. 526). "Of all the technological constraints," Chandler concludes, "the lack of coal was probably the most significant in holding back the spread of the factory in the United States" (1977, p. 76).

Americans first recognized the value of the anthracite fields of eastern Pennsylvania under the trade restrictions imposed by the War of 1812. Owners of these coal lands began to extend canals into the region in the 1820s. From virtually zero output before 1825, when the first canals became operational, annual production of anthracite exceeded 290,000 tons by 1830. Although this coal was first used to heat buildings, its increasing availability and declining price—from nearly $10 to less than $5 a ton by the mid-1830s—began to stimulate industrial applications, especially in New England, where coal became an attractive alternative to the limited sources of water power. Using anthracite to generate steam, Samuel Slater, who in 1789 had been first to bring a British design for a water-powered spinning mule to America, began in 1828 to operate the country's first steam-powered textile mill in Providence, Rhode Island (Ware 1931, chaps. 2, 5). Four years later nearly sixteen hundred sailing ships moved 158,000 tons of Pennsylvania's finest anthracite from Philadelphia to Boston to help feed new steam-powered textile, metal-making, and metalworking factories in southern New England.

As late as 1832, however, most American factories continued to run on water power. Outside of Pittsburgh, only four of 250 of the largest

U.S. manufacturers relied solely on steam power in that year, even though four-fifths of these firms made textiles; three other firms used steam to supplement water power (Chandler 1972, pp. 143–145). Although among them the 250 companies had a total of some one hundred steam engines, most were low-horsepower auxiliary engines (Temin 1966, p. 189). Indeed, more of the large firms relied on wind and mule power than on steam (Chandler 1977, pp. 61–62). Clearly steam-driven manufacturing, despite the steady supply of cheap coal by the late 1820s, did not precede the application of steam to rail transportation in the late 1830s.

Much the same can be said of iron. From mines to manufacturers, the entire industry continued to burn wood rather than coal and to rely on water power rather than steam until the coming of the railroads. Even in metal-making and metalworking, among the first American industries to exploit the growing domestic coal production to increase their own output, little change occurred before the 1840s. Peter Temin, specialist on early nineteenth-century manufacturing, noted that "before the 1840s . . . the iron industry of the United States was expanding only slightly more rapidly than the population . . . It was still using the old techniques and the traditional fuel, charcoal, inherited from previous centuries, much as the economy was largely concentrated in its familiar location" (1964, p. 15).

Well into the 1830s American domestic iron continued to come from the rural "iron plantations," little changed from colonial times, where adequate ore, trees to make charcoal for a blast furnace, and water power to run the forges all happened to be found in proximity. As colonial sites became depleted, plantations moved inland; furnaces to refine the ore into pig iron usually followed the miners. Because forges depended on water power and not steam to hammer pigs into wrought iron, however, these primitive operations often stayed behind at eastern sites, thereby increasing transportation costs. Output of both pig and wrought iron remained irregular: both furnaces and forges had to halt operations in extreme heat or cold or when a freeze or drought disrupted water supplies (Chandler 1972, p. 147). Even when they did operate, powered by several score indentured servants or occasionally by slaves, plantations produced at best thirty tons of pig iron per week (Hunter 1951, p. 178).

Small wonder, then, considering the low output, high cost, and irregular supply of domestic iron, that the large ironworks of New England and the Delaware Valley purchased 70 percent of their raw material

abroad, even after the stiff tariff—the so-called Tariff of Abomina-
tions—in 1828. As late as 1832, 97 percent of Maine blacksmiths, the
sector that consumed most wrought iron in the agricultural economy,
still depended *entirely* on foreign supplies; the other 3 percent used
only small amounts of American iron (Chandler 1972, pp. 145–146).

Metalworking itself remained nearly as primitive as domestic mining. Despite the increasing availability and decreasing cost of domestic
anthracite by the late 1820s, most wrought iron processors continued
to use water power and not steam to draw out wire and nails and to
hammer out sheets and fittings. Even New England's larger ironworks, source of most nails, barrel hoops, axes, and shovels for the
developing economy, continued to rely on charcoal and not anthracite
to work their metal, despite the mounting costs of wood as local
forests became depleted.

Table 5.5 summarizes selected events in the development of the
American anthracite and iron industries. The two industries developed
independently until the mid-1830s, when coal became plentiful and
cheap and anthracite reverberating furnaces first began to replace
charcoal-heated, water-driven forges. Owing to the demand for iron
created by the application of steam power to transportation, however,
by the mid-1840s anthracite, iron, and steam power began to develop
almost as one. By the late 1850s anthracite-powered iron rolling mills,
struggling to produce the material infrastructure for the steam-
powered railroads, constituted some of the largest industrial enterprises in the United States.

Thus did the primary, secondary, and tertiary sectors coevolve,
driven by the new inanimate power of steam. Until the 1840s, however,
despite growth in the coal, iron, and railroad industries, the industrial
factory—with complex, steam-driven machinery—had yet to appear
in the United States. With the exception of a few hundred textile mills
and a few dozen metalworking establishments like the U.S. Army's
Armory at Springfield, Massachusetts, almost all of which remained
water-powered, even multiunit factories, with division of labor and
integration of the several processes in a single establishment, had not
yet appeared. Alfred Chandler has pointed out that "before the mid-
1830s, when coal became available in quantity for industrial purposes,
nearly all production was carried on in small shops or at home. American manufacturing was still seasonal and rural. Workers were
recruited when they were needed from the local farm population and
paid in kind as well as wages. There was as yet only a tiny indus-

Table 5.5. Selected events in the development of the American anthracite and iron industries, 1789–1859

Year	Anthracite	Iron
1791	Anthracite discovered in Carbon County, Pa.	
1808	Anthracite burned experimentally	
1812	Anthracite used commercially	
1816		Boiler plate rolled at iron mill
1817		Iron mill puddles, rolls iron
1819		Angle iron rolled in Pittsburgh mill
1824	Large anthracite mining tunnel begun using black powder	
1826		Malleable iron castings produced
Late 1820s	Owners of anthracite fields in eastern Pennsylvania begin to extend canals into region	
1835	Coke demonstrated to be a successful blast furnace fuel	
Mid-1830s	Coal becomes plentiful and cheap as far north as Boston	
	Anthracite reverberating furnaces begin to replace charcoal-heated, water-driven forges	
1837	Domestic coal production exceeds a million tons per year	
	Anthracite used to smelt iron ore	
1839	Iron blast furnace adopts anthracite, produces 28 tons per week	
1840	Anthracite blast furnace begins production of pig iron	
1842		Hammered iron produced with water-powered trip hammers
1844		Rolling mill produces iron rails

Year	Anthracite	Iron
Mid-1840s	Pig iron producers adopt anthracite blast furnace; replaces rural iron plantations	
1846		Cast steel made, used in plows
1854		Wrought iron beams rolled, used in place of cast iron beams
1850s	Anthracite surpasses charcoal, bituminous in iron production	
		With large rail mills, railroad products consume 20 percent of domestic pig iron
Late 1850s		Four rail mills—each with more than $1 million in capital—become the largest enterprises in the iron industry

trial proletariat and a minuscule class of industrial managers" (1977, p. 245).

According to several of the so-called new economic historians (Fogel and Engerman 1971b), industrialization of the United States came in response to rising demand, at least prior to the American Civil War. Although demand for high-volume production existed before the 1840s in both America and abroad, at least for textiles (Zevin 1971, pp. 125–137) and iron products (Fogel and Engerman 1971a), the steam-driven factories of England and Europe continued to satisfy this demand. In short, the American Industrial Revolution cannot be said to have begun before the 1840s and the major application of steam power in the tertiary sector: the railroad. Not the steam engine itself, nor the knowledge of steam-powered machinery, nor established manufacturing in textiles and metals, nor the growing supply of domestic coal and iron— not even together did these factors bring rapid industrialization to the United States. The explanation, once again, involves speed.

As concrete open processing systems, societies cannot produce goods faster than they can move, process, and distribute throughputs to production and still maintain control of the system. We have already reviewed, in the previous three sections, early differentiation and specialization of the essential infrastructure for transportation, communication, and control. Until this system moved at the speed of steam power, however, steam could not be extensively applied to factory production—speeds in all parts of the system had to increase more or less in parallel.

As we have seen, a hundred steam engines could be found among the 250 largest manufacturers in 1832. Coal had become plentiful and cheap even in Boston by the mid-1830s, when the new anthracite furnace promised increasing supplies of wrought iron. Still, steam-powered factory production on a large scale awaited a means to move the correspondingly large volume of throughputs to the processing system— also at the speed of steam. Factories could sustain high processing speed and volume only if raw materials flowed in and finished goods could be distributed with equal volume and speed. For this to occur, the national network of canals and turnpikes would have to be replaced by a similar network of steam railroads. At the beginning of 1840, the United States—despite three thousand miles of track—still awaited completion of its first interregional railroad link. Contrary to the Clark-Bell sequence of industrialization, development of the economy's secondary sector awaited further development of its tertiary one.

Steam Speeds the Tertiary Sector

As in England, the first railroads in the United States employed horses to draw wagonloads of ore along cast-iron rails. The first such railroad, built in 1826, connected Quincy, Massachusetts, to the Neponset River; the following year a second linked the coal mines in Carbon County, Pennsylvania, with the Lehigh River. Although the first two locomotives arrived from England in 1829, they proved too heavy for existing tracks and trestles. The Baltimore and Ohio Railroad, begun in 1828 with horse-drawn cars, used steam power after 1830, when Peter Cooper, partner in a Baltimore ironworks, successfully tested his new locomotive, the *Tom Thumb*. By this time, steam-powered passenger and freight service on England's Stockton and Darlington Railway had been in full operation for five years.

Table 5.6 summarizes selected events before 1860 in the development of the second American infrastructure of transportation, one based not

Table 5.6.   Selected events in the development of a second national
transportation infrastructure based on steam power, 1807–1859

Year	Development
1807	Fulton, Livingston establish steamboat line
1809	Steamboat navigates ocean from New York to Philadelphia
1811	Steam ferryboat begins operations between New York and Hoboken, N.J. Steamboat travels from Pittsburgh to New Orleans
1816	First steamboat launched on the Great Lakes (Ontario)
1819	Steamboat (*Savannah*) built in United States crosses Atlantic
1825	Sheet-iron steamboat—weighing 5 tons—launched on Susquehanna Steam tugboat built to tow vessels at New York City dry dock
1829	Two locomotives arrive from England; too heavy for U.S. tracks
1830	Baltimore and Ohio Railroad adopts steam power after Peter Cooper successfully tests *Tom Thumb*
1831	Practical U.S. locomotive (*York*) introduced that burns coal, has coupled wheels, double pair of drivers
1834	Railroad tunnel—901 feet long—completed through Allegheny Mountains
1838	Regular transatlantic steamship service begins Annual steamboat inspection, federal inspection service established by Congress Steam shovel invented, first used on Western Railroad
1839	Express delivery service between New York and Boston organized using railroad and steamboat
1830s	Some three thousand miles of railroad track go into operation
1841	Western Railroad links Worcester, Mass., and Albany, N.Y., to become first interregional line
1844	Iron steamship (*Bangor*) built for transatlantic service Rolling mill produces iron rails
Early 1840s	Railroads move three to five times as much freight as canals for the same cost
1847	U.S. passenger steamship line to Europe established
1849	Regular steamboat service from New York to California via Cape Horn inaugurated
1840s	Six thousand miles of new railroad line built, double the amount laid in the previous decade

Table 5.6 (*cont.*)

Year	Development
1853	Railroad begins operations between New York and Chicago Union passenger station (Indianapolis) opened for trains of five companies
Early 1850s	Four great railroad trunk lines—Erie, Baltimore and Ohio, New York Central, Pennsylvania—link East and West
1855	Steam dredge built, used in Charleston harbor
1856	Railway bridge across Mississippi completed
1858	Overland mail service begun—twice weekly—to Pacific Coast
1850s	Rail network east of the Mississippi essentially completed Network of waterways, canals largely abandoned for railroads even in the Old Northwest; Mississippi Valley river boat lines also lose much traffic to railroads ′

on animate and wind power but on steam. Unlike the first infrastructure, which developed steadily from the 1790s through the 1820s and declined somewhat thereafter, the infrastructure of steam-powered transportation did not develop before 1830 and Cooper's *Tom Thumb*. Even the application of steam power to water transportation, as Table 5.6 shows, did not advance much beyond Fulton before the late 1830s and 1840s, when steamships first made their mark on the world commercial system with a regular transatlantic service (1838), a U.S. passenger line to Europe (1847), and regular service from New York around Cape Horn to California (1849).

On land, steam powered some three thousand miles of railroad by 1840 (Taylor 1951, p. 79; U.S. Bureau of the Census 1975, p. 731). Despite this rapid growth, railroads did not begin to stimulate widespread industrialization until America's first railroad boom in the mid-1840s. Until that time, lines remained short—most less than fifty miles—except in the West, where they carried fewer passengers and less freight. Most early lines connected existing commercial centers and therefore merely supplemented turnpike and water routes; no lines connected East and West until the 1850s. By 1850, indeed, only one railroad—the Western, completed in 1841 between Worcester, Massachusetts, and Albany, New York—linked major regions of the country.

The first decade of the railroad did, however, stimulate innovation and growth in iron and coal. By the mid-1830s the anthracite reverberating furnace had begun to replace the charcoal-heated, water-driven forge in the manufacture of wrought iron bars, sheets, and rods; the country's first anthracite blast furnace began production of pig iron in 1840. As both cause and effect of the innovations in iron-making, domestic coal production rose from 290,000 tons in 1830 to a million tons by 1837 and, despite a long economic depression, to two million tons by the mid-1840s; the price fell from $5 to $3 a ton. During the five years between 1831 and 1836 the number of vessels carrying anthracite from Philadelphia to Boston increased from 563 to almost 3,300 (573 percent); total shipments increased from 56,000 to 345,000 tons (616 percent) (Chandler 1972, p. 156).

After the protracted economic depression of the early 1840s, American railroad building boomed, first in New England and later in the South and West. The decade brought nearly six thousand miles of new track, double the amount laid in the railroad's first decade, compared to less than four hundred miles of new canals; America had as much rail as canal mileage by 1840, almost two-and-a-half times as much by 1850 (Taylor 1951, p. 79). Stagecoach-type rail cars gave way to rectangular "long cars" seating sixty in the 1840s; freight cars came to resemble smaller versions of the boxcars, cattle cars, and lumber cars used on American railroads to this day.

Between 1851 and 1854 four great trunk lines connecting East and West—run by the Erie, Baltimore and Ohio, New York Central, and Pennsylvania railroads—all began operations. The five states of the old Northwest, despite the region's natural transportation system of lakes and rivers, had by 1860 largely abandoned their network of waterways and canals for nine thousand miles of railroad. More canals were abandoned than built in the 1850s, when the country added twenty-one thousand more miles of track, thereby completing the essential rail network east of the Mississippi. Even the river boat lines had lost much of the Mississippi Valley traffic to the railroads by 1860 (Hunter 1949, pp. 484–488). As Chandler (1977, p. 86) observes, "Never before had one form of transportation so quickly replaced another" (Fig. 5.2).

In terms of costs alone, success of the railroads over canals depended on three factors: building, maintenance, and shipping costs. First, railroads cost considerably less to build over rugged terrain and, although canals cost somewhat less per mile on flat land, railroads could follow more direct routes (Taylor 1951, p. 53). Second, canals had much higher maintenance costs, particularly because of freshets, the sudden overflows caused by a thaw or heavy rain (Sanderlin 1946, pp. 191–

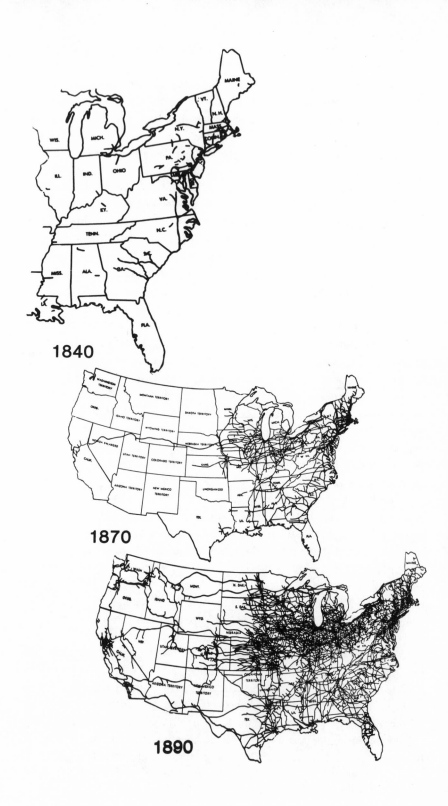

1840

1870

1890

193; Taylor 1951, p. 43). Third, because canals did not operate for much of the winter, even the early railroads could carry fifty times as much freight per mile annually, thereby greatly reducing unit shipping costs. As early as 1840 railroads could move at least four times as much freight as canals for the same cost (Lebergott 1966, p. 444).

In addition to these purely economic considerations, railroads also greatly surpassed canals in speed, regularity, and predictability—what we have already found to be the key factors increasing the need for control. Even the earliest railroads could increase the speed of a shipment tenfold: the canal and river route from Boston to Concord, New Hampshire, took five days upstream and four down, for example; the trip took only four hours each way on a rail link opened in 1842. "For the first time in history," Chandler (1977, p. 83) observed of the railroad, "freight and passengers could be carried overland at a speed faster than that of a horse."

Steam power made the American processing system move roughly three to ten times faster in the railroad's first quarter-century alone: between 1830 and 1857, according to data compiled by Paullin (1932, pp. 133–134, plates 138B, C), movement from New York City to New Orleans speeded up threefold (from two weeks to five days), to Cincinnati fivefold (from one week to one-and-a-half days), to St. Louis sevenfold (from two-and-a-half weeks to as many days), to Chicago tenfold (from three weeks to two days).

In addition to speed, steam power made transportation more regular, dependable, and predictable than that powered by animals, wind, or water currents. Although fraught with accidents and delays, railroad transportation, hindered little by either storm or calm, drought or flood, almost immediately proved more dependable than canals or sailing ships. Unlike the canals, railroads quickly developed the technology to continue operations year-round, even during the heaviest snowfalls. From the first, railroads ran on strict schedules: as early as 1835, for example, on the forty-four mile Boston and Worcester line, passenger trains departed each end of the single track at precisely the same times—6 A.M., noon, and 4 P.M.—with freights following

Figure 5.2. A half-century's growth in the U.S. rail network. During the period 1840–1890 the total mileage of completed railroad lines increased from 2,808 (1840) to 52,922 (1870) to 163,597 (1890); the system peaked in 1916 with a total of 254,000 miles of road, compared with about 200,000 miles today. (Courtesy of the Association of American Railroads.)

immediately behind the morning trains (Salsbury 1967, p. 114–115).

With the development of a year-round, dependable, and predictable transportation system that could move throughputs at the speed of steam, after 1840, the Industrial Revolution—grounded in similarly fast, steam-powered factory production—could at last take hold in the United States, nearly a century after its beginning in Great Britain. As domestic anthracite provided the energy, pig iron the matter, and steam-driven machinery the material processing for industrialization, the rapidly expanding rail network provided the infrastructure to move throughputs on an interregional, national, and finally continental scale. As Chandler (1977, p. 245) writes of the railroad, "Without a steady, all-weather flow of goods into and out of their establishments, manufacturers would have had difficulty in maintaining a permanent working force and in keeping their expensive machinery and equipment operating profitably."

Once the tertiary sector had established sufficient speed of movement, the primary sector provided the inputs of matter and energy. By the mid-1840s, with the price of domestic coal down from $10 to $3 a ton and annual production up to two million tons from virtually zero twenty years earlier, pig iron producers quickly adopted the new anthracite blast furnace. In 1849 sixty such furnaces had gone into blast; four years later the number had doubled. During the same decade the use of anthracite surpassed that of charcoal (in 1855) and charcoal and bituminous coal combined (1856) in U.S. pig iron production (Temin 1964, p. 266). For the first time, as a result of the railroad, American manufacturers could count on domestic sources for energy in abundance. Not only did anthracite generate the high steady heat needed in both furnace and foundary, thereby fueling a modern American iron industry and (indirectly) new metalworking industries, but it also powered the machines they produced.

Steam Moves to the Secondary Sector

As in the British Industrial Revolution, textiles paced industrialization of America's secondary sector. Aside from the production of firearms for the U.S. Army, as we have seen, the textile industry maintained the only substantial number of large factories—harnessed to the power of major rivers—before the 1840s. These so-called *integrated* mills, with all of the activities for both spinning and weaving under one roof, had appeared in America after 1815; the first integrated steam mill—Slater's Providence factory—began operations in 1828 (Ware 1931, chap. 2). With the increasing availability of coal for steam in the 1830s, the textile industry relocated from rivers to a string of integrated steam mills stretching along the New England coast from New London, Connecticut, to Portsmouth, New Hampshire.

Most of the major innovations in the new steam-driven textile machinery came before 1840 (Jeremy 1973), after which date the industrialization of American textiles can be said to have been essentially completed. Although productivity continued to increase dramatically in the decades immediately before and after the Civil War, these gains resulted less from further industrialization than from improvements in the processing skills of workers and in control technology—what economists call "learning by doing" (Arrow 1962; David 1970). In 1828, for example, Paul Moody applied belt drive to textile production, thereby overcoming the mechanical difficulties of the English gear drive and greatly raising the operating speeds potentially attainable (Gibb 1950, pp. 79–80). Greater productivity came through increased control learned through "experimental determination of optimum plant layouts, or work organization, or operating speeds and machinery maintenance schedules, or similar best-practice information" (David 1970, p. 537).

Abundant, inexpensive coal and iron brought the steam-powered integrated factory to the metalworking industries in the 1840s, and with it the so-called *American System*, manufacture through the processing and assembly of interchangeable parts. Pocket watches, clocks, locks, safes, and scales all came to be produced by the new method in large departmentalized factories. By 1850 manufacturers had begun to apply the American System to the production of newly invented machines of the Industrial Revolution: reapers, harvesters, and sewing machines. The need for still more specialized machinery to produce these machines, in turn, gave birth to a new machine tool industry—the technology of interchangeability applied to the production of parts for machines that would make parts for still other machines (Roe 1916, chap. 11).

Railroads themselves became major new markets for the metalworking industries. With completion of the first large rail mills, railroad products consumed more than 20 percent of American pig iron by the late 1850s, up from only 4.7 percent as late as 1840–1845; the rerolling of worn rails provided the mills a second substantial business (Fishlow 1965, p. 142). Because one rail mill could consume the total output of two or three blast furnaces, the total iron-making process—split up by the westward migration of the iron plantations—became quickly reintegrated in a single works with both furnaces and final shaping mills. By 1860 four rail mills, each with more than $1 million in capital, had become the four largest enterprises in the iron industry; the largest had nearly three thousand employees. Several rail mills had by this time branched out into the production of bar iron, beams, and wire (Temin 1964, pp. 109–111).

After the Civil War the large rail mills nurtured a new American steel industry. Sir Henry Bessemer's process for the manufacture of steel from molten pig iron, the first means of making the metal on an industrial scale, received its American trial in 1865. In 1866 the Pennsylvania Steel Company, rail-making subsidiary of the mammoth Pennsylvania Railroad, adopted the Bessemer process; by 1873 eight American mills had installed the new converters. In the decade that followed, large rail mills also pioneered the adoption of Sir William Siemens's open-hearth methods, which used both iron ore and pig iron, to produce steel on a mass scale (Temin 1964, pp. 170–172).

Not only did the railroad serve as a means to move throughputs to the industrial system, therefore; it also provided major markets for the products of that system. Because most locomotives burned coal, railroads stimulated both production and innovation in the coal industry—so vital to metals and manufacturing—while also serving as its major means of transportation. Because railroads consumed so much iron in their rails, wheels, and spikes, they also stimulated development and innovation in metal-making and metalworking, also industries vital to manufacturing.

Table 5.7 compares selected innovations in the application of steam power to transportation as compared to other sectors. American inventors seemed to think first of transportation when considering applications of the new inanimate power. Most of the major applications of steam to transportation, including a wide assortment of boats, ships, and locomotives, had been well developed by the early 1830s. The spate of innovations in the other sectors appeared to come in the late 1830s and 1840s.

This finding, however tenuous, does support our earlier conclusion about the relationship between transportation and the primary and secondary sectors. Despite the tertiary sector's role as a market for industrial production, its more likely major role was to move throughputs to the industrial system. As Chandler (1977, p. 245) concludes, "Of far more importance to the expansion of the factory system was the reliability and speed of the new transportation."

Table 5.7. Selected innovations in the application of steam, 1788–1870

Year	Transportation	Other sectors
1786		
88	Boat engine	
1790		
92		
94		
96		Practical engine
98		
1800		High-pressure engine
02		
04	Twin-screw boat	
06	Amphibious scow	
08	Successful boat	
1810	Seagoing boat	
12	Ferryboat	
14	Frigate	
16	Double-decked boat	Boiler plate
18	Transatlantic ship	
1820		
22		Printing press
24	Model locomotive	
26	Tugboat	
28		Textile mill
1830	Locomotive	
32	Practical coal locomotive	
34		
36		Coin mint
38		Thresher; shovel and crane
1840		Independent boiler pump
42	Engine for screw ship	Fire engine; grain elevator
44	Seagoing frigate, iron ship	Building heating system
46		Hotel, factory heating systems
48		Pile driver
1850		Percussion rock drill
52		
54		
56	Seagoing dredge	Calliope
58		Portable sawmill engine
1860		
62		
64		
66	Whaler; automobile	Suspended elevator
68	Motorcycle	
1870	Elevated railroad	

On the Eve of Crisis: Steam Power and Speed

This chapter has outlined a sequence of technological and economic innovations—most before 1850—in the material processing system that sustains the American society, the result of which was the harnessing of steam power throughout the economy, from extraction and production to distribution and consumption. Applied first to transportation, to railroads and to maritime shipping, steam power provided the means to move a large volume of throughputs to the processing system. With the development of this tertiary infrastructure and the quaternary and quinary infrastructure for its control, steam-powered factory production began to sustain the high processing speeds and volume that would usher in America's Industrial Revolution.

Speeding up the entire societal processing system, as the next chapter shows, put unprecedented strain on the quaternary and quinary sectors—on all of the technological and economic means by which a society controls throughputs to its material economy. Never before in history had it been necessary to control processes and movements at speeds faster than those of wind, water, and animal power—rarely more than a few miles per hour. Almost overnight, with the application of steam, economies confronted growing crises of control throughout the society. The continuing resolution of these crises, which began in America during the 1840s and reached a climax in the 1870s and 1880s, constituted nothing less than a revolution in control technology. Today the Control Revolution continues, engine of the emerging Information Society.

Industrial Revolution and the Crisis of Control

The most vital trait of the spontaneous organization of the industrial order is that its goal, and its exclusive goal, is to increase the control of man over things.

—Emile Durkheim, *Le socialisme*

INCREASING the speed of an entire societal processing system, from extraction and production to distribution and consumption, was not achieved without cost. Throughout all previous history material goods had moved down roadways and canals with the speed of draft animals; for centuries they had moved across the seas at the whim of the winds. Suddenly, in a matter of decades, goods began to move faster than even the winds themselves, reliably and in mounting volume, through factories, across continents, and around the world. For the first time in history, by the mid-nineteenth century the social processing of material flows threatened to exceed in both volume and speed the system's capacity to contain them. Thus was born the crisis of control, one that would eventually reach, by the end of the century, the most aggregate levels of America's material economy.

Problems began with a crisis of safety on the railroads, first with the intersectional lines in the early 1840s and then in the early 1850s with the great trunk lines connecting East and West. By the 1860s this safety crisis had given way to the railroads' continuing struggle to control their vast systems to maximum efficiency. As late as the 1870s, as we shall see, railroad companies actually delayed building large systems because they lacked the means to control them.

Meanwhile, the control crisis spread to distribution. With the proliferating network of grain elevators and warehouses in the 1850s, mercantile firms and other transporters had increasing difficulty in keeping track of individual shipments of wheat, corn, and cotton and in controlling the growing commerce in these commodities. In the 1860s this crisis in the control of distribution also began to affect the movement of goods in the opposite direction: from manufacturers westward to consumers. Commission merchants found it increasingly difficult to handle the distribution of mass-produced consumer goods; wholesalers struggled to integrate the movement of goods and cash among hundreds of manufacturers and thousands of retailers. By the late 1860s, with the rise of department stores and other large retailers and wholesalers, the control crisis had become one of maintaining high rates of "stock turn" in inventory.

At about the same time, the crisis of control also reached the producers themselves. Rail mills adopting the Bessemer process in the late 1860s struggled to control increasing speeds in the production of steel. Producers of basic materials—iron, copper, zinc, and glass— also worked with difficulty to maintain competitively fast throughputs within their plants. This crisis of production had by the 1880s reached even the metalworking industries, where companies engaged in the manufacture of everything from castings and screws to sewing machines, typewriters, and electric motors struggled to keep up with the volume and speed of their metal-producing suppliers.

Finally, in the 1880s the control crisis reached the area of consumption. Even in the 1860s petroleum companies adopting continuous-processing technologies, which increased output three to six times while halving unit costs, had confronted the need to stimulate consumption, differentiate products, and build brand loyalty. In 1882 a single miller adopting the same continuous-processing technologies to oatmeal produced at twice the national rate of consumption. Clearly the need to create new markets—and to stimulate and control consumption—had reached a crisis level in oatmeal. Over the same decade similar needs confronted producers of flour, soap, cigarettes, matches, canned foods, and photographic film as they began to adopt continuous-processing technology.

As this crisis of control spread through the material economy from the 1840s to the 1880s, it inspired a stream of innovations in information processing, bureaucratic control, and communications. This continuing innovation, effected by transporters, producers, distributors, and mass marketers alike, reached a climax in the 1870s and 1880s; by the turn of the century the crisis of control had largely been contained. Only through the dynamic tension between crisis and control, with each success at control generating still new crises, has the revolution in technology continued into the twentieth century and into the emerging Information Society.

Control Crisis in Transportation

Dramatic problems of control first appeared, as might be expected, on the railroads, the first part of the material processing system to harness the speed of steam power on a large scale. Because early railroads operated for most of their length on only a single track at unprecedented speeds of up to thirty miles per hour, they faced the problem of especially dangerous and costly head-on collisions. Lacking modern communication and control technology, most railroads adopted one of two solutions. On longer, lightly traveled roads, all trains ran one way one day and the other way the next. This solution did not prove economical or convenient enough for shorter, busier routes, however, where the first of two trains scheduled to meet running in opposite directions would wait at a midpoint station or siding until the other had passed. Without the technologies of centralized bureaucratic control, telegraphic communication, and formalized operating procedures along the line, however, and lacking even standardized signals, timetables, and synchronized watches aboard each train, many accidents did occur (Fig. 6.1).

For the Western Railroad, America's first intersectional rail link, the inability to control even a half-dozen trains quickly ended in tragedy. Because the 156-mile trip between Worcester and Albany took more than nine hours, the single-track Western could not adopt the Boston & Worcester method—waiting until one run had ended before starting another in the opposite direction—and complete more than one run daily in daylight. Instead, the Western scheduled two passenger trains and one freight a day each way between Worcester and Albany, a plan that required nine daily "meets," times when trains going in opposite directions had to pass each other. "Even assuming that there were no extra movements or work trains on the line," notes Stephen Salsbury, the Western's historian, "there was a scheduling problem on a single-track, unsignaled mountain system" (1967, p. 183), a system with much curved track often shrouded in fog.

On October 5, 1841, between Worcester and the Massachusetts state line, on a section of the road opened only the previous day, disaster struck: two Western passenger trains collided head-on, killing two, seriously maiming eight, and less critically injuring nine others. The public outcry, including an investigation by the Massachusetts legislature, reflected the fact that people were not yet used to travel at the speed of inanimate energy—certainly not to the Western's oper-

MOTHERS LOOK OUT FOR YOUR CHILDREN!

ARTISANS, MECHANICS, CITIZENS!

When you leave your family in health, must you be hurried home to mourn a

DREADFUL CASUALITY!

PHILADELPHIANS, your RIGHTS are being invaded! regardless of your interests, or the LIVES OF YOUR LITTLE ONES. THE CAMDEN AND AMBOY, with the assistance of other companies, without a Charter, and in VIOLATION OF LAW, as decreed by your Courts, are laying a

LOCOMOTIVE RAIL ROAD!

Through your most Beautiful Streets, to the RUIN of your TRADE, annihilation of your RIGHTS, and regardless of your PROSPERITY and COMFORT. **Will you permit this?** or do you consent to be a

SUBURB OF NEW YORK!!

Rails are now being laid on BROAD STREET to CONNECT the TRENTON RAIL ROAD with the WILMINGTON and BALTIMORE ROAD, under the pretence of constructing a City Passenger Railway from the Navy Yard to Fairmount!!! This is done under the auspices of the CAMDEN AND AMBOY MONOPOLY!

RALLY PEOPLE in the Majesty of your Strength and forbid THIS

OUTRAGE!

Figure 6.1. The crisis of safety on the railroads helped rival canal and turnpike interests to incite local resistance to the development of intersectional lines. In 1840 rioters actually tore up a Philadelphia rail line, which had to be abandoned. Despite this 1839 broadside intended to rally Philadelphians against becoming "a suburb of New York," that city had already been linked by a chain of railroads to Washington, D.C., by January 1838. (Courtesy of Metro-North Commuter Railroad, New York.)

ating speeds of up to thirty miles per hour. Travel on the shorter regional lines, although occasionally as fast, had been relatively free of accidents through the 1830s, so that the public had come to regard the railroad as a safe means of transportation. Salsbury sets the historical context for the great Western collision of 1841:

> The twentieth century has become blasé about disasters. Although train, aircraft, and even automobile wrecks are headline news, there is a general acceptance of the maxim that accidents are the price of progress. Few newspaper subscribers are surprised to learn of several major disasters on the same day. That was not the case in Boston in the 1840s. Although ship mishaps were considered normal, no tradition prepared people for spectacular land wrecks. True, stages often overturned or smashed, but for the most part such accidents usually resulted in injuries rather than death and involved only a few people. By contrast, a single train carried hundreds, at speeds up to 30 miles an hour. The railroad disaster, with its potential to kill or maim scores, if not hundreds, held a special terror. (1967, p. 183)

Investigation of the Western collision laid the blame on a failure of programming and communication. The company's management, aware that running six or more trains simultaneously on more than 150 miles of track required special information technologies, had settled on two: precise scheduling and a strict written program that defined procedures for various contingencies. The eastbound train had been late in arriving at the Chester Village siding, according to an investigation by the Western's board of directors, who reported of the conductor in charge of the train: "In conformity with the general order with which he had been furnished and the time sheet which he had then in his possession and which he had consulted on the route, he should have remained at Chester Village until the arrival of the westbound train. He must have known by examining his timetable that if the westbound was then acting in conformity to the same order, the trains would most certainly meet between Chester and Westfield" (Salsbury 1967, pp. 185–186).

Explanation for the behavior of the conductor, who himself died in the wreck, seemed to lie in a general failure of control. Programming for contingencies like delays lacked precision, detail, and integration among various workers and functions. As a result, the chain of command among Western employees could be ambiguous. The company's own investigation found "laxness in distributing copies of new orders to the train crews" and concluded that the "general control of the trains was too loose" (Salsbury 1967, p. 186).

As a result of this fatal failure of control, the Western management instituted a wide range of innovations in bureaucratic organization, programming, information processing, and communication. Alfred Chandler has hailed the result as an early milestone in bureaucratic control, "the first modern, carefully defined, internal organizational structure used by an American business enterprise" (1977, p. 97).

Control of the entire Western line became centralized in a new Springfield, Massachusetts, headquarters, linked to three regional offices by what Salsbury (1967, p. 187) describes as "solid lines of authority and command." One chain of command, headed by the chief engineer in Springfield, stretched through three regional "roadmasters" and controlled track, roadbed, bridges, and buildings. A separate chain of command, headed by the master of transportation at Springfield, stretched through three divisional masters and the various station agents and controlled all passenger and freight traffic. A third chain of command, headed by the master mechanic at the Western's major shops in Springfield, although normally under the master of transportation, stretched through deputy mechanics at each terminal and roundhouse and controlled all engines and rolling stock.

Certainly the most modern aspect of the organizational structure instituted by Western and eventually adopted by other railroads was the company's particular attention to regularity in data collection, to formalization of information processing and decision rules, and to standardization of communication with feedback. Responsibility for updating the three roadmasters on the condition of track and structures fell to conductors, enginemen, stationmasters, and other subordinates who passed up to the regional hierarchies a continual flow of data. The Western required each roadmaster, in turn, to keep a "journal of his operations" and to make a formal monthly report to the chief engineer in Springfield. The company also specified that "no alteration in the time of running or mode of meeting and passing of trains shall take effect until after positive knowledge shall have been received at the office of the superintendent that orders for such change have been received and are understood by all concerned" (Salsbury 1967, p. 186).

Although the Western left much to the discretion of central, regional, and even local administrators, the company's directors programmed its operating workers with "careful and explicit rules." Enginemen, for example, became little more than programmable operators, dutifully following rules like "in descending grades higher than 60 feet per mile passenger trains are not to exceed 18 miles per hour and merchandise trains not over 10 miles per hour." In transit the engineman obeyed the conductor, who told him even when to start and stop; upon arrival, however, enginemen fell under the supervision of the terminal's master mechanic. The Western directors specifically charged conductors with the responsibility of reporting "any disobedience of the engineman" directly to the superintendent in Springfield (Salsbury 1967, pp. 187–188).

Control of each train became centralized in its conductor, who had standardized detailed programs for responding to delays, breakdowns, and other contingencies, who carried a watch synchronized with all others on the line, and who moved his train according to precise timetables. The conductor controlled all operations between origin and destination, including those of the engineman and the brakeman on each car, from his platform outside the first car of the train. He controlled the brake of this car, and he alone—except in emergencies—determined when and where to stop and when to start, signaling his decisions by pulling a cord connected to the engine bell.

To describe the conductors on the reorganized Western line as "programmed" might at first seem anachronistic, a needless intrusion of contemporary jargon into the early nineteenth century. The fact remains, however, that in their control of trains the Western conductors might have been replaced in many of their functions by on-board microcomputers or, given modern telecommunications, by a more centralized means of computer control. Seen in this way, the Western conductors take on new significance: they are possibly the first persons in history to be used as programmable, distributed decision makers in the control of fast-moving flows through a system whose scale and speeds precluded control by more centralized structures. This use of human beings, not for their strength or agility, nor for their knowledge or intelligence, but for the more objective capacity of their brains to store and process information, would become over the next century a dominant feature of employment in the Information Society.

The directors of the Western Railroad labored quite consciously to program and reprogram the entire system. In investigating the company's operations, the Massachusetts legislature reported that "the directors have been at great pains to collect and compare their rules with those of other similar companies in this country and in England, with a view to adoption of those which would produce the greatest security." The legislators also found that the directors reassessed an employee's programming after an accident and used the experience to determine new rules that might make for still safer operations (Salsbury 1967, p. 189).

Because the Western was the first enterprise to extend beyond the span of a single manager's close personal contacts, a distance Chandler (1962, p. 21) sets for early railroads at roughly a hundred miles, and because the company attempted to control multiple units operating at the new speed of steam power, especially in opposite directions on the same track, it is perhaps not surprising that a crisis of control would arise in the first days of its operations or that its organizational and informational solutions to the crisis would serve as the earliest models for control by business well into this century. "As the first private enterprises in the United States with modern administrative structures," Chandler (1962, p. 23) finds, "the railroads provided industrialists with useful precedents for organization building when the industrial enterprises grew to be of comparable size and complexity." And, as we shall see, when they came to control movements at comparable speeds.

From Safety Crisis to Control for Efficiency

With the rapid diffusion of the telegraph, after Morse's successful demonstration in 1844, and the adoption and refinement of the Western's organizational innovations, the danger of collisions no longer ranked as the railroad's major control problem by the 1850s. In the first half of that decade, which brought the first four trunk lines—the Erie (1851), Baltimore and Ohio (1852), New York Central (1853), and Pennsylvania (1854)—connecting East and West, the control crisis of the railroads shifted from safety to efficiency in keeping track of trains, cars, and personnel in increasingly large, complex, and busy systems.

The history of the Erie Railroad, America's first great trunk line, illustrates this shift. In 1841, the year of the Western collision and the Erie's first in operation as a regional railroad, the company ran five locomotives, six passenger cars, and three freight cars, at an average speed of twelve miles per hour, on forty-six miles of track; this kept 112 people employed. Only a decade later, in its first year as a trunk line, the Erie ran 123 locomotives, 68 passenger cars, and 1,373 freight and baggage cars—at average speeds ranging from twenty-four up to twenty-nine miles per hour (for the express)—on 445 miles of track; this operation employed 1,325 people (Mott 1901, p. 483). Compared to the largest interregional railroad (the Western) in the previous year, the Erie in 1851 had three times as much track and moved about as much freight (a quarter million tons) and half again as many passengers—nearly 690,000.

In short, over the decade of the 1840s the Erie's control crisis had become how to keep track of 450 times as many freight cars moving at twice the speed over ten times as much track. Needless to say, traffic on the Erie and the other great trunk lines quickly overburdened the control technologies of the day. Imagine the manager who attempted to sustain efficient movement of several thousand rolling stock over hundreds of miles of track using the mercantile style of management. In 1854 Henry Varnum Poor, editor of the *American Railroad Journal*, wrote that "the utmost confusion prevailed" in the Erie system during its early years, "so much so, that in the greatest press of business, cars in perfect order have stood for months upon switches without being put to the least service, and without its being known where they were" (Chandler 1956, p. 147).

Although efficiency had come to overshadow safety, the Erie's accident rate also reflected its control problem: no one was killed in its first two years of operations, but twenty-six people were killed in 1851 alone. One possible way to attempt to regain control, of course, was to hire more people: The number of Erie employees increased nearly twelve times between 1841 and 1851, doubled again to twenty-six hundred in 1853, and again to fifty-five hundred by 1862. Employees had themselves to be controlled, however; 167 were killed in the Erie's first decade as a trunk line, a rate that in one year (1852) exceeded 1 per 100 employees (Mott 1901, p. 483).

Without innovations in organization and bureaucratic control, simply hiring more employees contributed to the control crisis as much as to its solution. As a result, railroads found that, contrary to anticipated economies of scale, as their systems grew larger, per-mile operating costs actually increased. As early as 1856, however, the Erie's superintendent, Daniel C. McCallum, saw the problem arising not from increasing scale per se but rather from the resulting decrease in ability to control operations efficiently:

A Superintendent of a road fifty miles in length can give its business his personal attention and may be constantly on the line engaged in the direction of its details; each person is personally known to him, and all questions in relation to its business are at once presented and acted upon; and any system however imperfect may under such circumstances prove comparatively successful. In the government of a road five hundred miles

in length a very different state exists. Any system which might be ap-
plicable to the business and extent of a short road would be found entirely
inadequate to the wants of a long one; and I am fully convinced that in
the want of system perfect in its details, properly adapted and vigilantly
enforced, lies the true secret of their [the large roads'] failure; and that
this disparity of cost per mile in operating long and short roads, is not
produced by a *difference in length*, but is in proportion to the perfection
of the system adopted.   (Chandler 1956, p. 146; 1965b, p. 101)

By virtue of such insights, Daniel McCallum must be considered
among the first to appreciate the breakdown of control that results
when a system exceeds the grasp of any one individual. As we saw in
Chapter 1, Emile Durkheim would reach much the same conclusion
for markets generally in his *Division of Labor in Society* (1893).
McCallum's report also made clear that, even though the railroads first
experienced loss of control as accidental loss of lives and equipment,
by the 1850s their control crisis had become one of mounting operating
costs and loss of business. McCallum's Erie Railroad, for example,
faced increasing competition from the short lines along the Erie Canal,
which in 1853 consolidated to form the New York Central, thereby
making that route more attractive for through traffic. In response to
the New York Central, the Erie promoted McCallum, a formally trained
civil engineer with experience in bridge building, to general superintendent of all five of its geographically separate operating divisions.

To complement the increasing regularity and speed in the movement
of matter and energy then possible on the Erie, McCallum sought
greater control over the railroad through greater regularity and speed
in the movement of information—greater regularity and speed of *communication*. This he accomplished by means of a new hierarchical
system of information gathering, processing, and communication designed to return control to the superintendent's office.

Among a half-dozen "general principles of organization and administration" McCallum placed major emphasis on intelligence gathering,
hierarchical communication, feedback, and error detection. Although
organizational control demanded that responsibility be formally divided
and power distributed, McCallum believed, the general superintendent
retained responsibility for system-wide control, which he saw as dependent on four informational capabilities: (1) "The means of knowing
whether such responsibilities are faithfully executed"; (2) "great
promptness in the report of all derelictions of duty, that evils may be
at once corrected"; (3) "such information, to be obtained through a

system of daily reports and checks that will not embarrass principal officers, nor lessen their influence with their subordinates"; and (4) "the adoption of a system, as a whole, which will not only enable the General Superintendent to detect errors immediately, but will also point out the delinquent" (Chandler 1965a, pp. 28–29). Clearly here, as in the subsequent implementation of these ideas, McCallum placed greater emphasis on communication from subordinates to their superiors than vice versa, that is, not on lines of command so much as on lines of feedback and control.

To illustrate these lines of communication and authority among the Erie's various offices and employees, McCallum drew up a detailed diagram that Chandler (1965a, p. 30) calls "certainly one of the earliest organizational charts of an American business enterprise." As Chandler (1956, p. 148) describes it, "the design of the chart was a tree whose roots represented the president and the board of directors; the branches were the five operating divisions and the service departments, engine repairs, car, bridge, telegraph, printing, and the treasurer's and the secretary's offices; while the leaves indicated the various local ticket, freight, and forwarding agents, subordinate superintendents, train crews, foremen, and so forth." Although McCallum intended this chart to be used for the Erie's internal purposes only, Henry Varnum Poor had it lithographed and offered copies to his *American Railroad Journal* readers for $1 each.

As if the formal hierarchy and lines of communication were not enough to unite the Erie's three thousand employees into a single information processor capable of controlling the growing system, McCallum also introduced the idea that all employees would wear a prescribed uniform indicating the wearer's particular subdivision and grade in the organizational hierarchy. By thus making manifest the lines of authority drawn in the General Superintendent's chart, the uniforms served to preprocess the crucial information that employees moving around in the far-flung Erie system would need in interacting with one another. As the general public came to recognize the various insignia, the uniforms also served to preprocess the Erie's organizational information for its passengers and clients. The revolutionary nature of this innovation is evident from the fact that it generated considerable controversy: Henry Varnum Poor, for example, defended the idea against the *Railroad Record* of Cincinnati, which considered it unbefitting a democracy.

To implement the flows of data that McCallum had outlined in his "general principles" and had drawn into his organizational chart, he required that three types of reports—hourly, daily, and monthly— be sent to the General Superintendent's office. Conductors and station agents began to report hourly, via telegraph, on the location of trains and the reasons for any delays, accidents, or breakdowns. Processing these data required "a very considerable amount of extra trouble and expense," the *American Railroad Journal* reported, including "the maintenance of a large office with eight active clerks" (Chandler 1956, p. 148). As McCallum himself described the data processing in his office, "The information being edited as fast as received, on convenient tabular forms, shows, at a glance, the position and progress of trains, in both directions on every Division of the Road" (Chandler 1965a, p. 30). Henry Varnum Poor, reporting triumphantly to his readers that "the superintendent can tell at any hour in the day the precise location of every car and engine on the line of the road, and the duty it is performing," also reflected the urgency of the control crisis: "All these reforms," he added, "are being steadily carried out as fast as the ground gained can be held" (Chandler 1956, p. 147).

So urgently were the reforms needed that many—including the organizational plan, the reporting system, and the use of the telegraph— had been initiated in part by the Erie even before McCallum became general superintendent in 1854. The first use of the telegraph in railroading, for example, came three years earlier when Charles Minot, then a superintendent on the Erie, wired fourteen miles to Goshen, New York, to delay a train so that his own would not have to wait (Mott 1901, p. 420).

McCallum systematized the Erie's initial plans, however, and integrated them into a comprehensive control system. His major contribution, according to Chandler (1965a, p. 30), lay in this larger understanding of communication and control: "McCallum's use of the telegraph brought universal praise from the railroad world both in this country and abroad. What impressed other railroad managers was that McCallum saw at once that the telegraph was more than merely a means to make train movements safe, but also a device to improve better coordination and better administration through this extremely efficient new technique of communication."

Data from the daily and monthly reports submitted to McCallum's office, although not dependent on the rapid new telegraphic communication, did prove useful—often in unexpected ways—to control variables not arising from the increasing distances, speed, and volume of the Erie's operations. When the railroad raised its rates, for example, it found that the resulting decrease in traffic actually reduced revenues. "To guard against such a result," McCallum argued, "and to establish the mean, between such rates as are unremunerative and such as are prohibitory, requires an accurate knowledge of the cost of transport of the various products, both for long and short distances" (Chandler 1965a, p. 31). Continuing time series data, in other words, could be used to control rate structure and thereby maintain maximum revenues despite continually changing conditions—an application of data collection to control that would be adapted to industrial production in the latter nineteenth century.

Even the monthly reports required of all heads of the Erie's various service departments, recorded and filed by McCallum's office in statistical format, proved useful to maintain the type of rational administrative control that would become known in the 1890s as "scientific management." Comparison of data from two different sources—conductors and station agents—on the loading and movement of freight, for example, served as a reliable check on the honesty and efficiency of these employees. Comparative analysis of the monthly engine reports revealed the engine best suited to even complex tasks (involving various loads, speeds, and grades) and the enginemen who operated their locomotives most efficiently. Even the *Railroad Advocate*, which claimed to speak for engineers and other skilled railroad laborers, favored such analysis: "Although, perhaps, for a purely selfish purpose, the monthly reports acknowledge the full doings of each engineer, they still serve as an honorable stimulant to exertion" (Chandler 1956, pp. 149, 321). And, we might add, as a means by which the engineer's exertion might be better controlled from the central office.

From this office, using data based on the daily and monthly reports, McCallum even proposed a crude type of *operations research*, what one text defines as "a scientific method of providing executive departments with a quantitative basis for decisions regarding the operations under their control" (Kimball and Morse 1951, p. 1). Because of the problem of unused capacity, for example (a control problem that continues to plague railroads to this day, according to a *New York Times* editorial, "Boxcar Follies," in 1983), McCallum proposed that his statistics be used to analyze traffic flow patterns so that prices might be "fixed with reference to securing, as far as possible, such a balance of traffic in both directions as to reduce the proportion of 'dead weight' carried." As Chandler (1965a, pp. 31–32) adds, "unused or excess capacity on a return trip warranted lowering prices for goods going that way."

Many historians find the origins of operations research in World War II or, occasionally, in World War I (Trefethen 1954, p. 4); Daniel Bell (1973, pp. 29–33) has associated such statistical control techniques with the postindustrial society. To describe McCallum's early use of systematically collected quantitative data to inform decision making and control as operations research may not seem anachronistic, however, when we consider that the term has been applied to the still earlier work of Charles Babbage in England, especially to his 1827 study of the British postal system, to his *On the Economy of Machinery and Manufactures* (1832), and to his series of studies of the Great Western Railway in 1839 (Halacy 1970, pp. 75–77; Dubbey 1978, pp. 221–224; Hyman 1982, pp. 158–163). McCallum's ideas, although they followed Babbage's by a quarter-century, stemmed from similar observations and experience with the new steam-powered systems—and they were more successful than Babbage's in effecting immediate control.

Despite the success of McCallum's innovations, the control crisis had not yet ended for America's railroads. Control problems with the first interregional trunk lines like the Erie in the early 1850s gave way to a mounting crisis in the control of national through traffic as the railroads pushed westward to Chicago (1853), crossed the Mississippi at Rock Island (1856), and—with the driving of the golden spike at Promontory Point, Utah, on May 10, 1869—connected to West Coast lines. Meanwhile McCallum had left the Erie in 1857 to return to his bridge-building business, although during the Civil War he served as "military director and superintendent" of the Union railroads (he spent $42 million to build or rebuild 2,745 miles of track and twenty-six bridges) and was made Brigadier General for helping to save Grant at Chancellorsville (Mott 1901, p. 434). Outside of the Union Army, McCallum's hierarchical system of information gathering, processing, and control continued to be tested and elaborated in other interregional trunk lines, most notably by the Pennsylvania Railroad.

By the late 1870s the control crisis of the railroads had shifted once again, this time to the maintenance and extension of vast multiregional systems. The first of these, the Pennsylvania, which by 1874 connected New York City and Washington, D.C., with Chicago and St. Louis, confronted the problem of controlling $400 million worth of capital and six thousand miles of track—more than any national system except for those in Britain and France. Until the 1880s most railroad companies delayed building giant systems like that of the Pennsylvania because they lacked adequate control technology. "Managers opposed expansion," according to Chandler, "because they considered any road much over five hundred miles in length to be too large and complex to manage" (1977, p. 136). As late as 1898, the Erie's express trains ran at an average speed of only thirty-five miles per hour, the same speed they had run under McCallum in 1854; ordinary passenger trains ran at twenty-five miles per hour—an actual *decrease* of five miles per hour over the forty-three years (Mott 1901, p. 483).

Control of even transcontinental rail systems gradually became feasible through a steady progression of innovations intended to facilitate control: the through bill of lading (1853), standardization of cars (1867), adoption of a uniform standard time (1883) and standard gauge of track (1886), regulation by an Interstate Commerce Commission (1887), and required standardized automatic couplers and air brakes (1893). By the 1890s, owing to these and many other innovations in control technology, the rail network of the United States had essentially been completed (Taylor and Neu 1956; Kirkland 1961, pp. 46–51; Stover 1961, chap. 6). Freight that in 1849 required nine transshipments between Philadelphia and Chicago could by the late 1880s move from coast to coast without a single one (Chandler 1977, pp. 122–123).

Table 6.1 summarizes the major innovations in information-processing and communication technology for the control of U.S. transportation and distribution from the 1830s—and the first applications of steam power—to the establishment of a standardized, federally con-

Table 6.1. Selected innovations in information processing and communication for control of transportation and distribution, 1830–1889

Year	Innovation
1830s	Wagon lines carrying freight between rural towns and ports begin to operate on regular schedules
1837	Telegraph demonstrated, patented
1839	Express delivery service between New York and Boston organized using railroad and steamboat
1840s	Freight forwarders operate large fleets on canals, offer regular through-freight arrangements with other lines
1842	Railroad (Western) defines organizational structure for control
1844	Congress appropriates funds for telegraph linking Washington and Baltimore; messages transmitted

Table 6.1 (*cont.*)

Year	Innovation
1847	Telegraph used commercially
1851	Telegraph used by railroad (Erie) First-class mail rates reduced 40–50 percent
1852	Post Office makes widespread use of postage stamps
1853	Trunk-line railroad (Erie) institutes a hierarchical system of information gathering, processing, and telegraphic communication to centralize control in the superintendent's office Through bill of lading introduced
1855	Registered mail authorized, system put into operation First-class mail rates reduced—a second time—40 percent
1858	Transatlantic telegraph cable links America and Europe, service terminates after two weeks Overland mail service begins—twice weekly—to Pacific Coast
1862	Federal government issues paper money, makes it legal tender
1863	Free home delivery of mail established in 49 largest cities
1864	Railroad postal service begins using special mail car Postal money order system established to insure transfer of funds
1866	Telegraph service resumes between America and Europe "Big Three" telegraph companies merge in single nationwide multiunit company (Western Union), first in United States
1867	Railroad cars standardized Automatic electric block signal system introduced in railroads
1874	Interlocking signal and switching machine, controlled from a central location, installed by railroad (New York Central)
1876	Telephone demonstrated, patented
1878	Commercial telephone switchboards and exchanges established, public directories issued
1881	Refrigerated railroad car introduced to deliver Chicago-dressed meat to Eastern butchers
1883	Uniform standard time adopted by United States on initiation of American Railway Association
1884	Long-distance telephone service begins
1885	Post Office establishes special delivery service
1886	Railroad track gauges standardized
1887	Interstate Commerce Act sets up uniform accounting procedures for railroads, imposes control by Interstate Commerce Commission

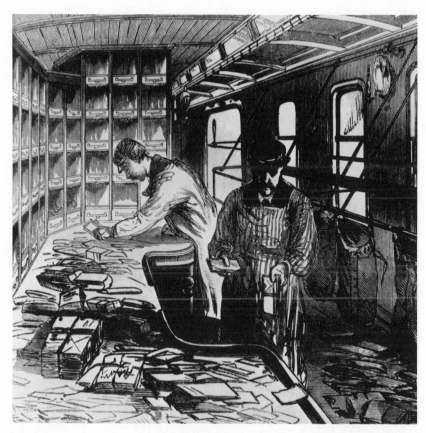

Figure 6.2. Integration of information-processing and communication systems: the mail car, instituted in England in 1837, adopted by the U.S. Post Office in 1864. By 1888 the New York and Chicago "Fast Mail" employed nineteen postal workers in five cars to process mail bags seized by means of iron catch-arms from more than a hundred stations en route. (Hornung 1959, p. 135; courtesy of A. S. Barnes.)

trolled transcontinental rail network by the late 1880s (Fig. 6.2). The list of innovations illustrates a shift from basic organizational technologies like scheduling, multisystem coordination, and centralization of bureaucracy—in response to the control crises of the 1840s and early 1850s—to what Daniel Boorstin (1973, pt. 5) has called "leveling times and places," for example, through the development of refrigerated railroad cars (1881) and uniform standard time (1883). Control of the speed and volume of material flows, at first largely a matter of bureaucratic coordination, increasingly came to be built into the transportation infrastructure itself through automatic signals (1867) and switches (1874) and through standardization of cars (1867) and track (1886).

The entire system of transportation and distribution derived progressive coordination and integration—as Table 6.1 suggests—from a succession of new communication systems involving both infrastructure or common carriers and generalized media of exchange. The common carriers, which developed increasingly apart from the development of the railroads, included the telegraph, postal, and telephone systems—all point-to-point networks well suited to control transportation and distribution. New generalized media of exchange—and therefore of communication and control—included postage stamps (1852), the through bill of lading (1853), federal paper currency (1862), and postal money orders (1864), among many others.

That the degree of control enjoyed by the directors of the Western, Erie, Pennsylvania, and other interregional and national rail systems was truly new and revolutionary in the mid-nineteenth century is reflected in the fact that few of their innovations drew on previous experience in large business, government, or military bureaucracies. Although a few such bureaucracies already existed and controlled the processing of materials and information, none did so over hundreds of miles at the speed of steam. Alfred Chandler describes the difference:

> The management of such enterprises did not require the constant, almost minute-to-minute supervision . . . Such constant coordination and control were, however, fundamental to the management of the railroads . . . Without the building of a managerial staff, without the design of internal administrative structures and procedures, and without communicating internal information, a high volume of traffic could not be carried safely and efficiently . . . No other business enterprise, or for that matter few other nonbusiness institutions, had ever required the coordination and control of so many different types of units carrying out so great a variety of tasks that demanded such close scheduling.   (1977, p. 94)

Most of the early innovators of the Control Revolution were civil engineers with experience in the construction of roadbeds and bridges, hardly an obvious preparation for the control of dynamic systems. "All evidence indicates that their answers came in response to immediate and pressing operational problems," Chandler (1977, p. 95) finds. "They responded to these in much the same rational, analytical way as they solved the mechanical problems of building a bridge or laying down a railroad." Just as the early pioneers of genetic biology were information specialists from physics and mathematics, as we saw in Chapter 2, and the pioneers of computer control came from mathematics and engineering, so too did the early leaders of organizational control—as typified by Charles Babbage, mathematician, and David McCallum, engineer—represent abstract and analytic rather than practical experience with information processing and decision. In times of true crisis, it would seem, experience with the old technologies provides little help in devising revolutionary new ones—more theoretical and general disciplines better fill that need.

Similar crises of control have persisted to this day in newer transportation and distribution systems. As recently as January 1984, for example, *Newsweek* magazine focused public debate on a control crisis in transportation in a fashion that recalled the outcry following the Western Railroad tragedy of 1841. "Can We Keep the Skies Safe?" asked the bold headline on the *Newsweek* cover, calling attention to a possible deterioration of the country's air traffic control (Beck 1984). Like the directors of the Western, Erie, and other railroads who helped launch the Control Revolution, the Federal Aviation Administration planned to strengthen control of its system through improvements in data processing and communications based not on the telegraph and human clerks but on computers and microprocessors. Boeing's 757 and 767 jets already come equipped with 130 microprocessors and six computer screens—two replacing the flight engineer of the older jets. Much as the Western's conductors took control of its trains from their enginemen, Boeing's designers now call pilots "flight managers," according to *Newsweek*, because "their jobs involve much less hands-on flying and much more monitoring of sophisticated information systems" (Beck 1984, p. 30).

Control Crisis in Production

Application of steam power to production created a crisis of control much as it did in transportation and for much the same reason: increased speed. By dramatically increasing the speed and reliability with which raw materials arrived at a factory and the speed and volume with which finished goods could be distributed, the railroad created a niche in the societal processing system for manufacturing at comparable speeds. A steadily growing array of steam-powered machinery by the 1860s promised to fill the new niche with a processing capability that might match the railroad in volume and speed. Incentive came from economic theory, which suggested that unit costs would decline and profits rise as a direct function of the speed of throughputs to any given amount of machinery and labor. Before producers could realize these potential gains in output, productivity, and profits, however, they had to confront the problem of controlling the increased speed and volume of flows through their factories.

Crises first arose in the rail mills that attempted to adopt the Bessemer process of mass steel production in the late 1860s. By 1890 the weekly output of a single blast furnace had risen from seventy tons to more than a thousand; Andrew Carnegie's furnaces, largest in the world, each produced two thousand tons per week in the late 1890s. By the century's end a modern rolling mill processed as much throughput in a day—three thousand tons—as a mid-century mill had in a year (Temin 1964, pp. 159, 165). Throughout this period, according to historian Peter Temin, "the speed at which steel was made was continually rising, and new innovations were constantly being introduced to speed it further" (1964, pp. 164–165).

Maintaining control of these accelerating throughputs proved to be a continuing struggle. Unlike the material flows in most other processing industries, each step in the production of steel involved many different activities. Melting and converting operations had to run concurrently, their outputs synchronized with rolling and the various rail, beam, wire, and other finishing operations. Even though integration of these processes in single factories had helped to improve overall control somewhat, economies of scale resulting from shared energy, lighting, and maintenance remained small compared with those achieved by increasing the volume of throughputs to men or machines or, for a given capacity, by increasing flow velocities. Coordination and control of production therefore became no less important than the innovations in equipment or more intensive applications of energy that made possible the continual increases in volume and speed. Producers with the best control technologies could maintain the greatest speeds, produce at the lowest costs, and thereby enjoy an important edge on competitors.

Solutions to this aspect of the control crisis, the competition among steel producers for control of ever faster processing, centered on two general information technologies, formal organization and preprocessing. The essential insight: to construct steel works as processors *explicitly*, with structures and procedures so determinate that information extraneous to any particular function would be eliminated on the designer's table, so to speak, and hence pose no further need for control. Because metal-making proved more difficult to control than any other major manufacturing process, steel producers became the first to design, build, and maintain—on a scale that involved thousands of individuals—the concrete open processors that distinguish all living systems.

The first such system, Carnegie's Edgar Thomson Steel Works in Pittsburgh, began operations in 1875. Unlike all previous manufacturing establishments, including Bessemer installations in England, the E.T. (as it came to be called) had been explicitly designed to facilitate throughputs. Located beside the Monongahela River at the confluence of three major railroads, the E.T. Works were, in the words of their designer, engineer Alexander Lyman Holley, "laid out not with a view of making buildings artistically parallel with the existing roads or with each other, but of laying down convenient railroads with easy curves; the buildings were made to fit the transportation" (Chandler 1977, p. 262).

Indeed, throughputs to the E.T. Works only occasionally left the rails. Carloads of coke and pig iron passed directly to the company stockyard and from there to blast furnaces and cast houses by means of an internal, narrow-gauge rail network. Trains loaded at nearby mines could dump their coal directly onto the floors of producer and boiler houses. Wide-gauge tracks lined both sides of the finishing mill so that the E.T.'s outputs of rails might be distributed as quickly as its inputs had arrived. Chandler reports that the plant and its operations were "designed to assure as continuous a flow as possible from the suppliers of the raw material through the processes of production to the shipment of the finished goods to the customers" (1977, p. 262).

Thus Carnegie and the other steel companies eventually resolved their control crisis and managed to coordinate upwards of three thousand workers in the complex processing of pig iron, coal, and coke into rails, beams, and wire with continually greater speeds of throughput and corresponding increases in output per unit of input. Control rested largely on the fact that material processing had been made as specific as possible, so that men and machinery, by virtue of their concrete and functional organization, could quite literally do little else. This recalls John von Neumann's observation, cited in Chapter 2, that end-directedness varies directly with organization (Pittendrigh 1970, p. 392). Traditional nineteenth-century factories like the Springfield Armory, with buildings well ordered with respect both to roads and to one another, stood in dramatic contrast to the sprawling new E.T. Works—much as we earlier saw that a crystal differs from the amoeba. Although both the Springfield Armory and the crystal are undeniably better *ordered*, both the E.T. Works and the amoeba are better *organized* to effect control.

Like the amoeba, the E.T. and similar steel works depended largely on organization and preprocessing to maintain control rather than on a centralized capability for information processing, programming, and decision, the control structure more characteristic of higher vertebrates that would come to characterize modern factories until the decentralization made possible by microprocessors. As late as the 1880s only a handful of managers and a small staff kept the E.T. running: Carnegie himself made the occasional management decision (Bridge 1903, chap. 18), three engineers maintained the plant and equipment, and a single chemist provided all quality control (Livesay 1975, pp. 99–111).

Carnegie did pioneer one centralized decision and control technology: the so-called voucher system for the collection and processing of quantitive data (Wood 1895). Each subunit of a Carnegie steel works tallied the amount of its own throughputs and the costs of all materials and labor added to them. These data, processed by accountants in the general manager's office, served as the basis for detailed cost statements that landed daily on Carnegie's desk. Soon they became his primary means of control not only of the quality and mix of material throughputs but of departmental operations and the performances of managers and foremen. Obsessed with costs even more than with profits, Carnegie continually asked his employees to account for changes in unit inputs (Wall 1970, p. 342). In the words of one, "The men felt and often remarked that the eyes of the company were always on them through the books" (Bridge 1903, p. 85). Better testimony to the importance of information in the control of large-scale production would be difficult to imagine.

As Carnegie and other steel makers came to resolve the crisis of control in their own companies by the 1880s, similar methods began to appear in the production of other basic materials: iron, copper, zinc, and glass. In each of these industries, according to Alfred Chandler, "expansion of output came more from increasing the velocity of throughputs within the plant than from increasing the size of the establishment in terms of area covered and workers employed" (1977, p. 269). By dramatically increasing the speed and reliability of basic material flows, however, the metal makers merely shifted the control crisis to the metalworking sectors. From producers of simple castings, moldings, nails, and screws to the manufacturers of the new repeating rifles, typewriters, and electric motors, the metalworking industries struggled to match the production of basic materials in volume and speed. As with their suppliers, the metalworkers realized, unit costs would decline and profits rise as a direct function of throughput speeds. Before these industries could hope to achieve the same gains as metal producers, however, they had to solve similar problems of controlling flows through their factories at increasing speeds.

Toward Control of Processes as Flows

The new control crisis varied directly with the number of steps in a production process. In the case of simpler fabricated products like castings and cutlery, manufacturers merely lined up men and machinery in processing order: first stamping forges, then welding and tempering furnaces, grinding and polishing machines, and so on. Additional stages for the assembly of interchangeable parts were necessary for more complex manufacturing like that of plows, stoves, scales, and harvesters. In all of these metalworking industries reasonable throughput speeds could be achieved simply by locating each stage in the process as close as physically possible to the preceding stage. Such manufacturing experienced little if any of the larger crisis of control.

Control problems did increase with the numbers and complexity of parts to be assembled, however, especially in the manufacture of smaller items like locks, clocks, and watches. These problems grew to crisis proportions as parts also increased in variety, both in size and materials, as they did in a rapid succession of new manufactures: sewing machines (1850s), breechloading and repeating firearms (1860s), typewriters (1870s), electric motors (1880s), and—in the 1890s—the first American automobiles. The manufacture of sewing machines, for example, required hundreds of different types of materials, including pig- and malleable iron and steel in various bar, sheet, and wire forms, as well as several kinds of woods, varnishes, and machine supplies.

Crises arose from the need to coordinate hundreds and sometimes thousands of workers in several simultaneous processing lines assembling such wide varieties of materials and parts. Sewing machine manufacturers, for example, had to control a major flow of metals from foundry through tumbling, annealing, japaning (enameling and lacquering), drilling, turning, milling, grinding, polishing, ornamenting, varnishing, adjusting, and testing. Meanwhile, these processes had to be coordinated with parallel work lines producing metal attachments, needles, and tools, while woodworking and cabinet-making operations—among the most complicated in the mass production of furniture in the nineteenth century—kept pace in still other departments. Outputs from these various production lines had to be coordinated in a final line that completed the complex tasks of assembly, gauging, inspection, final testing, and preparation for shipment.

Factories producing sewing machines and other complex metal goods managed to coordinate their internal operations, at least until the late 1870s, by means of a loose system of distributed control called "inside contracting." Especially in the metalworking operations requiring the most careful control, like grinding and polishing, skilled foremen often contracted to produce a minimum number of parts in a given time using the manufacturer's factory, tools, patterns, and materials. Such a foreman had total control over his own department—he hired, promoted, and fired workers and had responsibility for sustaining the production levels of his contract. Although paid a minimum foreman's wage by the manufacturer, he depended for the bulk of his earnings on the value of his department's output that passed inspection, minus debits to his account for wages, energy, and tools (Buttrick 1952; Williamson 1952, pp. 85–91).

Inside contracting provided greater control of a work force—even a large one engaged in complex manufacturing—than had prevailed in much simpler factory operations a half-century earlier. The system did not provide control of material flows among the various contractors and departments, however, nor the cost controls of Andrew Carnegie's voucher system. By the 1880s this lack of higher-level factory control hampered production in the metalworking industries, which were expanding again after the economic depression of the previous decade. The owner and manager of a factory with fourteen hundred workers described his own crisis of control: "The trouble is . . . in constantly running over the back track to see that nothing ordered has been overlooked, and in settling disputes as to whether such and such an order was or was not actually given and received . . . I spend so much of my time in 'shooing' along my orders like a flock of sheep that I have but little left for the serious duties of my position" (Chandler 1977, p. 273).

These words were recorded by Captain Henry Metcalfe, superintendent of several federal arsenals, and used to justify the first book on cost control of factories (Metcalfe 1885). Among the earliest to describe management as a problem of information processing, Metcalfe sought to control the flow of materials through a factory by means of what he called a "shop-order system of accounts." Under Metcalfe's system the superintendent's office assigned an identifying number and prepared routing slips for each order it accepted. These slips, which indicated the sequence of departments through which each order would progress and the parts to be processed or assembled at each stage, accompanied all movements of materials in the factory. On the same slips foremen had to account for all use of machinery and workers in their departments and for the time, materials, and wages expended on each order. By continuously tabulating this moving record, a factory's central office could coordinate the material flows among its various departments and thereby resolve the major control crisis of the inside contracting system. Using the same data, a superintendent could evaluate his departments and contractors in terms of their operating costs and contributions to profits, monitor the performances of individual workers and machinery, and maintain weekly and even daily factory-wide cost sheets for each order.

Similar methods had begun to appear in metalworking factories after the mid-1870s, nearly a decade before Metcalfe's influential book (Garner 1954; Litterer 1963). Frederick Winslow Taylor, who in the late 1890s would come to be known as the father of scientific management, had by 1880 introduced a similar shop-order control system at Midvale Steel, manufacturer of specialized heavy machinery and machine tools; by 1884 Taylor had provided Midvale with a separate "rate-fixing department" (Copley 1923; Wren 1972, chap. 6; Nelson 1974). Two years earlier the *Tenth U.S. Census* described a shop-order control system in use by the Wilson Sewing Machine Company; another factory with twelve separate departments and foremen employed a similar system by 1886. Because inside contractors had little incentive to fill out order slips properly, by the 1890s factories employed timekeepers and specialized clerks for the purpose—the first "staff" employees in many metalworking companies (Chandler 1977, p. 274).

Outside of metalworking and metal making, industrialization did not create a crisis in the control of production. In the non-heat-using and hence less energy-intensive industries like textiles, clothing, leather, and wood products, where steam-driven machinery simply replaced human labor, production quickly reached natural bounds on the speed of essential processes like cutting, shaping, tanning, and sewing; control requirements therefore remained relatively simple. Even in the newer mechanical industries like processed flour milling, canning, and the production of bar soap, cigarettes, matches, and photographic film, where continuous-process machinery greatly increased both throughput speeds and the volume of outputs, the technology itself integrated the various processes of production, so that additional control was not required—workers did little more than watch over the machinery. At least among the functions that it integrates, continuous-process machinery is itself a control technology.

In the case of cigarette making, for example, by the late 1880s a single machine could produce 120,000 cigarettes in a ten-hour day, the equivalent output of forty highly skilled hand workers; thirty such machines could have met the total U.S. demand for cigarettes in 1885. Far from creating a crisis in the control of this unprecedented volume and speed of factory throughputs, however, the same machinery integrated the various stages of production—from sweeping and compressing the raw tobacco to packaging the finished cigarettes—in one continuous process (Tennant 1950, chap. 1). Similarly with the automatic, all-roller, gradual reduction flour-milling machinery, which increased the average output of Minneapolis mills sevenfold in the fifteen years after its introduction in 1879, the mill became quite literally an automatic factory into which raw wheat entered on the ground floor and quality patent and bakers' flour exited from the third (Kuhlmann 1951; Storck and Teague 1952, chap. 16).

Even in the heat-using and hence energy-intensive industries like sugar and petroleum refining, the processing of cotton and linseed oil, distilling of alcohol and liquor, brewing of beer and ale, and production of sulphuric and other acids, paints, and solvents, no control crisis developed. These and other refining and distilling industries, because they involved chemical rather than mechanical processes and because of the inherent ease with which flows of the liquids might be integrated by means of pipes, hoses, and troughs, required little additional control technology, even though throughputs increased fivefold (in petroleum) to a hundredfold (beer) through continually more intensive applications of energy (Cochran 1948; Williamson and Daum 1959). In Alfred Chandler's words,

> Mass production came in much the same way in the refining and distilling industries as in continuous-process mechanical industries, though in a less dramatic manner . . . It appeared earlier because of the ease in integrating the flow of liquids through the processes of production and because

the chemical nature of these processes permitted the application of more intense heat to expand the volume of throughput from a set of facilities . . . But precisely because of the ease of controlling and coordinating throughput, their operation had only a little more impact on the development of modern systematic or scientific management methods than did the supervision of the processes of production in the non-heat-using mechanical industries.   (1977, pp. 253–254)

The fact that liquids of all kinds—from acids to beer—rank among the most easily controlled industrial outputs serves to bolster the argument that maintenance of *flow* is the essential problem in control of production.

Table 6.2 places the early American innovations in the control of production in broader historical and comparative context. Ironically enough, two technologies that would be central to much wider control of material economies in the twentieth century—namely, Hollerith punch cards and operations research—emerged early in the Industrial Revolution in England and France for the control of steam-driven mechanical production (Fig. 6.3). During the comparable stage of U.S. industrialization control of production—the so-called American system—was achieved mostly through innovations in preprocessing to facilitate flows: interchangeable parts, standardization of sizes and pro-

Table 6.2.   Selected innovations in the control of production, 1800–1889

Year	Innovation
1801	French industrialist Joseph-Marie Jacquard develops a loom programmed by a continuous belt of punched paper cards that automatically controls the patterns woven
c. 1815	Jacquard loom perfected for industrial use, becomes foundation of the silk industry in Lyons
1832	Charles Babbage publishes *On the Economy of Machinery and Manufactures*, a pioneering treatment of operations research
	United States
1838	Factory established solely to manufacture machinists' tools Machine introduced to manufacture solid-headed pins; one worker tending two machines produces 10,000 pins per hour
1840	Bolt factory established; 6 operators produce 500 bolts per day

Table 6.2 (*cont.*)

Year	Innovation
1840s	Metalworking industries adopt steam-power, integrated factory, American System of manufacture by processing, assembly of interchangeable parts General principles of machine tools developed; tools themselves become substantially what they are today
1844	Rolling mill produces iron rails
1849	Wire gauge of "V" type introduced to standardize sizes
1850s	Outside consultants commissioned for industrial research
1853	Paper folding machine introduced; enables 3 workers to produce 2,500 envelopes per hour
1854	Wrought iron beams rolled, used in place of cast iron beams in New York City building
1860s	Petroleum producers adopt continuous-processing technologies, increase output three to six times while halving unit costs
Late 1860s	Rail mills adopt Bessemer process, increase speed of steel production
1870s	Producers of basic materials—iron, copper, zinc, glass—adopt continuous-processing technologies, increase speed of throughputs
Mid-1870s	Shop-order system of accounts—based on routing slips—developed for control of material flows through factories
1875	Plant (Carnegie's Edgar Thomson Steel Works) explicitly designed to facilitate throughputs
1882	Henry Crowell adopts continuous-processing technology to oatmeal, produces twice national consumption from a single mill
1884	Factory (Midvale Steel) incorporates rate-fixing department
1885	Book (Metcalfe 1885) on cost control of factories published
1887	Time recorders ("autograph type") introduced in production Industrial accident reports required by state (Massachusetts)
1888	Employee time recorder (paper tape) introduced
Late 1880s	Producers of flour, soap, cigarettes, matches, canned foods, and photographic film adopt continuous-processing technologies

Figure 6.3. Origin of programmable control of production: Joseph-Marie Jacquard's loom, invented in 1801, in which a continuous belt of punched paper cards automatically controlled the patterns woven. The idea may have come from late eighteenth-century musical instruments programmed to perform automatically under the control of rolls of punched paper. (Courtesy of IBM Corporation.)

cesses (as effected by machine tools), integration of processes by fitting outputs to inputs (as perfected in continuous-processing technologies), and factory design to facilitate throughputs. Advances in actual information processing—as opposed to information reduction or preprocessing—included organizational specialization (as in rate-fixing departments, expert consultants, and governmental regulation), intraorganizational communication (shop-order systems based on routing slips), and programmed control (automatic recording devices, cost control of factories).

To summarize this section, rapid industrialization created a crisis of control in only those industries—primarily metal making and later metalworking—where progressively more intense applications of heat brought corresponding increases in the volume of production. There crisis resulted for much the same reason that it did in applications of steam power to transportation: increases in the speed and reliability of throughputs out-paced the development of information-processing and communication technologies adequate to control the larger systems. In other words, the wider crisis in control of societal systems, born of the Industrial Revolution, made itself felt in all but three types of production: where more intensive energy could not be exploited, where continuous-processing machinery maintained control by integrating entire systems, and where the liquidity of flows facilitated their continued control even at vastly increased volumes and speeds. Whether or not a particular industry required new technology to control its throughput processing, however, all sectors, because of the increasing speed and volume of industrial production, experienced crisis in controlling the distribution of outputs on a comparable scale.

Control Crisis in Distribution

Because genetic programming severely constrains the speed with which plants can be grown, industrializing countries experience no crisis nor even much acceleration in the movement of materials through the various stages of agricultural production: tilling, sowing, cultivation, harvest. Even though various technological innovations—new machinery, fertilizers, strains of crops—may increase the speed or volume of throughputs at any one of these stages, the relatively fixed length of the growing season prevents integration of these improvements into a much faster system; small family farms remained the basic unit of agricultural production in the United States until well into the twentieth century. Once crops had been harvested, however, they increasingly entered a high-speed, national distribution system based on rail transportation and telegraphic control. Even without a control crisis in agriculture itself, therefore, America confronted—after completion of the East-West trunk lines in the 1850s—a growing crisis in the distribution of wheat, corn, and cotton.

In essence, the problem of distribution became one of coordinating and controlling the movement of these commodities from several million farmers scattered throughout the South and West to thousands of processors in northern and European cities. Central to the solution were specialized commodity dealers and brokers who purchased directly from farmers, sold directly to processors, and thereby eliminated most problems of coordinating intermediary transactions. The mercantile firm, which had served the function for half a millennium, disappeared from American agricultural markets in the 1850s and 1860s and was supplanted by new distributional structures made possible by rail transportation and telegraphic control. The key innovation in social technology was the commodity exchange, based on the telegraph and later on telephone exchanges, which permitted crops to be sold in transit and even before harvest and allowed the exploitation of even minute-by-minute changes in prices.

Commodity exchanges accompanied diffusion of the telegraph, which was launched in 1844 and in eight years comprised a continental telecommunications network of some twenty-three thousand miles (Fig. 6.4). The Chicago Board of Trade, established in 1848 as a merchant exchange, had by the early 1850s become a modern commodity exchange; the Merchants Exchange of St. Louis underwent a similar transformation in 1854. Meanwhile, wholly new commodity exchanges opened in other large cities: New York (1850), Philadelphia and Buffalo (1854), Milwaukee and Kansas City (1860), Toledo, Omaha, Minneapolis, and Duluth by the 1880s (Huebner 1911); cotton exchanges began operations in New York in 1870 and in New Orleans the following year (Woodman 1968, pp. 289-294).

Evidence that these exchanges functioned to control distribution can be found in the list of agricultural products *not* traded there: tobacco, meat, sugar, cacao, and other imports, the only foodstuffs processed by mass producers, who quickly replaced commodity dealers and brokers to integrate even further (vertically) the distribution of farm products. Coffee, the single import to have an exchange, was the only one not processed domestically; it reached retailers in the same bags

Figure 6.4. Rapid growth of a continental telecommunications network in the late 1840s required that cable be spun in the field and strung across high masts at river crossings. Submarine cables appeared after gutta percha proved to be a good underwater insulator (Shaffner's Telegraph Manual).

packed on Brazilian plantations. Only when processing did not become concentrated in a few companies, as with domestic grains and cotton, did products continue to be distributed by commodity dealers and brokers using an exchange.

Because the growing networks of railroad and telegraph greatly increased both the potential speed and predictability with which agricultural products could be delivered, an elaborate material infrastructure—including spur lines and sidings, grain elevators and warehouses, and cotton compresses—developed to control this distribution. Although the first grain elevator appeared in Buffalo in 1841, three years before the telegraph, the second did not come until 1847; only in the 1850s did they begin to be constructed in any numbers.

With mounting demand for storage (Chicago had at least fifteen elevators by 1860), it became increasingly difficult to ship grain "in separate units as numerous as there were owners" (Clark 1966, p. 259).
This control crisis eased only with various innovations in preprocessing:
standardized methods of sorting, grading, weighing, and inspecting.
The people hired to preprocess grain in this way became some of the
first of a new wave of information workers created by the Control
Revolution, at a time when information still accounted for less than 5
percent of the U.S. labor force (see Table 1.2).

Along with increased preprocessing came innovations in the actual
processing of commodities through distribution. In the mid-1850s fast-
freight and express companies—many founded a decade earlier for
local deliveries—began to make alliances with railroads for the shipping of goods across the lines of several companies. A congressional
committee reported that by 1874 fast-freight lines carried substantially
all of the rail shipments in the country (Taylor and Neu 1956, p. 72).
The success of these cooperative arrangements for coordinating the
movement of through freight depended on two additional innovations
in information processing and communication: the car accountant office
and the through bill of lading.

Railroads developed the first of these innovations, the car accountant
office, to monitor the location and mileage of "foreign" cars—including
those of other railroads, diners and sleepers of the Pullman Company,
and the tankers and coal cars owned by manufacturers—on the rail
company's own tracks. Because the same office also kept track of cars
owned by the company on other roads, dual control could be maintained
over much rolling stock. By the 1890s accountant offices had largely
supplanted joint fast-freight and express companies (Stover 1961, p.
156), which placed more centralized control in the hands of the railroad
managers themselves.

The second innovation, the through bill of lading or *waybill*, first
appeared in 1853 for shipments between Cincinnati and Atlantic ports
(Taylor and Neu 1956, pp. 74–75, 97). Perfected by the 1860s, the
waybill detailed goods shipped, the route covered, and charges accrued; the shipper, intermediary carriers and stationmasters, and the
receiver all retained copies. Copies also went to the carriers' auditors,
who used them to credit the sender and bill the next receiver. When
checked against the stationmasters' daily logs of bills processed, the
auditors' accounts provided ultimate control over all through freight
shipped in the country as well as over each carrier's finances. After

the fast-freight lines guaranteed the accuracy of their waybill listings by the 1870s, the bills quickly became negotiable paper, as did elevator and storage receipts (Chandler 1977, pp. 129, 212). These are two more examples of information-processing systems, like earlier ones based on promissory notes and bills of exchange, where the communications medium itself acquired the formal characteristics of specie in more generalized commercial exchange.

Successful coordination of through freight traffic, combined with the growing network of elevators, warehouses, and other storage facilities increasingly accessible by telegraphic communication, meant that the flow of agricultural commodities, despite its unprecedented velocity, could be regulated with precision. Deliveries could be scheduled for times when manufacturers would be ready to process or when retail inventories would likely be depleted, and trade in agricultural commodities could be carried on throughout the year. This more precise control of distribution led to yet another innovation in information-processing technology: the "to arrive" or futures contract, purchased with cash, which stated the amount, quality, and price of a commodity to be delivered on a prespecified date. In contrast to the long-established "consignment" contracts which they quickly replaced, the futures contracts' prespecified prices served to stabilize markets, reduced risks and hence credit costs, and thus speeded financing (Odle 1964). So successful did the new contracts become that their trading quickly found institutionalization in the modern futures market and commodity dealers could shift to speculators what little credit costs remained in the distribution of agricultural outputs (Rothstein 1965, pp. 68–71).

In short, the newly specialized dealers and brokers, by exploiting the growing railroad and telegraph systems, managed to control national and even worldwide networks for the distribution of agricultural products with little labor and even less capital (Fig. 6.5). Often a long chain of middlemen and advances could be replaced by a few managers operating out of a single central office. This not only lowered the credit costs in moving goods; it also, through the resulting improvements in information processing and communication, more closely integrated agricultural supply with demand. By increasing the productivity and efficiency of American agriculture in this way, and by increasing the regularity and speed with which its output could be processed, the modern commodity dealer hastened the advent of the Information Society—even as he helped to ease the crisis in controlling the mass distribution of products from millions of farms scattered throughout the South and West.

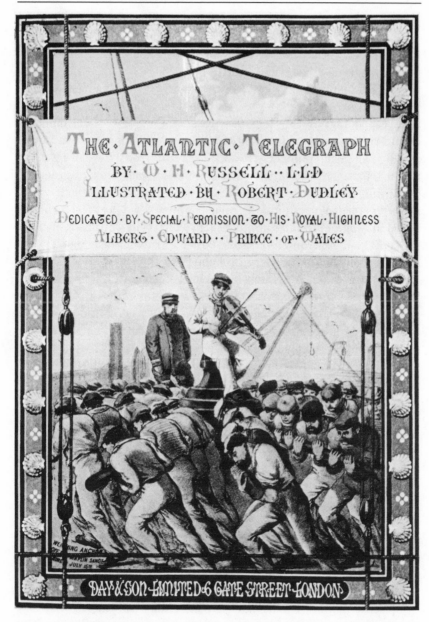

Figure 6.5. Completion of the Atlantic Telegraph in 1866 greatly increased control of worldwide networks for the distribution of agricultural products and manufactured goods. Title page from a book by William Howard Russell (1866), the only newspaper reporter aboard the cable-laying Great Eastern. At nearly seven hundred feet and a displacement of 22,500 tons, it was five times the size of any other ship afloat, the world's largest ship until the launching of the Lusitania in 1906. Painting by Robert Dudley, also on board.

Much the same distributional crisis confronted the manufacturers of consumer goods, including many—clothing and other dry goods, drugs, hardware, groceries, candy, liquor, stationery, tobacco, furniture, jewelry, and shoes—that still dominate shopping malls more than a century later. Maintaining control of the speed and volume of industrial production of these goods was one thing; distributing the outputs on a correspondingly mass scale was quite another. As increasing numbers of enterprises produced in continually greater volume for even more rapidly increasing numbers of consumers scattered ever more widely, the distributional system based on commission merchant transactions began to be overburdened. No longer could the southern factors and western storekeepers on their regular trips to the northeastern seaboard fully accommodate the mounting volume of industrial production. After East-West trunk lines made possible truly national distribution with unprecedented speed and reliability in the 1850s, new pressure arose to fine-tune the system—to enhance profits through inventory reduction and turnover at more rapid rates.

Just as specialized commodity dealers appeared in the 1840s and 1850s to integrate the distribution of agricultural output, so too did the distribution of consumer goods come to be consolidated about the same time, as we saw in Chapter 4, in a new economic actor, the wholesale jobber. Jobbers bought directly from manufacturers and sold to specialized retailers, which eliminated many intermediate links in the chain of middlemen. The resulting drop in distribution costs brought lower prices and an increased volume of business for the wholesalers, who, like the new grain and cotton dealers, did not sell on commission but actually took title to goods, thus enabling themselves to keep more than the usual 2 to 5 percent markup. So great an advantage did this prove to be that by the 1880s America had about five hundred wholesalers of dry goods, almost as many in hardware, and some two hundred for drugs alone. The commission merchant, for centuries the mainstay of worldwide distribution, had all but disappeared from the United States by the 1870s—replaced in a single generation by a distributional system based on rail transportation and telegraphic control.

With the two new technologies jobbers could drastically reduce their inventories; now they relied both on the telegraph to order goods as needed and on the railroads to ensure prompt delivery. This allowed wholesalers to move goods with greater speed and regularity, which reduced both inventories and unit costs, and at increased volume, which improved cash flow and reduced credit needs; it also meant that jobbers could service much broader areas. Geographical expansion depended, in turn, on marketing networks sufficient to reach rural general stores as well as more specialized urban retailers. The "drummers" used by wholesalers to solicit trade from rural merchants during their visits to the North gave way, in the late 1860s, to salesmen who themselves journeyed by rail and buggy to even the most isolated country store—and into American folklore and humor as the archetypal traveling salesman. Quite apart from their fabled exploits with farmers' daughters, these new information workers promoted products, recorded orders, assisted storekeepers with inventory control, accounting, and merchandise displays, and provided their wholesale companies with a constant flow of feedback necessary to maintain control of distribution. In addition to credit evaluations of local businessmen, this feedback included data on economic conditions, tastes and fashions, the relative success of the company's lines, and other factors likely to affect future demand (Clark 1944, chap. 6).

Toward Bureaucratic Control and Vertical Integration

Extensions of marketing operations required a corresponding increase in purchasing organization, the networks through which wholesalers obtained goods directly from manufacturers both at home and abroad. For example, Alexander T. Stewart, America's foremost dry goods distributor, had by 1873 a branch purchasing office in every major textile and apparel center in Great Britain and Western Europe (Resseguie 1965, p. 316). A new control crisis arose as wholesalers attempted to integrate the movement of goods from hundreds of manufacturers and purchasing offices to thousands of retailers with cash flows in the opposite directions. A single hardware jobbing firm, for example, might need to control distribution of six thousand different products from a thousand manufacturers to perhaps ten thousand customers.

This crisis led to innovation in bureaucratic structure, particularly the progressive subdivision of operating units, the whole controlled by a growing hierarchy of salaried managers. As we have seen, railroad and telegraph companies had begun developing such multiunit business structures in the 1850s; by the late 1860s they had been adopted by large wholesale houses attempting to control unprecedented volumes of trade made possible by the new technologies. As late as the 1840s, the staff of one of America's largest importers consisted of the owner, his son, two or three clerks, and a porter; together they handled perhaps a quarter-million dollars in annual sales (Tooker 1955, pp. 64–65, 225). A quarter-century later A. T. Stewart had two thousand employees and annual sales of $50 million (Resseguie 1965). By 1866 Chicago alone had fifty-nine jobbers with annual sales of more than $1 million each (Twyman 1954, p. 31). Alfred Chandler reports that, more generally, the annual sales of wholesale firms jumped from "tens and hundreds of thousands of dollars to tens of millions of dollars" following the spread of the railroad and telegraph (1977, p. 218).

In addition to sales and purchasing operations, large wholesale houses normally differentiated a half-dozen or more other departments or offices: advertising, orders, traffic and shipping, credit and collections, accounting. Each of these divisions filled particular information-processing, communication, and control functions. Purchasing developed the specifications for manufactured goods, determined both the purchase and selling prices and volume, and maintained contact with manufacturers and foreign commission agents; the sales organization, as we have seen, provided similar feedback from retailers in addition to recording orders and drumming up new trade. The general sales department included an advertising office, usually small, which prepared the catalogs carried by traveling salesmen and mailed to regular customers; occasionally it advertised to retailers through local newspapers. The orders department had the responsibility for filling orders promptly, the traffic department concentrated on scheduling larger shipments from suppliers to warehouses or from there to retailers, and the shipping department handled the details of moving the goods by rail. Credit and collections sustained cash flow—and thereby minimized the cost of credit per unit—by continuously processing credit information supplied by salesmen and outside credit agencies and by responding quickly to extend credit or to close delinquent accounts. Accounting recorded the receipts and expenditures of all funds, separately for each department and office as well as each supplier, retailer, and shipper, and continuously compiled management information like the rates of inventory turnover or stock turn.

Together these separate departments and offices, each of which maintained some specialized functions of information processing and control, comprised—in the great wholesale houses of the late 1870s— some of the largest and most differentiated bureaucratic structures to appear before the twentieth century. Often each of a house's specialized units kept its own books and accounts, so that each could have at least theoretically operated as a largely independent enterprise. At Chicago's Marshall Field and Company, for example, the head buyer of each major product line ran his department, in the words of the company's historian, as would "a merchant completely and independently responsible for the results within his own separate department or store," which "was run as though it were an independent business firm" (Twyman 1954, p. 65). More of the informational functions might also have been taken over by separate businesses, as with the Mercantile Agency, formed in 1841 as the first American credit reporting firm, which by the 1870s under R. G. Dun—later to form Dun and Bradstreet—employed ten thousand investigators in sixty-nine branch offices to process some five thousand queries daily (Madison 1974).

Counterpoised to these tendencies toward structural differentiation and increasing independence of specialized units is the fact that the functionally distinct but interdependent elements of any information processor—no less bureaucratic ones—need themselves to be controlled. Only by maintaining the separate operating units under a single structure of control could the wholesale houses of the late nineteenth century precisely monitor and coordinate the various activities of large-scale mass distribution, with resulting increases in productivity, decreases in costs, and consequently higher profits. Similar arguments for internal integration within a single firm have been developed by Coase (1937), Arrow (1964), and Williamson (1970), among others.

The key to such bureaucratic integration proved to be another innovation in organizational control: managerial hierarchy. Top managers replaced the informational mechanism of the market in allocating resources among a firm's future business activities; middle management, in turn, its work thus coordinated from above, supervised that of still other managers. Because middle managers had not yet appeared anywhere in the United States as late as the 1840s, the enterprise of multiple units coordinated by a hierarchy of salaried managers, as pioneered by the large wholesale houses, must be considered a response to the control crisis of the 1860s and 1870s—yet another innovation in information processing and control technology that marked the origins of the Control Revolution.

Each of the specialized units of the large wholesale houses had its own administrative office and full-time salaried managers. What generalized control developed in middle management usually devolved to

the general merchandising manager, who not only oversaw ware-
housing—including unpacking, labeling, branding, packaging, and re-
packing—and whatever manufacturing operations the firm might have
acquired, but also supervised the buyers in each product department.
In many cases, as we have seen, the senior buyers were themselves
important executives, salaried managers of large purchasing organizations who made careers of their specialties. The general sales manager,
perhaps next in importance in the control of wholesale jobbing operations, had responsibility for directing, monitoring, and evaluating the
entire sales force, sometimes with the help of assistant sales managers
for different regions, and also for supervising the advertising office.

Managers in the other departments had more direct responsibility
for control of a concrete open processing system: the physical movement of goods from manufacturer to retailer and the flow of cash or
credit—the information that controlled the system within the larger
market economy—in the opposite direction. Maintaining the greatest
possible speed, volume, and regularity of these two flows at the lowest
possible operating cost directly affected a wholesale house's unit costs
and thus its profits. Traffic managers maintained their own shipping
office for this reason and bargained continuously with the various railroads to get the lowest possible classifications and rates for their shipments and the largest possible rebates for the firm and its customers.
Because credit terms are also reflected in unit costs, as we have seen,
credit managers sought the shortest-term and most tightly controlled
credit possible, subject to competition and the longer-range advantages
of helping new retailers and others with temporary cash flow problems.
Alfred Chandler describes this situation at Marshall Field's:

> The granting of credit was of such importance that it became an almost
> full-time responsibility for one partner, Levi Leiter. Leiter's abilities in
> this field made it possible for the enterprise to carry out most of its
> business on a cash basis. "With their carefully selected customers dis-
> counting their bills as regularly as a group of faithful employees punching
> a time clock, the two partners had little capital tied up in delinquent
> accounts, knew with reasonable certainty how much money was coming
> in each month, and were subsequently able to maintain an unsurpassed
> reputation themselves for prompt payment." The resulting steady cash
> flow reduced the cost of credit per unit of merchandise obtained to a new
> low.   (1977, p. 222; quote from Twyman 1954, p. 36)

Much the same account might be given for the need to fine-tune the
processing of material flows in the opposite direction—from manufac-

turers to warehouses and then on to customers. Here senior executives of wholesale houses relied on accounting data to evaluate the performances of their operating managers. Most important of these were monthly reports on stock turn, defined formally as the number of times that a supply sold out and had to be replaced in a specified period, usually a year (Beckman et al. 1937, chap. 19). Like cash flow, stock turn reflected the speed with which a wholesaler processed the exchange of goods between manufacturers and retailers. The faster a house managed to distribute goods, for a fixed amount of capital and labor, the lower would be its unit costs and the greater its productivity and hence profits.

Not surprisingly, since stock turn reflects speed and efficiency in the processing of material flows, the concept did not appear before the railroads and the resulting crisis of control. Chandler (1977, p. 223) finds "no example of a prerailroad merchant using the term." By 1870, however, the exhortation to "keep one's stocks 'turning' rapidly" had become Marshall Field's most repeated message to his managers. Between 1878 and 1883 his firm's average stock turn hovered above 5, a good rate even by modern standards (Twyman 1954, pp. 50–51, 118–119, 175–176). Field urged the same goals on retailers and at least one, Rowland Macy's New York store, achieved for six months in 1887 a stock turn of 12, at least double the rate for department stores in this century (Hower 1943, pp. 185–188).

As suggested by the comparative stock-turn rates for Macy's and Field's in the 1880s, this decade saw the wholesalers challenged by new mass retailers—department and chain stores and mail-order houses—that purchased from manufacturers directly and thereby integrated still further the processes of distribution and marketing. Although the total number of wholesalers continued to grow into this century, increasing six- to eightfold between the 1880s and 1925, their market share began to decline in the early 1880s. Between 1869 and 1879 the ratio of wholesale to direct sales rose to 2.40 from 2.11, with only $1 billion worth of goods passing directly from manufacturers to retailers in the latter year, while some $2.4 billion worth went by way of wholesalers. After 1889, however, when wholesaling's predominance had already declined slightly to 2.33, the ratio began to fall ever more sharply: to 2.15 in 1899, to 1.90 in 1909, and to 1.16 by 1929 (Barger 1955, pp. 69–71).

Table 6.3 summarizes the major developments in market and retail control of distribution in the American economy from early industrial-

Table 6.3.   Selected events in the development of market and retail control of distribution, 1830–1889

Year	Development
1830s	Jobbers begin to purchase directly from domestic and foreign manufacturing agents; auctions decline Specialized New York City jobbers adopt fixed prices, publish catalogs, mail to customers
1840	Auctions fall to only one-eighth of U.S. import sales Nearly three-quarters of New York merchant establishments and almost all in New Orleans operate on commission
1842	Credit protection group formed by importers, commission houses
1845	Benjamin Babbitt markets waste shavings of soap in 1.5- to 2-pound boxes—an instant success with laundries, hotels
1850	New York City directory lists more jobbers than importers in a half-dozen specialty goods
Early 1850s	Commodity exchanges opened; standardized methods adopted for sorting, grading, weighing, inspecting
1850s	Specialized jobbing begins to spread from large coastal ports Warehouse receipts with uniform grades serve as exchange media "To arrive" or futures contracts, futures markets appear
1853	Trade association (American Brass Association) organized to regulate prices
1857	Weighing scale for use on railroad tracks introduced
1859	First store in Great American Tea Company chain—to become A&P in ten years—opens in New York City
1862	Fixed-price policy adopted by major store (Stewart's of New York)
1867	Macy's department store remains open until midnight Christmas Eve, sets a one-day record of $6,000 in receipts
Late 1860s	Large wholesale houses develop organizational structures with a half-dozen or more operating departments controlled by a hierarchy of salaried managers. Traveling salesmen employed Large wholesalers and retailers like department stores begin to monitor rate of stock turn
c. 1870	Christmas makes December retail sales more than twice those of any other month

Year	Development
1872	Mail-order house (Montgomery Ward) established
1874	Labor union labels attached to products (cigars)
1876	Great Atlantic and Pacific Tea Company (A&P) has 67-store chain
1879	Cash register demonstrated, patented Five-cent store (Woolworth's) opens
1880	Book (Ryan 1880) of measurements of human body for various ages and sizes becomes guide for standardizing ready-made clothing
1880s	Cheap paper bags speed up retail sales, especially of groceries
1884	Montgomery Ward issues a 240-page illustrated catalog with 10,000 items representing $500,000 in stock
1885	Self-service restaurant opened
1886	F. W. Woolworth controls a chain of 7 five-and-ten-cent stores Richard W. Sears begins a mail-order watch company

ization to the late 1880s. As with the control of production (Table 6.2), early control of distribution came mostly through innovations in preprocessing, in this case to faciliate *market transactions*, the processing equivalent of *throughputs* to production. These marketing innovations included—within a fifty-year period—standardized methods of sorting, grading, weighing, and inspecting, packaging in containers of fixed sizes and weights, fixed prices, standardized sizes, and periodic presentation to consumers via catalog. New types of retailers (like the five-and-ten) came to be specialized by price as well as by commodities sold, with general purpose containers (like grocery bags) and self-service introduced to cut transaction processing costs while at the same time speeding sales.

As Table 6.3 indicates, innovations also occurred in transaction processing as distinct from either information reduction or preprocessing. Processing innovations included the use of warehouse receipts and "to arrive" or futures contracts as generalized media of exchange, differentiation and specialization of bureaucratic control structures, monitoring of transaction rates via stock turn and other quantitative indicators, and the introduction of cash registers to record and control sales. Progressive development of an organizational infrastructure for control of geographical distribution via marketing over the same early industrial period included direct routinized contact of jobbers with manufacturing agents (1830s), use of the postal system to distribute catalogs (1840s), and the growth of commodity exchanges (1850s), chain stores and traveling salesmen (1860s), and mail-order houses (1870s) (Fig. 6.6). As with the control of production, in short, early market and retail control of distribution included innovations at all levels of control: preprocessing, organization, programming, processing, and communication.

During the three decades of the 1860s–1880s, for example, America's great wholesale houses like A. T. Stewart's in New York and Marshall Field's in Chicago developed all of the essential features of modern bureaucracy: subdivision into distinct operating units, distinguished by particular information-processing, communication, and control functions, which are themselves coordinated by an administrative hierarchy. This structure could not have replaced smaller, more traditional ones, at least not in a market economy, until bureaucratic control could yield lower costs, greater productivity, and higher profits than did control by the market. Not until steam power had sufficiently increased the speed and volume of material processing and the resulting increase in outputs could be distributed widely with the precision made possible by coevolving networks of railroad and telegraph did bureaucratic control become more efficient and more profitable than coordination by the market.

Such conditions first appeared in the distribution of manufactured consumer goods by the 1850s; the multiunit firm—one of the first of many information-processing innovations that would mark the Control Revolution—came largely in response to the resulting control crisis in distribution. The resulting transformation of the market by increasing administrative control—what Alfred Chandler aptly termed a progressive replacement of Adam Smith's "invisible hand" by management's visible one (1977, p. 1)—gained impetus from continued increases in the speed and regularity of production and distribution, developments that for the first time in history made communication through market mechanisms too slow and control by bureaucracy relatively more profitable. Market control could not have been transcended, however, without major innovations in information-processing and control technology: a new processor in the form of modern bureaucratic struc-

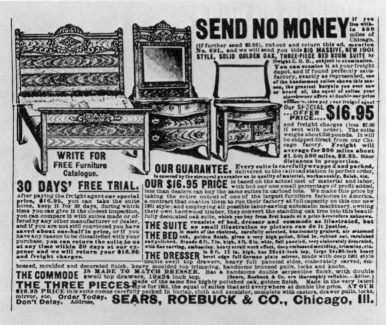

Figure 6.6. Innovations in market and retail control of distribution included the use of the postal system to distribute catalogs (1840s) and the resulting rise of large mail-order houses (1870s). By the 1890s Sears, Roebuck of Chicago ran crowded advertisements offering a "30 days' free trial," a "binding guarantee," and various free catalogs or "an immense catalogue of everything" for fifteen cents. (Courtesy of Sears, Roebuck & Company.)

ture, new internalized control in the hierarchy of salaried managers, unprecedented preprocessing through the increasing rationalization of society, and a third level of programming—in addition to genes and culture—in bureaucracy's formal sets of rules governing institutionalized decisions and responses. Bureaucratic control of production and distribution would have mattered little, however, without corresponding control over final consumption.

Control Crisis in Consumption

Crisis in the control of consumer demand did not arise until the early 1880s, when new continuous-process technologies began to be applied within a short span of years to a wide range of industries: flour milling and soap making (1879), cigarette rolling and match manufacture (1881), food canning (1883), and photographic film processing (1884). So well did the new continuous-process technologies control production through mechanical integration of the various stages that overnight they made more traditional industries highly capital-intensive, often increasing the ratio of output to workers several hundredfold. As production became more capital-intensive, assurance of adequate return on investment required large, steady, and predictable demand for products in order to keep the new plants and machinery running at peak efficiency, even as it forced concentration of production in only a few locations and companies. This sudden need of largely local and regional firms for vastly greater demand, often available only through control of national and even global markets, led to the late nineteenth-century crisis in the control of consumption.

In the words of Harry Tipper, advertising manager of the Texas Company (later Texaco), with "the continued improvement in the machinery of production, transportation, communication, etc. . . . the problem of disposing of goods became, consequently, more important" (Tipper et al. 1915, pp. 4–6). Especially with the adoption of new continuous-processing technologies like those of the petroleum industry, which increased its throughputs three- to six-fold while halving unit costs during the 1860s alone (Williamson and Daum 1959, pp. 282–285), producers had to teach consumers, Tipper argued, "to use more than they formerly had used, and to discriminate between different sellers or sections in order to control the market" (Tipper 1914, p. 13).

Tipper's view, that consumer demand had to be stimulated and controlled in response to sharply improved production technologies and corresponding increases in output, seems to belie current economic wisdom. The so-called new economic historians find that demand led industrial expansion in the first half of the nineteenth century (Fogel and Engerman 1971b), as we have seen; Jacob Schmookler (1966) argues that technological innovation comes in response to rising demand. Counter arguments, however, abound in economic and business studies. Alfred Chandler, for example, notes that "the precise timing of innovations in production, like the organizational innovations in marketing, can be related more closely to the new speed and volume at which materials and goods *could* flow through the economy than to any change in demand resulting from an obvious shift upward in the rate of growth of population and income" (1977, p. 253; emphasis added). Economists following the lead of John Kenneth Galbraith (1967, esp. chap. 18) have begun to consider mass communication and market feedback technologies in the control by modern corporations of consumer demand.

Some of the earliest and perhaps clearest cases of production-generated crises in the control of consumption came with the first continuous processing of agricultural commodities in the 1880s. This resulted in nothing less than a thoroughgoing transformation—in a matter of decades—of even the most basic eating habits of the American people (Cummings 1941).

Three years after completion of the first automatic, all-roller, gradual reduction flour mill in 1879, Henry P. Crowell adopted comparable technology to the production of oatmeal. His plant literally received raw oats at one end and shipped cartons of packaged oatmeal out of the other: it has been described as "the first in the world to maintain under one roof operations to grade, clean, hull, cut, package, and ship oatmeal to interstate markets in a continuous process that in some aspects anticipated the modern assembly line" (Marquette 1967, p. 33). When Crowell's plant began operations in 1882, however, most Americans scorned oats as fodder for horses and associated oatmeal with invalids and a few Scottish immigrants whose taste for the cereal was thought to reflect their dour personalities. As a result, Crowell soon produced twice as much oatmeal as the market could absorb. In Chandler's judgement, "A new market had to be found if the great volume of output from the new machines was to be sold" (1977, p. 294).

Crowell addressed this crisis with a revolutionary new technology for the control of consumption: national advertising of a brand name product directly to the mass household market. By repackaging his bulk meal in convenient twenty-four-ounce boxes, which he marketed under the now-familar brand label of the black-coated Quaker, Crowell managed to dispose of surpluses created by the control revolutions in production and distribution by inventing not only the modern breakfast food industry but breakfast cereal itself—a product then almost entirely new to American tastes.

Crowell's innovations in advertising included many of the fundamental techniques and gimmicks still used today: scientific endorsements, testimonials, prizes, box-top premiums, and the like. In 1889, a year after the country's seven largest mills had merged to form the American Cereal Company, Crowell introduced another mass marketing innovation, the first prepared mix, Aunt Jemima Ready-Mix (Kelley 1954, p. 104). Two years later, Crowell ran a fifteen-car freight train from his Cedar Rapids, Iowa, plant to Portland, Oregon, in perhaps the first national publicity stunt to promote commercial products. The train included not only public exhibits of breakfast foods but a professional actor dressed as a Quaker to attract and entertain spectators. In Portland every household received a half-ounce sample of Quaker Oats, probably the first use of free samples distributed door-to-door (Marquette 1967). Seven years later the advertising trade journal *Printers' Ink*, founded in the same year as the American Cereal Company (1888), reported that "one result of its extensive advertising of Quaker Oats is that exceedingly few people now buy oatmeal in bulk" (Pope 1983, p. 55), early proof that mass consumption patterns could be created, altered, and controlled by means of national advertising to the new mass markets.

Key to Crowell's success were the brand name label and trademark, which themselves constituted a new technology for the control of consumption (Fig. 6.7). When Abraham Lincoln tended store at Old Salem, Illinois, in 1833, on the threshold of the railroad and the industrial age, only one packaged, branded product graced his shelves—Walter Baker's Chocolate. In the words of Daniel Pope, historian of advertising,

Shopkeepers in more settled areas might carry a few more items, but only a sprinkling of canned specialty goods, an occasional import, and some patent medicine concoctions carried a manufacturer's brand. Manufacturing in pre-Civil War America did not, for the most part, produce goods suitable for national advertising, nor, of course, was there a network to distribute those goods throughout the nation. The most important industries in mid-19th century America—flour and grain milling, lumber and saw milling, even the relatively advanced textile and footwear indus-

Figure 6.7. New technologies for control of mass distribution, including consumer-packaged and brand-labeled products, encouraged imitators. The original Smith Brothers cough drop package (upper left), first registered in 1877, is one of the oldest and best-known trademarks in America. Its arrangement of graphic elements inspired the popular misconception that the brothers, William and Andrew, were named Trade and Mark, respectively. The label also inspired competition from other Smith brothers (and sisters) and packages with bearded pairs, including Presidents Lincoln and Garfield. Some competing labels even warned buyers to "beware of imitators." (Courtesy of Smith Brothers, F & F Laboratories, Chicago.)

tries—were, for the most part, processing the nation's abundant raw materials which consumers themselves would then fashion into items of utility—bread from flour, buildings from boards, clothing from textiles . . . It would be hard to conceive of developing a consumer allegiance to the flour of a local grist mill or the lumber of a particular saw mill. (Pope 1983, pp. 31–32)

Although truly national advertising did not appear before Crowell and other continuous-process manufacturers in the 1880s, a separate advertising industry did begin to differentiate itself with the advent of intersectional railroads in the early 1840s. The country's first advertising agency emerged from the Philadelphia real estate and coal business of Volney B. Palmer in 1841. According to his own advertising, Palmer provided access to "newspapers in Pennsylvania and New Jersey and in many of the principal cities and towns throughout the United States" for some of the first businesses to advertise beyond their own localities, including "merchants, mechanics, professional men, hotel and boardinghouse keepers, railroad, insurance and transportation companies."

Palmer also offered to assume the information-processing burdens of advertising, "the trouble of perplexing and fruitless inquiries, the expense and labor of letter writing, the risk of making enclosures of money, &c., &c.," and a convenient method of payment: his receipts provided evidence that an advertiser had paid his bill (Holland 1974, p. 359). In exchange, he probably received about 25 percent of the gross cost of the newspaper space, a commission method of compensation that survives to this day throughout the advertising industry (Pope 1983, p. 116). So great did demand for Palmer's service prove to be that by 1849 he had opened additional offices in Boston, New York, and Baltimore and claimed to have exclusive rights to sell advertising for some 1,300 newspapers.

Despite Palmer's initial success, which came largely in response to the expanding rail network, most advertising agencies found their business—until the 1880s—not in national markets but in regional and especially in local ones. Because the palatial retail houses, which appeared in the late 1840s, and the new department stores, a product of the 1860s, sold directly to final consumers, they needed to allocate more money and attention to advertising than did even the largest wholesale houses, and they soon became the advertising agencies' major clients. In Philadelphia, for example, John Wanamaker's store helped to make N. W. Ayer and Son the nation's largest advertising agency within thirty years of its founding in 1869 (Hower 1949, pp. 58, 214).

Well into the 1870s only books, journals, and patent medicines were advertised nationally; most manufacturers left advertising to the wholesalers who marketed their output in particular regions or localities.

The first federal trademark legislation, enacted in 1870, attracted no registrants for more than three months and only 121 during the year. When in 1880, on the advent of continuous-process manufacturing, the law was declared unconstitutional as a restraint on commerce within states, Congress quickly enacted new legislation; Henry Crowell's Quaker became one of the first trademarks registered under the new law (Smith 1923; Lambert 1941; Barach 1971). The power of the new technology to control consumption became apparent in the same decade when Henry L. Pierce, who had purchased the Walter Baker Chocolate Company, agreed to pay a royalty of $10,000 a year to Baker's widow for rights to the family name and trademark—the only one in Lincoln's store only a half-century earlier. By 1905 *Printers' Ink* estimated the value of Royal Baking Powder's trade name to be $5 million, "a million dollars a letter" (Pope 1983, p. 69).

Trademarks helped to ease crisis in the control of consumption of a wide range of products: not only oatmeal, but flour, cereal, cigarettes, matches, various canned goods, soap products, and photographic film. All of these products could suddenly be produced far in excess of demand during the late 1880s, owing to the vastly greater control of manufacturing brought about by more integrated continuous-processing machinery. Just as almost simultaneous innovations in this technology across scores of industries resolved mounting crises in the control of production, the resulting crises of consumption could also be met by the still newer control technology pioneered by Henry Crowell: mass marketing of trademarked, consumer-packaged, and brand-labeled products through national advertising. Many of today's best-known brand names—Gold Medal and Pillsbury flour, Kellogg's corn-flakes, American Tobacco, Diamond matches, Borden and Carnation condensed milk, Campbell Soup, Heinz 57 Varieties, Procter and Gamble soap, Kodak film—began as trademarks for the fruits of new continuous-processing machinery in the 1880s. As a result of massive national advertising campaigns, all had become household words by 1900 (Fig. 6.8).

Rapid diffusion of the automatic all-roller, gradual reduction flour mill, after successful completion of C. C. Washburn's experimental mill in Minneapolis in 1879, brought overproduction not only in the milling of oats but of wheat, barley, rye, and other grains as well (Kuhlmann 1951). Over the next decade a half-dozen pioneers of the new milling

Figure 6.8. Mass marketing of trademarked, brand-labeled products through national advertising made many manufacturers of continuous-process goods household names by the turn of the century. Other businesses exploited this new element of mass culture—for example, a sheet music publisher marketed a popular new song, "Signs We See as We Pass Along." (Courtesy of the Union Music Company, Studio City, California.)

technology scrambled to control consumption by means of trademarks and national advertising. Washburn's firm introduced Gold Medal Flour in 1880; his leading rivals, the Pillsbury brothers, quickly countered with a flour bearing the family name. After 1890, when Minneapolis flour production had almost tripled to some seven million bushels annually in less than a decade, advertising alone could not control demand sufficiently and prices fell. As a result, both the Washburn and Pillsbury firms adopted the strategy Crowell had developed for Quaker Oats—packaging rather than selling in bulk, stepped-up national advertising, and vertical integration through networks of buyers and sellers—in order better to control distribution and consumption (Storck and Teague 1952; Gray 1954, chap. 4).

Other entrepreneurs copied Crowell more directly by inventing new breakfast foods. In 1893 Henry Perky and William Ford patented a machine that made wheat into filaments or "shreds" and thereby introduced shredded wheat biscuits into the morning meal. C. W. Post, considered by many to be the father of ready-to-eat cereal, introduced "Grape Nuts" in 1896 and "Post Toasties" in 1915; he also began promoting his earlier "Postum" as a hot beverage that might substitute for coffee, which he attacked in advertising as bad for the nervous system. W. K. Kellogg, of Battle Creek, Michigan, who began producing fifteen-cent boxes of "breakfast food" in 1896, had in ten years established Kellogg's Corn Flakes on the American breakfast table (Cummings 1941).

Table 6.4 summarizes the major developments in U.S. advertising and mass communication technology from early industrialization through the 1880s. As this list of innovations illustrates, increasing control of consumption in response to the resolution of crisis in mass production and distribution came through the coevolution of mass media and their messages to attract, hold, and imprint the mass attention: short slogans endlessly repeated (1856), secular symbols of Christmas (early 1860s) and other "festivals of consumption" (Boorstin 1973, chap. 18), commercial premiums (1865), trademarks (1870), patented package labels (1874), and the multiple elements of national advertising and publicity campaigns (1889).

Power-driven printing distributed by rail, the major mass medium before broadcasting, improved rapidly—parallel to the developing crisis in control of consumption—though a spate of innovations: the first electric press (1839) and rotary printing (1846), wood pulp and rag paper and the curved stereotype plate (1854), paper-folding machines (1856), the mechanical typesetter (1857), high-speed printing and folding press (1875), and linotype (1886). Utilizing these purely mechanical inventions, mass publishers developed a wide range of new organizational and social innovations to improve control: the penny newspaper (1833) to expand mass readership, a press association (1848) and

Table 6.4.   Selected innovations in advertising and mass communication technology for control of consumption, 1830–1889

Year	Innovation
1833	Penny newspaper (*New York Sun*) opens way for mass press
1839	Printing press run by electricity Photographs (daguerrotypes) produced
1841	Advertising agency (V. B. Palmer) established
1842	Design patents authorized, issued to typeface, other designs Illustrated weekly (*Brother Jonathan*) published
1846	Double cylinder rotary printing press (Hoe) adopted by *Philadelphia Ledger*, produces 8,000 sheets per hour
1848	Newspaper press association formed Periodical index (Poole's) published
1854	Wood pulp and rag paper introduced for printing Curved stereotype plate for Hoe rotary press cast, used
1855	Professional printing magazine (*Typographic Advertiser*) published
1856	Advertiser (*New York Ledger*) establishes "iteration copy" by purchasing full-page newspaper ad, repeating same line 600 times Machine to fold paper for books, newspapers installed
1857	Typesetting machine demonstrated
Early 1860s	Display type introduced, ending requirements of agate type and single columns in newspaper advertising Christmas begins to be promoted commercially
1865	Advertising monthly (*Advertising Agency Circular*) published Premiums (lithographed pictures) given for coupons on manufactured product (Babbitt Soap)
1869	Book (Rowell 1869) published for advertisers listing all U.S. newspapers with accurate estimates of circulation
1870	First federal trademark law passed, 121 registered Human-interest illustrations appear in newspaper advertising
1873	Successful illustrated daily (*New York Daily Graphic*) begun
1874	Label patents authorized, first issued to breakfast hominy label Macy's offers window display devoted exclusively to Christmas

Year	Innovation
1875	Advertising weekly (*Advertiser's Gazette*) issued "Open contract" makes advertising firm sole agent for advertiser High-speed newspaper printing and folding machine—producing 400 four-page sheets per minute—installed at *Philadelphia Times*
1878	Full-page newspaper advertising introduced
1879	N. W. Ayer & Son surveys grain market for an advertising client
1881	Second federal trademark law passed
1883	Joseph Pulitzer takes over the *New York World*, makes it America's first modern mass-circulation daily
1884	Newspaper syndicate (McClure) organized
1885	Daily railroad delivery of newspapers begun
1886	Linotype machine used commercially (*New York Tribune*) American Newspaper Publishers Association organized
1888	Trade journal (*Printers' Ink*) established for advertising industry
1889	National publicity stunt used to promote commercial product (Quaker Oats)

professional printing magazine (1855), mass-circulation dailies (1883), newspaper syndicates (1884), daily railroad deliveries (1885), and a publishers association (1886).

Parallel to the combined mechanical and organizational infrastructure of mass publishing, a new infrastructure developed for mass control of consumption via national advertising and market feedback. An increasingly specialized advertising sector met the control crisis in consumption, following the spread of industrialization through the material economy, with an array of innovations no less impressive than the better-known ones in communications technology, including advertising agencies (1841), indexes of periodicals (1848), a monthly trade journal (1865), book of newspaper circulation (1869), weekly trade paper and sole-agent contracts (1875), market surveys for clients (1879), and an industry-wide journal (1888). Almost a century before the advent of television, which John Kenneth Galbraith (1967, p. 342) declared "essential for effective management of demand," nineteenth-century advertisers pioneered the use of visual media and techniques of communication to stimulate and control consumption: daguerrotypes (1839), patented typefaces and illustrated magazines (1842), iteration copy (1856), display type (early 1860s), national consumption symbols like Thomas Nast's cartoon Santa Claus (1863), human-interest illustrations (1870), illustrated daily newspapers (1873), thematic store window displays (1874), and full-page advertisements set-off by "white space" (1878) (Fig. 6.9).

So important did such mass advertising innovations become for controlling consumption of the output of the new continuous-process mills that as late as 1913 (the earliest year for which statistics are available), despite new consumer products like the automobile, phonograph, and electrical appliances, foodstuffs from the mills still dominated national magazines, the most important advertising medium of the age. Of the top thirty-five advertisers in that year, six marketed processed grain products, including two of the three leading accounts: Quaker Oats (ranked second behind Proctor and Gamble), Postum (third), Kellogg (eleventh), Cream of Wheat (seventeenth), National Biscuit (twenty-second), and Washburn-Crosby, producer of Gold Medal flour (thirty-third). None of these companies ranked among even the top seventy-five industrials in assets, however, and only two—National Biscuit (at seventy-sixth) and Quaker Oats (one hundred and thirty-fourth)—ranked in the top two hundred and fifty (Pope 1983, pp. 41–45). Among all consumer packaged foodstuffs produced by continuous-process technology in 1917, the leading manufacturers of each type had begun operations before 1900—many before 1890; all had made early use of national advertising to establish their trademarks as household names: Borden's Condensed Milk, Fleischmann's yeast, Royal Baking Powder, Coca Cola, Heinz, Wrigley's gum (all firms that ranked among the top two hundred and fifty U.S. industrials in 1917 assets) (Navin 1970).

Apart from the producers of breakfast foods and packaged flour, who prospered from the application of integrated continuous-processing machinery to milling, the only other crisis in the consumption of

---

Figure 6.9. Increased control of consumption came through the coevolution of mass media and messages to attract, hold, and imprint the mass attention, including the secularization of Christmas and other festivals of consumption in the 1860s. By 1873 the illustrated periodical press promoted Easter as an occasion to display the latest fashions in men's and women's clothing (Harper's Weekly, *April 26, 1873*).

Easter Sunday, 1873

foodstuffs resulted from use of this same new technology—and the similar threat of overproduction—in canning. Although the tin can had been patented in England in 1810, a good tinsmith could cut, mold, and solder only about sixty cans per day. Even after the introduction of machine-stamped cans in 1847, canned items sold only in specialty shops in small quantities and at high prices. Indeed, cans could only be opened through intensive work with hammer and chisel until 1865, when the introduction of tins of thinner steel brought the first can opener; the so-called key opener followed in 1866 (Collins 1924; Woodcock and Lewis 1938). Crisis in the control of consumption did not come until after 1883, however, when the brothers Edwin and O. W. Norton put into operation the first "automatic line" canning factory, with machinery so arranged that cans could be soldered at the rate of fifty per minute and tops and bottoms at the rates of forty to seventy per minute—the output of roughly five hundred skilled tinsmiths only forty years earlier (May 1938, pp. 350–351).

So important did this continuous-process technology become that the first firms to adopt it on a year-round basis—Campbell Soup, Heinz vegetable products, Borden, Carnation, and Pet condensed milk, Libby canned meats—still rank among the largest and best-known canners in the country. Where canning remained seasonal, however, as with most vegetables, fruits, and fish, large canners did not appear. Instead, local and regional canneries bought cans and canning equipment from two large can makers: American Can, a 1901 merger that had Edwin Norton as its first president, and Continental Can, formed in 1906 (Bitting 1916; McKie 1959, pp. 103–107). By 1917 both companies ranked (thirty-first and two hundred twenty-second, respectively) among the largest U.S. industrials (Navin 1970), as did Borden's (103), Libby (220), Heinz (226), and Campbell (442). These latter food-processing giants all emerged from older family firms that had controlled the distribution and consumption of outputs from the new line technologies with their already familiar brand labels, national advertising, and backward and forward integration into purchasing and sales.

Gail Borden had received both American and English patents for his process of evaporating milk in a vacuum in 1856; two years later he launched the New York Condensed Milk Company with a large-scale plant about a hundred miles north of the city. During the Civil War the Union Army commandeered Borden's entire output of canned milk for field rations. Even at sixteen thousand quarts a day the New York plant could not keep up with government orders, and Borden had to license several new operations. The war gave the first taste of canned foods to many soldiers who carried the experience back to homes throughout the country in the early 1860s. Over that decade alone, demand for cans of food rose from five to thirty million per year. During the 1880s, as European milk processors like the Anglo-Swiss Condensed Milk Company (precursor of Nestle) and Helvetia (later to split into two American firms, Carnation and Pet) set up U.S. plants and sales organizations, the Borden Company—now directed by Borden's oldest son—adopted continuous-process technology and expanded its purchasing and marketing organizations (Frantz 1951). By 1914 Borden ranked among the top fifty advertisers in national magazines (Pope 1983, p. 43).

Henry John Heinz of Pittsburgh, a small processor of pickles and relishes for local consumption, had gone bankrupt in 1876. In the early 1880s Heinz adopted continuous-process methods of canning and bottling and quickly built a national network of offices to advertise and distribute the massive outputs (Alberts 1973). When Heinz died in 1919, the company employed 6,323—including 952 salesmen—in twenty-five branch factories, including one each in Canada and Spain, plus eighty-seven raw product stations, eighty-five pickle salting stations, and fifty-five branch offices and warehouses. The company also owned and operated its own can, bottle, and box factory, its own seed farm, and 258 railroad cars, which in that year hauled 17,011 carloads of goods (McCafferty 1923, pp. 106–107).

Although much less information survives on the Joseph Campbell Soup Company of Camden, New Jersey, it seems to have appeared at about the same time as H. J. Heinz and to have adopted continuous-process canning technology in much the same way (May 1938, pp. 341–346). Because of its considerable reputation in the Philadelphia area, Campbell did not advertise on a large scale until 1899, when the company's secretary is said to have remarked to its treasurer, "Well, we've kissed that money goodbye!" The money must have bought something useful for Campbell Soup, however, because five years later the company launched its now-famous Pork and Beans with a massive ad campaign, and by 1915 it ranked ninth in the country in national advertising (Pope 1983, pp. 43, 61).

The modern canned meat industry began in the 1870s when a Chicagoan, J. A. Wilson, designed a can with the familiar "truncated pyramid" shape. When the can was opened at the larger end and tapped on the smaller, as the instructions explained, the contents would "slide out in one piece so as to be readily sliced." By 1878 a Chicago factory turned out such cans by the thousands. Four years later Wilson and another small Chicago meatpacker, Libby, McNeil and Libby, began the line canning of meats (Bitting 1916; Collins 1924; May 1938; Woodcock and Lewis 1938; McKie 1959); within decades both ranked among the top industrials in the country (Navin 1970).

In short, all successful continuous-line food processors in the 1880s and 1890s resorted to consumer packaging, brand labeling, and mass advertising to control consumption. National advertising proved particularly important to these processors, not only because breakfast food, packaged crackers, and canned milk, soup, and meat remained unfamiliar products to many consumers at the turn of the century, despite the popularity of canned rations following the Civil War, but also because low unit prices—usually five or ten cents—made demand inelastic; producers had little room to control consumption by manipulating prices (Chandler 1977, p. 298). Small wonder that by 1900 processed foods accounted for the largest part of N. W. Ayer's volume (Boorstin 1973, p. 147) or that as late as 1913 twelve of the top forty advertisers in national magazines produced foodstuffs—double the number in any other industry (soaps and chemicals followed with six, automobiles with five, tires and rubber products with four; Pope 1983, pp. 43–45).

From Industrial to Control Revolution

The nineteenth-century revolution in information technology was predicated on if not directly caused by social changes associated with earlier innovations. Just as the Industrial Revolution presupposed a commercial system for capital allocations and the distribution of goods, as we saw in Chapters 4 and 5, the Control Revolution developed in response to problems arising out of advanced industrialization: a mounting crisis of control at the most aggregate level of national and international systems, levels that had had little practical relevance before the mass production, distribution, and consumption of factory goods.

Resolution of the crisis demanded new means of information processing and communication to control an economy shifting from local segmented markets to increasingly higher levels of organization—what might be seen as the growing "systemness" of society. This capacity to communicate and process information is one component of the problem of *integration*, the need for coordination of functions that accompanies differentiation and specialization in any system.

Increasingly confounding the need for integration of the structural division of labor were the corresponding increases in commodity flows through the system—flows driven by steam-powered factory production and mass distribution via national rail networks. Never before had the processing of material flows threatened to exceed—in both volume and speed—the capacity of technology to contain them. Suddenly, owing to the harnessing of steam power, goods could be moved at the full speed of industrial production, night and day and under virtually any conditions, not only from town to town but across entire continents and around the world.

To do this, however, required a system of manufacturers and distributors, central and branch offices, transportation lines and terminals, containers and cars, that grew increasingly staggering in its complexity. Even the logistics of nineteenth-century armies, then the most difficult problem in processing and control, came to be dwarfed in complexity by the material economy. Just as the problem of control reached crisis proportions, however, a series of new technological and social solutions began to contain the problem. This was the beginning of the Control Revolution.

Foremost among all the technological solutions to the crisis of control—in that it served to control most other technologies—was the rapid growth of formal bureaucracy. As we saw in Chapter 3, bureaucratic organization serves as the generalized means to control any large social system; it tends to appear wherever a collective activity needs to be coordinated by several people toward explicit and impersonal goals, that is, to be *controlled*. Because of the venerable history and pervasiveness of bureaucracy, historians have tended to overlook its role in the late nineteenth century as a major new control technology. Nevertheless, bureaucratic administration did not begin to achieve anything approximating its modern form until the Control Revolution.

Evidence for the magnitude of this change can be found in the rapid development of office technology during early industrialization. In 1780 a modern American office might contain printed business forms and file cabinets, communicate via mail, parcel post and courier, subscribe to various types of news publications, and hire financial and other professional information services. In general, informational goods and services—as well as media and content—were still sharply separated. Bureaucracy, where it could be said to exist at all, lacked structural differentiation and specialization of function. As John McLaughlin (1980, p. 10) notes of the "information business" in 1780: "Some individuals or companies engaged in both vertical and horizontal integration of economic enterprises during this period. Thus Benjamin Franklin worked as a writer, produced books, newspapers and magazines, developed printing equipment and sold printing services while serving as postmaster general of the colonies" (and, we might add, while conducting primary research on electricity).

A century later, in the midst of the revolution in generalized information processing and control, the modern American office had added a dozen major information technologies and services: telegraph and telephone, international record carriers and other local delivery services, newsletter, loose-leaf and directory subscriptions, news and advertising services, and differentiated security systems (including district telegraphs that could summon police or the fire brigade with the turn of a crank). By the 1890s typewriters, phonographs, and cash registers had also come into common use in American business (Fig. 6.10). The new office technologies and services, added since the advent of industrialization and the resulting need for increased control, reflected a trend toward integration of informational goods and services, media and content, that has continued unabated to this day. Loose-leaf subscriptions provided more of a service than a product compared to bound volumes, for example; news services constituted more conduit than content compared to books (McLaughlin 1980, chap. 2).

Table 6.5 summarizes the major developments in office technology and bureaucracy from early industrialization to the late 1880s. Notable in this list are the parallel developments of the various components of a generalized information processor, innovations not only in the processor itself (bureaucratic structure), but also to improve its information creation or inputs, recording (storage), programming, processing, communication, and control.

Innovations in bureaucratic structure included its first careful definition (1842), a hierarchical information-processing system to centralize control of rail movements (1853), the line-and-staff concept (1857), and by the late 1860s the emergence of bureaucracy in its modern form, complete with a half-dozen or more operating departments controlled by a hierarchy of salaried managers. Within forty years, as we shall see in the next chapter, one of the world's largest bureaucracies—rural free delivery or RFD—had been successfully incorporated within the U.S. postal system.

Concurrent with the development of formal bureaucratic structure

THE TYPE-WRITER.

WHAT "MARK TWAIN" SAYS ABOUT IT.

Hartford, March 19, 1875.

GENTLEMEN: Please do not use my name in any way.   Please do not even divulge the fact that I own a machine.   I have entirely stopped using the Type-Writer, for the reason that I never could write a letter with it to anybody without receiving a request by return mail that I would not only describe the machine, but state what progress I had made in the use of it, etc., etc. I don't like to write letters, and so I don't want people to know I own this curiosity-breeding little joker.   Yours truly,

SAML. L. CLEMENS.

Figure 6.10. Typewriters, phonographs, and cash registers came into common use in American business by the 1890s. The Adventures of Tom Sawyer *(1876) is reputed to be the first typewritten manuscript to be set into a book. Its author, Mark Twain, wrote a humorous testimonial for the new machine. By 1930 union printers refused to set manuscripts that had not been typed (Illustrated Phonographic World).*

Table 6.5.   Selected events in the development of office technology and
bureaucracy for generalized control, 1830–1889

Year	Development
1839	Mass-produced envelopes introduced
1842	Internal organizational structure in business (Western Railroad) carefully defined Business school (Eastman Commercial College) founded
1848	Shorthand magazine (*American Phonographic Journal*) published
1850s	Railroads come to employ more accountants and auditors than any government, federal or state
1853	Hierarchical system of information gathering, processing, and telegraphic communication instituted to control a trunk-line railroad (Erie) from the superintendent's office
1856	Domestic blotting paper—manufactured on Fourdrinier machine—introduced, replaces sand-boxes
1857	Business (Pennsylvania Railroad) enunciates line-and-staff concept: line managers direct workers in basic functions of organization, others (staff executives) set standards
1858	Pencil introduced with eraser attached Commercial manufacture of steel pens (Esterbrook) begun
1867	Telegraph ticker installed inside brokerage house
1868	"Type-Writer" patented, word enters the language as Americanism
1869	Earliest patent issued for carbon paper
Late 1860s	Modern bureaucracies, organizational structures with a half-dozen or more operating departments controlled by a hierarchy of salaried managers, emerge in large wholesale houses
1870	Stock ticker invented
1872	"Carbon Paper" designed for use in a typewriter patented, term enters the language as Americanism
1873	Typewriter with modern "QWERTY" keyboard marketed
1874	Record-keeping system introduced into hospital (Bellevue)
c. 1880	Modern offices contain paper business forms, file cabinets, directories and a telephone, subscribe to newspapers and loose-leaf services, use international record carriers
1880s	Information required to run a large business—including billing, sales analysis, inventory—grows rapidly in scope and complexity

Year	Development
1881	University business school (Wharton) founded
1882	Messenger news service (Dow Jones) begun on Wall Street
1883	Accounting firm (Barrow, Wade, Guthrie, & Co., New York City) founded
1884	Bonding company (American Surety) begins business Press clipping bureau (Romeike's) opens
1885	Dictating machine demonstrated, patented
1886	American Association of Public Accountants formed Desk telephone introduced
1887	Modern office calculating machine (Comptometer) with keys, multiple rows marketed
1889	Punch-card tabulating machine (Hollerith) introduced
c. 1889	Typewriters come into common use in U.S. offices

came a spate of basic inventions to improve the generation of information within bureaucracy. These innovations included blotting paper (1856), the pencil with eraser and steel (Esterbrook) pen (1858), carbon paper (1872), and modern keyboard typewriter (1873); all can be found in many offices to this day. The new tools for creating information could also be applied—by the 1870s—to the preprocessing of new informational inputs to the modern office: not only the loose-leaf and directory services already mentioned but also the stock ticker (1870), messenger news service (1882), and press clipping service (1884). Processing of numerical data came to be facilitated by two inventions—the keyboard calculator (1887) and punch-card tabulator (1889)—whose social implications would be felt well into the twentieth century (as we shall see in Chapter 8).

Other major components of the generalized information processor that the modern office and—by aggregation—emerging bureaucratic structure would become also appeared in the early decades of industrialization. Recording or information storage capabilities increased with the systematization of shorthand, including the first professional shorthand journal (1848), the systematization of office record keeping (early 1870s), and the dictating machine (1885). The programming of management with generalized bureaucratic practices improved with the organization of separate business schools after 1840, culminating in Wharton—the first university school—in 1881. Office communication improved with a continuous stream of innovations, from manufactured envelopes (1839) to the desk telephone (1886) (Fig. 6.11).

As we might predict, internal bureaucratic control—in the form of the comptroller or auditor—also developed rapidly during early industrialization and the resulting crises of control. By the 1850s U.S. railroads already employed more accountants and auditors than the federal government or any state. America's first independent accounting firm, Barrow, Wade, Guthrie of New York City, began operations in 1883; three years later the public accountants organized a national professional association.

With this rapid development of rationalization and bureaucracy came the succession of dramatic new information-processing and communication technologies that contained the continuing control crisis of in-

Figure 6.11. The telephone had become a standard item in modern offices by 1880, inspiring the introduction of a desk model in 1886. This sketch was used to promote the telephone to New York City businesses for both local and out-of-town calls. (Courtesy of New York Telephone.)

dustrial society in production, distribution, and consumption of goods and services. Control of production was achieved by the continuing organization and preprocessing of industrial operations. The resulting flood of mass-produced goods demanded comparable innovation in control of distribution. Growing infrastructures of transportation, including rail networks and steamship lines, depended for control on a corresponding infrastructure of information processing and telecommunications. Controlled by means of this infrastructure, an organizational system rapidly emerged for the distribution of mass production to national and world markets.

Mass production and distribution could not be completely controlled, however, without control of demand and consumption. Such control required a means to communicate information about goods and services to national audiences in order to stimulate or reinforce demand for these products, as well as a means to gather information on the preferences and behavior of this audience—reciprocal feedback to the controller from the controlled (a nascent consumers movement, inspired by the first muckraking journalists, demonstrated that information might also serve to control the controllers). Communication to a national audience of consumers developed with the first truly mass medium: power-driven, multiple-rotary printing and mass mailing by rail.

At the outset of the Industrial Revolution, most printing was still done on wooden handpresses that differed little from the one Gutenberg had used three centuries earlier. Steam power was first successfully applied to printing in 1810; by 1893, with an octuple rotary power press, 96,000 eight-page copies of Joseph Pulitzer's New York *World* were printed every hour (Fig. 6.12). The postal system also served as a new medium of mass communication through bulk mailings of mass-produced publications. By 1887, for example, Montgomery Ward mailed throughout the continent a 540-page catalog.

As we saw in the first chapter, many other communication technologies that we do not today associate with advertising were tried out—early in the Control Revolution—as means to influence the consumption of mass audiences. Popular books like the novels of Charles Dickens contained special advertising sections. Mass telephone systems in Britain and Hungary carried advertisements interspersed among music and news, although application of telephony to mass communication was undoubtedly stifled by the rapid development of broadcast media beginning with Guglielmo Marconi's demonstration of long-wave te-

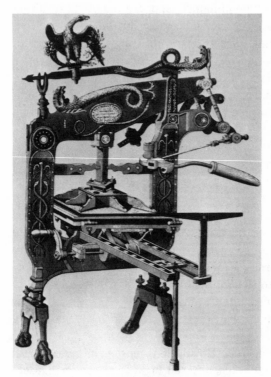

Figure 6.12. Printing speed increased three-hundred-fold during the period 1827–1893. Top: *The Columbian press, manufactured in Philadelphia in 1816, included an ornate eagle and serpents; otherwise it differed little from the hand-press used by Gutenberg three centuries earlier.* Bottom: *The Hoe Web Printing Machine of 1875 was the result of a thirty-year search for a high-speed cylindrical press; it printed 25,000 sheets per hour, a tenfold increase in speed over other power presses. By 1893 octuple rotary power presses could print up to 96,000 eight-page copies per hour. (Courtesy of the Free Library of Philadelphia.)*

legraphy in 1895. With the development by Edison of the "motion picture" after 1891, advertisers had yet another mass medium to exploit.

Mass media were not sufficient to effect true control, however, without a means of feedback from potential consumers to advertisers, a mechanism that would restore to the emerging national and world markets an essential relationship of the earlier segmental markets: communication from consumer to producer. As we shall see in the next chapter, the development of mass communication led to *mass feedback* technologies: market research (the idea first appeared as "commercial research" in 1911), including questionnaire surveys of magazine readership, the Audit Bureau of Circulation (1914), and house-to-house interviewing (1916).

Although most of the new information technologies originated in the private sector, their potential for controlling systems at the national and world level was not overlooked by government. Control is control, it would seem, and technologies originally developed to control production, distribution, and consumption of goods and services would increasingly be adopted by states encompassing the entire range of political ideologies to control their citizens. As corporate bureaucracy came to control increasingly wider markets by the turn of this century, its power came increasingly to be checked by a parallel growth in state bureaucracy.

Because the activities of information processing, programming, decision, and communication are inseparable components of the control function, a society's ability to maintain control at all levels—from interpersonal to international relations—will be directly proportional to the development of its information technologies. Because technology defines the limits on what a society *can* do, technological innovation might be expected to be a major impetus to social change, as indeed it proved to be in the Control Revolution. The Industrial Revolution, which brought about the nineteenth-century crisis of control, began with greatly increased use of coal and steam power; the Control Revolution that eventually resulted was achieved by innovation at a most fundamental level of technology—that of information processing and communication.

III

Toward an Information Society: From Control Crisis to Control Revolution

7

Revolution in Control of Mass Production and Distribution

Are we not ourselves creating our successors . . . daily giving them greater skill and supplying more and more of that self-regulating, self-acting power which will be better than any intellect?

—Samuel Butler, *Erewhon* (1872)

IN ONLY recent years have the industrial economies of the United States and perhaps a dozen other advanced industrial nations appeared to give way to information societies. If this great societal transformation owes its origin to the Industrial Revolution and resulting crisis of control, as argued in the last chapter, why has the resolution of the crisis—the Control Revolution in information-processing and communication technology—continued unabated to this day, almost a century later?

The smooth transition from control crisis to Control Revolution in the 1880s and 1890s can be attributed to three primary dynamics, each of which has sustained the steady development of information societies through the twentieth century. First, control technologies have co-evolved with energy utilization and processing speeds in a positive spiral, advances in any one factor causing or at least enabling improvements in the others. The mighty dynamos of the Trocadero Exposition of 1900 that rounded out the nineteenth-century education of Henry Adams (1918), for example, led directly—as means to control the utilization of their power—to the development of analog computers ("network analyzers") by the 1920s (Hughes 1983, chap. 13). Further development of computers, themselves dependent on electrical power, made possible the control of systems of increasing speed and complexity—for example, air traffic control of jet transportation through major airports.

Second, increased control brought increased reliability and hence *predictability* of processes and flows, which in turn meant increasing economic returns on the application of information-processing technology. When ships depended on wind power, Atlantic crossings took from three weeks to three months; hence little could be gained from attempts to coordinate transatlantic shipping with production, distribution, or consumption flows. Steam power not only reduced the trip to ten days, the improvement most often noticed by scholars, but also made possible the prediction of arrivals to the day and even hour and hence increased profits through planning, scheduling, and coordination of material flows.

Increasing reliability and predictability of flows, in turn, sustained a continuing succession of planning technologies: centralized economic planning by national government (Soviet Union after 1920), state fiscal control (following Lord Keynes), national income accounting (after 1933), econometric forecasting (mid-1930s), input-output analysis (after 1936), linear programming and statistical decision theory (late 1930s), operations research and systems analysis (World War II), economic simulations via computer (late 1940s), cost-benefit analysis and planned programming and budgeting (1950s). Increasing reliance of both the modern state and the modern corporation on forecasting and planning technologies has been a central theme of recent books by Galbraith (1967), Bell (1973), and Chandler (1977).

Third, information processing and flows, increasing in response to the crisis in material control, needed themselves to be controlled, so that new control crises have appeared and been resolved throughout the twentieth century at levels of control increasingly removed from the processing of matter and energy. The shop-order system of accounts based on routing slips developed in the mid-1870s to control material flows through factories, but by the 1890s a growing hierarchy of timekeepers and specialized clerks had become necessary to control not these throughputs themselves but their controlling flows of information. By the 1960s the hierarchies began themselves to fall under computer control requiring still newer information workers.

This progressive layering of control levels, with resulting increases in the total amount of information processing and communication, has been described by James Grier Miller (1978, pp. 1–4) as "shred-out," a phenomenon he finds throughout the evolution of living systems. Such layering of control may also account for what Bell noted as the increasing importance of the quaternary and quinary sectors of advanced industrial economies: although the quaternary sector (finance, insurance, real estate) develops to control the extraction (primary sector), processing (secondary), and distribution (tertiary) of matter and energy, the quinary sector (law and government) develops to control the quaternary—that is, to control *control* itself at a still higher level.

In summary, three primary dynamics—the coevolution of energy utilization, processing speed, and control, the gains from control technologies that accrue through increasing reliability and predictability, and the increasing control required of control technologies themselves—account for the Control Revolution that has continued unabated from the 1880s to the present. The third part of this book, Chapters 7–9, traces the transition from crisis to sustained revolution in the control of material and energy flows in production, transportation, distribution, and consumption. This analysis for the transitional period of the 1890s through the 1930s completes the corresponding sections on the crises of control in Chapter 6. Contrary to prevailing views, which locate the origins of the Information Society in World War II (Wiener 1948, 1950) or in the commercial development of television (McLuhan 1964) or computers (Berkeley 1962; Martin and Norman 1970; Tomeski 1970; Hawkes 1971) during the 1950s, or computer-based telecommunications in the 1960s and early 1970s (Brzezinski 1970; Oettinger 1971; Hiltz and Turoff 1978; Martin 1978, 1981; Nora and Minc 1978), or microprocessing technology in the late 1970s (Evans 1979; Forester 1980; Laurie 1981), we shall see from this analysis that the basic societal transformation from Industrial to Information Society had been essentially completed by the late 1930s.

Control Revolution in Mass Production

Chapter 6 traced the crisis of control of production through various industries, from rail mills adopting the Bessemer process (late 1860s) to the producers of basic metals who struggled to maintain competitively fast throughputs within their plants (1870s) to the metalworking industries that scrambled to keep pace with the metal producers (1880s). These crises eased before a succession of innovations in preprocessing designed to facilitate flows: interchangeable parts, standardization of sizes and processes, integration of outputs and inputs (continuous-processing technology), and factories explicitly designed for throughput processing. Only in the late 1870s and 1880s did advances in actual information processing—as opposed to information reduction or preprocessing—come to industrial production. These innovations gave the late nineteenth-century factory the components basic to all information processors: internal communication and process control, as in the shop-order systems based on routing slips (mid-1870s); hierarchical differentiation and specialization for control, as in the rate-fixing departments (early 1880s); programmed control, as in the cost control of factories (1885); and data collection and storage, as with the new automatic recording devices (late 1880s).

By the early 1890s a sustained revolution in control technology had largely ended the crisis of control in production. Frederick Winslow Taylor, who had given Midvale Steel an effective shop-order control system by 1880 and a separate rate-fixing department in 1884, in the following decade introduced timekeepers and specialized clerks—the first "staff" employees at Midvale and many other metalworking companies—to ensure that shop-order slips would be filled out properly (inside contractors, as we have seen, had little incentive to do this). In 1895 Taylor delivered his first paper—before the American Society of Mechanical Engineers—on what by 1910 he would call "scientific management." By the turn of the century Taylor had begun his famous "time studies" that would provide the quantitative empirical basis for a more rationalized control of industrial production.

In essence, scientific management aimed to preprocess the activities of individual workers *qua* processors, much as earlier efforts at preprocessing in industrial production—interchangeable parts, standardization of sizes, integration of flows—had focused on the entire factory as a continuous processor. Scientific management sought to preprocess out of industrial operations the personal idiosyncrasies that distinguished workers as individuals. "In the past, the man has been first," Taylor declared in his *Principles of Scientific Management* (1911); "in the future the system must be first."

Taylor's carefully refined methods for subordinating the human processor to the industrial process can be found in his six-step prescription for a proper time study: (1) "Find, say 10 to 15 different men . . . especially skillful in doing the particular work"; (2) "study the exact series of elementary operations or motions which each of these men uses"; (3) "study with a stop watch the time required to make each of these elementary movements"; (4) "eliminate all false movements, slow movements, and useless movements"; (5) "collect into one series the quickest and best movements"; and (6) substitute "this new method [for the] inferior series which were formerly in use."

Steps 4 and 5, in effect the preprocessing of individual and idiosyncratic movements into the most rationalized or "efficient" method of processing, became the first steps toward elimination of human workers altogether. Here Taylor anticipated the idea of *automation*, a term introduced in 1936 by General Motors executive Delmar S. Harder for what he called "the automatic handling of parts between progressive production processes." By 1948 the term had found its way into popular print as a synonym for "automatic control" in general (Burchfield 1972, p. 159). More recently it was defined by a computer scientist (Dertouzos 1979, p. 38) as "the process of replacing human tasks by machine functions."

Taylor's Step 6 meant, in practice, the reprogramming of even the most basic human movements to conform to system-level rationality. After three-and-a-half years of work to improve the efficiency of shoveling at Bethlehem Steel, for example, Taylor had cut the cost of handling materials in half, reducing the number of shovelers from 600 to 140. Where shovel loads had once ranged from 3.5 pounds (for rice coal) to 38 pounds (for iron ore), Taylor brought all loads to what his studies had found to be an optimal 21.5 pounds using fifteen standard types of shovel (huge scoops for rice coal, small flat shovels for iron ore) (Fig. 7.1). Workers who might well have thought that they shoveled in their own "natural" styles found themselves lectured by Taylor's team of experts on "the science of shoveling" (this became a term of derision when adopted by his opponents). Although the shovelers who remained earned 60 percent more in wages, scientific management had detrimental effects on them. According to historian Daniel Boorstin,

> those rule-of-thumb ways of doing things which were anathema to Taylor had at least given a man on the job the feeling that he was doing what he should. To abolish the rule of thumb in factory work would excise a part of every worker's emotional investment and personal satisfaction. Could a worker now fail to feel that he was doing somebody else's job, or a job dictated by the machine? Rule of thumb was personal rule. Scientific management, which made the worker into a labor unit and judged his effectiveness by his ability to keep the technology flowing, had made the worker himself into an interchangeable part. (1973, p. 369)

From today's perspective of quality control circles and distributed control, Taylor's time study approach, despite its contribution to the rationalization of production, might seem to represent the worst of both extremes: centralized and yet highly fragmented control. Indeed, Taylor himself argued that, because no man could efficiently sustain

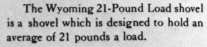

Figure 7.1. In a three-and-a-half year study of shoveling at Bethlehem Steel, Frederick Winslow Taylor found that shovel loads ranged from 3.5 to 38 pounds and that 21.5 pounds was the optimal load. As a result Taylor introduced fifteen different types of shovel, including large scoops for relatively light rice coal and small shovels for much denser iron ore. The Wyoming Shovel Works of Wyoming, Pennsylvania, brought out a comprehensive new line of products in accordance with Taylor's "science of shoveling" (Engineering and Mining Journal, courtesy of McGraw-Hill).

the versatility required of a shop foreman, workers ought to report to no fewer than eight bosses: route, instruction card, and cost-and-time clerks, gang, speed, and repair bosses, an inspector, and a shop disciplinarian. Despite such fragmented and specialized control at the lowest levels, however, Taylor had all shop bosses report to a planning department, a centralized controller responsible for the entire range of information-collection, processing, programming, and decision functions: review of orders received; establishment of the daily work plan, the schedule of flows, and the standards of output; monitoring of costs and expenses; analysis of jobs; hiring and laying off of workers; and the maintenance of a planning room, an information bureau, a rush-order department, and the factory's messenger and post office delivery systems (Taylor 1911b, pp. 95–120).

Because Taylor seemed to emphasize specialization at the expense of integration, no factory owner—even the many who hired him and his disciples as consultants—adopted his system without modification. A study of twenty-nine factories (Nelson 1975) that explicitly adopted Taylor's principles found that only six kept his functional foremen, and all of these had fewer than eight. Instead, many factories adopted the line-and-staff structure pioneered by the Pennsylvania Railroad in the late 1850s, first adapted to production by Yale and Towne Lock Company in 1905, and promoted by Harrington Emerson—a former railroad manager himself—in a series of articles in *Engineering News* (1908–1909) and in two major books (Emerson 1911, 1913). Under the new system shop foremen remained generalists on a line of authority running from the company president through a factory manager or superintendent who had responsibility for overall planning, coordination, and control. Even factories adopting the line-and-staff structure often realized partial benefits of Taylor's differentiation, however, in the manager's staff, a group of experts in the various specialties that Taylor distinguished in his planning department or among his functional foremen.

In no industry did the new ideas for control have greater initial impact than in the production of automobiles, to this day the most complex product of metalworking in great volume. Automobiles might have remained a luxury of the rich (as Princeton University President Woodrow Wilson among many other turn-of-the-century observers had predicted) had not the essential product, the internal-combustion vehicle developed by German engineers Karl Benz and Gottlieb Daimler, appeared just about the time (1885) the control crisis of energy-intensive mass production had finally been resolved. U.S. factory sales of passenger cars totaled four thousand in 1900, 3.6 million in 1923 (U.S. Bureau of the Census 1975, p. 716), a phenomenal growth that wrought major social and cultural changes as well as economic ones (Rae 1965). It resulted from, above all else, the moving assembly line, early embodiment of the Control Revolution and best-known symbol of modern mass production.

Preprocessing by Moving Assembly

Introduced in 1913 in Ford's Highland Park plant to produce the Model T (15 million by 1927, when 11.3 million could still be found on the road), the moving assembly line inspired Aldous Huxley to reckon time as "B.F." and "A.F." (before and after Ford) in his twenty-fifth-century *Brave New World* (1932). With the moving line system, according to Alfred Chandler, "the process of production in the metal mass production industries had become almost as continuous as those in petroleum and other refining industries" (1977, p. 280), an astounding feat when the complex operation of building even the simple Model T is compared to the relative ease with which liquids and gases can be processed through pipes.

Essential to the assembly line is the standardization and interchangeability of parts, ideas "in the air" almost from the beginnings of industrialization in the late eighteenth century (Giedion 1948, pp. 47–50) and the basis of the American System of manufacturing by the 1840s. Henry Leland, an American machinist who created both the Cadillac and Lincoln, elaborated the system of interchangeable parts for the country's automotive industry (Leland and Millbrook 1966). In a public demonstration in 1908, workers disassembled three Cadillacs, mixed the parts, then reassembled the vehicles and drove them away— a level of standardization in mass production that made moving assembly possible. By the following year Ford had decided to carry standardization to what may be its modern extreme: he would build *only* Model T's on his line and would use the same chassis for his runabouts, touring, town, and delivery cars. "Any customer can have a car painted any color he wants," Ford added, "so long as it's black" (Nevins and Hill 1954).

Thus did the imperatives of standardization, interchangeability, and the moving assembly line, reinforced by the logic of scientific management, bring increased control of production by preprocessing away much of the information contained in the final products themselves. Henceforth, industrial control would have its effect "not only on *how* anything was produced but also on *what*," as Boorstin (1973, p. 369) puts it. "Items to be manufactured were designed and selected for production according to how quickly and economically they could be produced. In place of the naive consumer, the 'scientific' system now made its own demands."

Because of the number and complexity of parts on even the simplest models, automobile production via moving assembly line also aggravated scientific management's inherent tendency toward increased specialization of function: the Model T required 7,882 distinct tasks of workers, of which—as Ford noted in his autobiographical *My Life and Work* (1923)—only 949 (barely 12 percent) required "strong, able-bodied, and practically physically perfect men"; of the rest, he disclosed with obvious pride, "we found that 670 could be filled by legless men, 2,637 by one-legged men, two by armless men, 715 by one-armed men and ten by blind men" (1923, pp. 108–109).

Ironically, Ford's idea for the assembly line owes its origin (in Siegfried Giedion's phrase) to the "disassembly line," the overhead trolley, introduced by Gustavus Swift in his Chicago meat-packing houses in the early 1880s, which Giedion (1948, p. 89) traces back to Cincinnati in the late 1860s (Fig. 7.2). Ford first realized the larger control implications of moving disassembly from Upton Sinclair's best-selling exposé *The Jungle* (1906, p. 42), which describes the system in brutally graphic detail. Much as Swift, Philip Armour, and other meatpackers sought to integrate the flow of meat from range to consumer using the railroads, disassembly lines, refrigerated cars, and warehouses, so too did Henry Ford dream of a nonstop flow—never quite achieved—from raw materials to finished product. "If Ford had succeeded perfectly," according to Boorstin (1973, p. 549), "a piece of iron would never have stopped moving, from the moment it was mined until it appeared in the dealer's showroom as part of a completed car," surely the ultimate dream in control of production.

As a result of Ford's dream, control of production flows extended well beyond his moving assembly lines by the 1920s. Much credit belongs to Albert Kahn, an architect who specialized in concrete construction, who served Henry Ford much as Alexander Holley—designer of the Edgar Thomson Steel Works—had served Andrew Carnegie a quarter-century earlier. In 1903 Kahn designed a new Packard Company works so that, like the E.T., throughputs could be moved with

Figure 7.2. Top: *the overhead trolley and disassembly line, introduced by Gustavus Swift in his Chicago meat-packing houses in the early 1880s* (Harper's Weekly, *September 6, 1873*). Bottom: *the disassembly line inspired Henry Ford's moving assembly line, introduced at his Highland Park automobile factory in 1913 (courtesy of the Motor Vehicles Manufacturers' Association).*

a minimum of carrying and hauling. Six years later Kahn helped Ford overcome burdensome rail-freight charges by designing a branch assembly (or "reassembly") plant beside the rail yards in Kansas City, Missouri, to which the manufacturer sent knocked-down Model T's—introduced the previous year—on regular freight cars and with great economy of space.

In the same year Kahn also began construction of the Highland Park plant to accommodate the world's first "line production system" based on a well-planned sequence of processes and machinery to be used for automobile manufacturing. After the moving assembly line had been perfected by 1917, Ford himself supervised construction of the much more ambitious River Rouge plant in which each building had been carefully designed to accommodate streams of production (Arnold and Faurote 1915; Nevins and Hill 1954). In both grandeur and pretension, River Rouge might be seen as the ultimate building *qua* processor—the rational culmination of a half-century's effort to preprocess away the informational burdens of complex production at the draftsman's table.

Automatic Control

Even as Ford completed River Rouge, physical control of industrial processes had begun to shift away from architecture to greater use of automatic control devices (Fig. 7.3). When continuous processing began to replace batch stills in the oil industry about 1910, for example, workers controlled most of the processes manually—aided by off-on controllers and pneumatic valve actuators—based on readings from local thermometers and pressure gauges. By the late 1920s pneumatic controllers had been developed with proportional, integral, and reset modes that could be tuned on the plant floor according to recorders mounted nearby. Such instrumentation of factory processes flourished in the early 1930s after pneumatic devices gained a derivative mode to become the so-called "PID" (proportional-integral-derivative) three-term controller. This innovation, developed in factories without benefit of supporting theory, brought automatic control to many industrial processes previously difficult to regulate at all, even manually. Development of pneumatic transmitters in the mid-1930s brought the still-familiar centralized control rooms with their heavily instrumented control panels by the end of the decade (Evans 1977).

Figure 7.3. Origin of the centralized control room with large control panel: the electrical load-dispatching center with systems operator's board, introduced by Consolidated Edison of New York in 1898. The board's diagram, photographed in 1902, shows lines among generators and stations; tags hung on pegs indicate the status of system components. (Courtesy of Consolidated Edison Company.)

Table 7.1 places the development of automatic control of industrial processes in its intellectual and historical context. The idea of feedback control dates at least from the waterclock of the third century B.C., as we saw in Chapter 5. No fewer than nineteen innovations in automatic control, listed in Table 5.1, appeared in the early industrial period, 1740–1830, culminating in Andrew Ure's generalization of the principle of negative feedback in the thermostat. Practical use of electricity in the 1830s made possible industrial applications of electromagnets, inspiring thousands of new feedback controllers, many to feed and adjust arc light carbons or to regulate the voltage and current of generators.

In 1852 French physicist Leon Foucault established a new self-regulating device and its modern name, *gyroscope*, meaning literally "to view the turning." Although Foucault predicted its use as a compass, its first notable application proved to be his own pendulum, which was used to demonstrate the earth's rotation (Hughes 1971, p. 131). A decade later, Foucault patented another controlling device, a governor incorporating a centrifugal pendulum but relying on an air brake in-

Table 7.1.   Selected innovations in automatic control, 1830–1939

Year	Innovation
1830	Andrew Ure generalizes the idea of the thermostat, an application of negative feedback to control
1852	Physicist Leon Foucault introduces the modern gyroscope; its first notable use is in a pendulum
1862	Foucault patents centrifugal weight-driven governor with pendulum and air brake, the first of his several new governors
1866	Charles Siemens invents a liquid governor with speed-sensor like the impeller of a centrifugal pump, alternative to pendulum
1868	William Thomson (Lord Kelvin) describes a new centrifugal weight-driven friction governor
	United States
1883	Elmer Sperry files for patent on dynamo-electric machine regulator, first of 11 patents in automatic control in next five years
1902	Air-conditioned factory (Brooklyn printing company) established with automatic temperature and humidity control
1905	New factory built that changes air five times per hour, automatically filtering and washing it and controlling humidity
1907	Sperry begins investigating gyroscopic control
1910	Automatic bread plant opens in Chicago; dough and loaves are untouched by humans except when placed on wrapping machine
1914	Electric substation with a rotary converter goes into service completely unattended in Detroit
1926	Photomaton Studios, based on fully automatic film developing machinery, launched on Broadway, New York City
Late 1920s	Pneumatic proportional controller is developed for industrial processes; integral reset mode can be tuned in the plant
1930	Colorscope, a photoelectric cell which reacts to colors more precisely than the human eye, is publicly demonstrated; it gives off electric currents capable of controlling machinery Windowless factory erected, has ultraviolet lighting, soundproofed cork walls, temperature and humidity control
Early 1930s	Adjustable controllers gain the derivative mode to become the three-term or PID (proportional-integral-derivative) controller, a major turning point in process instrumentation

Table 7.1 (*cont.*)

Year	Innovation
1931	Photoelectric cell ("magic eye") installed commercially, provides automatic control of swinging doors in restaurant
1933	H. L. Hazen develops a light-sensitive servomechanism that follows plotted line to alter signal controlling a machine Wolverine-Empire Refining Company completes a new oil-distilling plant run automatically by process control equipment
Mid-1930s	Pneumatic transmitters are developed for industrial process control, bringing within a decade centralized control rooms with large control panels
1936	Term *automation* introduced in automotive industry for the replacement of human tasks by machine functions

stead of mechanical friction for control. An alternative to the centrifugal pendulum, a "liquid governor" whose control depended on the depth of immersion of an impeller attached to its drive shaft, was described in an 1866 paper by British industrialist Charles William Siemens. Two years later Lord Kelvin introduced an extremely simple centrifugal governor that used the friction of its two weights pressing outward against a stationary ring to maintain control. All of these innovations, as Mayr (1976) has shown, influenced James Clerk Maxwell to write his famous paper "On Governors" (1868), generally considered the first theoretical analysis of control (Evans 1977).

Despite considerable initial publicity, Maxwell's paper had limited influence before Norbert Wiener resurrected it in his widely read *Cybernetics* (1948, pp. 11–12); Mayr (1976, p. 187) finds only a few references to "On Governors" up to World War I. Certainly American contributions to automatic control came independently and brought increasing numbers of patents by the 1880s, when Elmer Sperry began work on a regulator for dynamo-electric machines that exploited automatic control. Of Sperry's nineteen patent applications between 1883 and 1887, eleven included some form of automatic control, more than half involving closed-loop feedback (Hughes 1971, pp. 45–46). Although Sperry's inventions were hardly unique (the U.S. Patent Office granted protection to twenty-two generator regulators in 1884 alone), his early career does provide further evidence that information engineering, cybernetics, and even computer science trace their origins to the 1880s

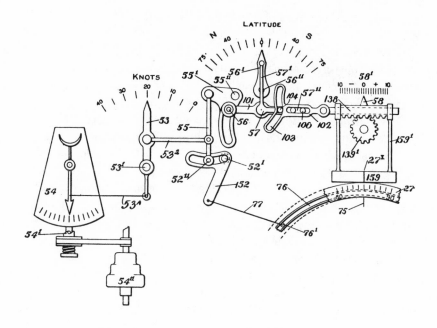

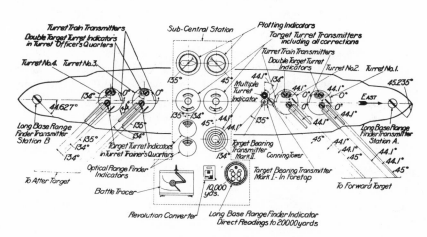

Figure 7.4. Elmer Sperry's early analog computers. Top: *his 1911 gyrocompass included a complex computer to make latitude, course, and speed corrections according to a trigonometric function, which it embodied mechanically (Diagram, U.S. Patent No. 1,255,480).* Bottom: *By 1916 Sperry had developed a system of gunfire control for battleships that included a "battle-tracer" that combined inputs from four electrical motors to plot both the ship's and its target's courses and speeds (Sperry Gyroscope Company,* The Sperry Fire Control System, *Bulletin 301 [1916]).*

and 1890s—the beginning of the Control Revolution—and not to World War II or to subsequent developments.

By 1896 Sperry had begun thinking about Foucault's gyroscope as a means of control (Eames and Eames 1973, p. 61); his notebook records his first professional interest early in 1907 and cites an 1874 article in a Smithsonian journal (Hughes 1971, p. 105). Soon Sperry had developed two new control devices, the gyrocompass and the gyrostabilizer, both successfully tested on U.S. Navy ships by 1911 (Fig. 7.4). Sperry's gyrocompass included a complex analog computer to make latitude, course, and speed corrections; by 1916 his target-bearing and turret-control system included the "battle tracer," another analog computer that combined inputs from four electric motors to plot both the ship's and its target's course and speed.

Nor did Sperry work in isolation from the development of modern computers. The young man whom he had helped to establish in the gyroscope field, Hannibal Ford, went on to develop range-elevation instruments for fire control by ships. Ford Instrument Company merged with Sperry Gyroscope in 1933, five years after the latter's sale by its founder (Hughes 1971, pp. 232–233). In 1955 the Sperry Corporation, as a natural complement to its specialties in instrumentation and control, merged with Remington Rand, owner since 1950 of the Univac line, the first commercial successor to the original ENIAC machine (1946) developed by its inventors, John Mauchly and J. Presper Eckert (Sperry Corporation kept the machine's memory alive in the name of its computer division, Sperry Univac, until 1982). In 1984 Sperry ranked as the world's second largest manufacturer of mainframe computers in revenues behind IBM (which also traces its origins to an 1880s idea: Herman Hollerith's punch-card sorter); Sperry also ranked fifth in microcomputers and seventh in data processing more generally (Archbold and Verity 1985).

Historian of technology Thomas Parke Hughes, recent biographer of Elmer Sperry, argues for his subject's place among the founders of control theory:

> Sperry never described himself, nor was called by his contemporaries, a control engineer or a pioneer in cybernetics, but his stress on control devices in the eighties and his great work in gyro controls several decades later justify his being so characterized. His consistent stress upon control devices as he shifted from one field of technology to another—from electric light, to electric streetcars, and to gyro applications—suggests that concern with control was his prime characteristic as inventor . . . His first

two important inventions, the generator and arc light, featured automatic controls; their success may have caught his interest and imagination. Or he may have been hoping to increase the sale of his lighting system by lowering its cost of operation; his automatic devices performed some of the complex tasks of the station attendant. Perhaps the idea of controlling large forces with delicate devices attracted him, for often in his lifetime he referred to getting control of the "brute," meaning a ponderous machine or vehicle.   (1971, p. 45)

As Table 7.1 suggests, countless other American inventors—including many industrial engineers whose names have gone unrecorded—shared with Sperry the task of developing automatic control in the early decades of this century. Their innovations included air-conditioned and humidity-controlled factories (1902), photoelectric improvements of hand-eye control (1930), and light-sensitive servomechanisms to program control of machinery by means of plotted graphs (1933). Increasingly the Control Revolution meant that industrial processes could be run with little or no human intervention: a bread plant (1910), an electric substation (1914), a photographic film-developing studio (1926), the doors of commercial establishments (1931), an entire oil refinery (1933). By 1936, as we have seen, the automotive industry had adopted the term *automation* for the replacement of human functions by machine control. Within a decade it would be in popular print as a synonym for automatic control.

Quality Control via Statistics

In taking control of material processing by the 1920s, American industries confronted a new, higher-order control problem: that the *quality* of goods might exceed the economically optimal level and thereby waste capital. As with the struggle to push throughputs past fixed capital as quickly as possible, the desire to control the quality of outputs arose not from technology itself but from the benefits of maximizing returns on investment. This goal of capitalist production—to minimize capital inputs per unit value of output—could be achieved not only by increasing throughput speed but also by decreasing output quality, ordinarily a direct function of nonfixed inputs per unit. For this reason, at least, the idea of "quality control" could hardly be called new to the 1920s, the decade in which developing technologies of the Control Revolution first made possible mass production of parts no stronger or more precise than they needed to be.

This new sense of quality control first appeared in 1922 in a pioneering treatise, *The Control of Quality in Manufacturing* by George S. Radford, an American factory engineer. Radford argued that mass production methods (what he called "Repetition Manufacturing") could never make parts truly identical but merely interchangeable within a certain allowable variation or "tolerance" (the word had acquired this technical meaning by 1909). "When we generalize that it is best to make things uniform," Radford wrote, "what we really mean is likeness, uniformity, or standardization of quality *within limits;* this, in a word, is why quality requires control" (Radford 1922).

Because factory engineers had to keep quality within tolerance limits to effect mass production in any case, Radford further reasoned, they might as well use this control of quality to minimize costs by striving to attain only minimal standards, possibly by minimizing the cost of quality control itself. Parts unlikely ever to be replaced need not even be interchangeable, Radford argued, but should approximate this ideal only as an economy of mass production, not as an end in itself. "We make things alike because it is cheaper rather than for the sake of having them alike," he concluded, and it would contradict this same economic logic to squander resources on precision for its own sake or to make a fetish of uniformity—especially when the same quality control techniques could be used to exploit *maximum* allowable variation or to approximate *minimum* standards.

To establish this last point, Radford proposed a system to sample statistically and to inspect "the flow of work in process." For his analysis, he drew on work in mathematical and applied statistics begun in England by Francis Galton, Karl Pearson, and their students around the turn of the century. To enforce his quality control, Radford proposed a new information worker: the quality control inspector, equipped with precision calipers, limit gauges, and other measuring devices, working continually to keep throughputs within tolerance. Through his quality control inspector, Radford hoped to transform factory statistics from a historical tool, used to keep inspection records and to evaluate finished products, into an active tool for control of ongoing industrial production.

Credit for effecting this transformation belongs to another American factory engineer, Walter A. Shewhart, who spent decades elaborating and testing what by World War II would be widely known as statistical quality control or SQC (Littauer 1950). Working at Western Electric in the early 1920s and later at Bell Laboratories, Shewhart established specific, quantitative, mathematically derived tolerances for each part of each manufactured product, then used mathematical and statistical techniques to derive quality control goals that would minimize production costs. Shewhart also introduced the concept of maximum acceptable defects, to be fixed in advance for each part as well as for larger components and for final products.

Eager to promote his ideas beyond his own company, Shewhart in 1930 taught the world's first college course on statistical quality control at Stevens Institute of Technology. The following year he published his definitive treatment, *Economic Control of Quality of Manufactured Product*. His 1938 address, "The Future of Statistics in Mass Production," published in the *Annals of Mathematical Statistics* (Shewhart 1939), drew citations for decades as engineers applied his ideas to statistical controls ranging from Hollerith card-punching verification to prediction of epidemics to consumer protection policy. Some ten thousand specialists in statistical quality control would be formally trained during World War II alone; they helped to form the American Society for Quality Control in 1946. Thirty years later, when U.S. manufacturing lost market share to other countries in many industries and widespread analysis of Japanese techniques revealed comparative inadequacies of American management, the response would be "back to basics," in the words of Harvard business historian Wickham Skinner (1984, pp. 47–48), "with the old tools of statistical quality control, process and control charts and quality assurance dusted off and newly reinvigorated."

Statistical Control via Market Feedback

In the same years that Radford and Shewhart pioneered statistical quality control, General Motors—then the fifth largest U.S. industrial enterprise—adopted and refined a new organizational control structure also based largely on statistical analysis. Here change came in response to the sharp recession following World War I, a sudden and continuing drop in demand from the summer of 1920 until the spring of 1922. The first downturn in the economy since the depression of the 1890s, the postwar recession left most American companies with inventories vastly overstocked; General Motors' write-downs in 1921 and 1922 exceeded $83 million. Even the meatpackers, who constantly coordinated supply with demand using telephone and telegraph, failed to adjust inventory fast enough—with disastrous results. The Armours, for example, lost control of their family firm to a managerial enterprise (Gras and Larson 1939, pp. 630–640). Suddenly and dramatically, the recession placed the goal of controlling production flows relative to demand at the top of most management agenda.

Within three months of the September 1920 crash in automobile sales, which left General Motors close to bankruptcy, its Board of Directors adopted a new reorganizational plan. Worked out by Alfred Sloan, then manager for parts and accessories, the plan remains to this day the corporation's basic structure. In sharp contrast to the centralized, functionally departmentalized organization pioneered by General Electric in the 1890s and perfected by DuPont in the early 1900s, Sloan chose a multidivisional decentralized structure—then also evolving at DuPont—with a general office and autonomous but integrated operating units. Because of the success of this structure, General Motors became, more than any other organization, the model for other large industrial enterprises in the 1920s and 1930s. According to Chandler (1977, p. 463), the structure remains today the basic organizational type for companies "manufacturing several lines for a number of product and regional markets."

Central to Sloan's plan was the idea that production flows and resource allocation be controlled by market feedback. His general office required each division to submit for each coming month and the following three a forecast of all inputs required for the anticipated output. These forecasts soon came to be tied to annual forecasts of demand provided by the new financial staff. Not until the general office had approved the monthly estimates could the separate divisions purchase their materials and equipment. In short, feedback from the market— current and future—served to integrate formerly independent divisions into a coordinated enterprise that optimized its response to changing consumer demand—thereby solving the most pressing industrial problem of the post–World War I economy.

Despite the early success of Sloan's structure, unanticipated demand for automobiles in the first half of 1923 resulted in what the General Motors Annual Report described as "the loss of sales and some dissatisfaction on the part of the Corporation's dealer organization on account of failure to make adequate delivery" (Chandler 1962, p. 149). As a consequence, the company kept production at capacity through the second half of 1923—to achieve its greatest annual sales ever— and into 1924. As the second quarter passed, however, Sloan began to suspect—contrary to the reports of his own organizational feedback system—that demand had fallen below supply. He decided to investigate by taking a trip west. Finding crowded dealer lots in St. Louis and Kansas City, Sloan immediately slashed production and began to institute a still more refined structure for controlling factory flows based on statistical feedback from dealers themselves (Dale 1956, p. 45).

Albert Bradley, a trained economist in the company's Statistical Department, later described Sloan's new plan to the American Management Association:

> The first and controlling principle in the establishment of General Motors production schedules is that they shall be based absolutely upon the ability of its distributors and dealers to sell cars to the public. Each car division now receives from its dealers every ten days the actual number of cars delivered to consumers, the number of new orders taken, the total orders on hand, and the number of new and used cars on hand. Each ten-day period the actual results are compared with the month's forecast, and each month, as these figures are received, the entire situation is carefully analyzed to see whether the original estimate was too high or too low . . . In other words, instead of attempting to lay down a hard and fast production program a year ahead and stick to it regardless of the retail demand, the Corporation now follows the policy of keeping production at all times under control and in correct alignment with the indicated annual retail demand.   (Chandler 1962, pp. 150–151)

Bradley's statement that "the actual results are compared with the month's forecast" recalls the original meaning of *control* as comparison "against the rolls." His use of the word *program* approximates the more recent computer science definition, and his desire to transcend "a hard and fast production program" can be seen as an appreciation of iteration with feedback to control a dynamic process with external contingencies toward the goal of "keeping production at all times under control," that is, "in correct alignment" with the prior programmed but modifiable goal, "the indicated annual retail demand." This goal, in turn, he maintained subject to four inputs about the most recent and current states of the processing system: deliveries, new orders, orders on hand, and inventory. In short, Sloan's new system extended control of production from the factory through his distributors and dealers to the consumer himself—toward the ideal that literally no automobile would be built unless a customer had already agreed to buy it.

Sloan obtained additional market feedback by commissioning regular reports from R. L. Polk and Company on new motor vehicle registrations. These provided the market share of each General Motors division—from Cadillac to Chevrolet to trucks—and how that share might be changing relative to those of competitors in the same market. Continual comparison of registration data, dealer reports, and market projections further served to control inventory, to calibrate internal flows, and to refine forecasting techniques. Because of General Motors' relatively heavy investment in fixed capital like factory equipment, its scheduling and related decisions coordinating product flow at higher organizational levels continued to be based on anticipated long-term trends rather than on current conditions, even though actual flows were adjusted every ten days according to consumer demand. Long-range market forecasting, increasingly used to control allocation of resources among divisions as well as daily use, had by 1925 proved essential to strategic planning at General Motors and had begun to be adopted by other large business enterprises (Chandler 1977, pp. 456–463).

Sloan's new industrial control structure served General Motors well. In the three years following his 1924 reforms, the company's share of the motor vehicle market rose from 18.8 to 43.3 percent, a leading position which it has never relinquished. Henry Ford, by contrast, had concentrated on controlling output without feedback from demand. Relying on a single model of automobile (first T, then A) built largely in one plant (River Rouge), he saw his market share drop rapidly, especially during the Great Depression, from 55.5 percent in 1921 to 18.9 (behind GM's 47.5 and Chrysler's 23.7) in 1940; Ford's profits also faltered after 1926. Alfred Chandler summarizes the comparative economic advantages of Sloan's new industrial control structure:

> By keeping its varied activities tied together closely and related directly to day-to-day changes in market demand, the corporation was able to make a real reduction in costs. Manufacturing, plant, and sales facilities could be utilized at a fairly even and regular capacity. Both the manufacturing and sales force were assured of steadier employment. Outside suppliers, too, could keep their plants operating more evenly. By reducing the amount of inventory needed, this careful coordination cut down the cost of working capital, lowered expenses of storing, carrying charges, and so forth. It is hard to see how expenses could not have been reduced by such rational control.   (1962, p. 153)

Despite the savings and increased profits, however, Sloan's success at General Motors belonged less to economics than to control systems engineering. An MIT graduate in electrical engineering, Sloan appreciated the value of communication and feedback for control. By exploiting the widest possible range of demand information, he managed to coordinate and sustain smooth flows of materials and products into and through his factories and onward to distributors, retailers, and consumers. At the same time, he managed to maintain reasonably steady use of plants, equipment, and personnel in an industry plagued by wild market fluctuations.

Perhaps the greatest triumph of Sloan's system of industrial control came during World War II, when his techniques—refined by protégés on the War Production Board—provided the basis for the Controlled Materials Plan or CMP (Novick et al. 1949; Smith 1959; Chandler 1967). Board economist Charles Hitch described CMP as "the basic mechanism for integrating strategy, production, and the flow of materials," declaring that "it attempts to organize all war plants in the United States (more than 100,000) in much the same way that a large, industrial corporation organizes its operating plants." Compared to the Board's previous organizational plans, auto executive Ernest Kanzler concluded enthusiastically in December 1942, the CMP was "the closest approach by far to automotive shop materials scheduling" (Cuff 1984, pp. 39–41).

CMP used vertical allocations for administrative coordination of production programming—effected by the Army, Navy, and other procurement agencies—with actual material flows. Control itself concentrated on three critical materials—steel, aluminum, and copper—which not only provided generalized media of exchange for trade-offs among production programs but also greatly facilitated accounting control. In the assessment of historian Robert Cuff (1984, p. 37), "CMP was, indeed still is, the most ambitious attempt ever mounted in the United States to allocate, coordinate and monitor the flow of scarce materials for the entire industrial economy." What more convincing evidence could we have that the late nineteenth-century crisis in the control of industrial production had become—a half-century later—truly a revolution in industrial control?

Industrial Control via a Science of Human Relations

The 1930s brought a final ironic twist to the effort initiated by Frederick Winslow Taylor nearly a half century earlier to preprocess the personal idiosyncrasies of workers out of industrial operations. The result, a new approach to industrial control via the worker, would come to be called a science of human relations. Its repercussions, including a host of new control techniques usually included under the general rubrics of industrial relations (Moore 1951) and personnel management (the latter term, tellingly, from Taylor's original lexicon), continue to be felt in American industry.

Interest in the human relations approach to industrial control stemmed from a series of experiments involving some twenty thousand workers conducted between 1924 and 1932 by Elton Mayo and his Harvard University research team at Western Electric's Hawthorne plant near Chicago. Originally intending to test the relationship between working conditions and employee productivity, Mayo manipulated factors like temperature, humidity, and lighting to assess their relationship to worker efficiency and fatigue. To his initial amazement, he discovered that, regardless of how his experiments altered the environment in a switchboard assembly room, productivity usually rose and continued to rise. Mayo's eventual conclusion, what would come to be known as the *Hawthorne effect*, was that the productivity of a work group increases as a direct result of *any* concern shown by outsiders in the group's activities, including concern communicated by researchers changing experimental conditions.

Industrial researchers have since concluded that, contrary to Mayo's initial results, negative Hawthorne effects occur more often than positive ones. More generally, however, the Hawthorne studies established that human factors may be more important than environmental ones in determining worker productivity. Even though each Hawthorne worker's pay would be determined by the number of components he or she assembled (following Taylor), for example, most work teams established informal norms about the range of socially acceptable productivity. Workers who exceeded ("rate busters") or fell below ("chiselers") the normative range would be harassed by other team members, either through good-natured but pointed kidding or by "binging," a series of punches to the upper arm (Roethlisberger and Dickson 1939).

Mayo's discovery of the informal work group and its interactions, which might be expected to undermine most control based on Taylor's scientific management approach, began a trend away from authority toward manipulation as the means of exerting control over workers. Industrial control quickly came to be reformulated as a science of human relations, in the phrase of Mayo (1945) "a new method of human control." The book *Management and the Worker* (1939), an influential report on the Hawthorne studies, gave further impetus to the human relations movement. By 1947 professionals in the field had established the Industrial Relations Research Association. In the decade that followed, industrial relations institutes and departments appeared on many campuses: Harvard, Yale, Cornell, and the University of California at Berkeley, among others. Today most major U.S. corporations have specialized departments of industrial relations as permanent parts of their management structures. Harvard Business School historian Wickham Skinner summarizes the lasting legacy of Elton Mayo and his colleagues in management theory:

> The famous Hawthorne experiments demonstrated that working conditions may be important but social expectations and personal feelings are even more so. Worker counseling, foremen training, human relations training for managers, sensitivity training, worker participation plans, profit sharing, gain-sharing plans (such as the Scanlon Plan) were given time, money and hope. Experiments of many types were tried in the late 1950's, often featuring such heresy as nonsupervised work groups and off-plant sessions with third parties to surface feelings and promote better understanding. Whether or not this wave of new human resource management concepts experiments worked is less the question than the fact that for the first time long-smoldering serious labor problems were seen in terms other than adversarial or paternal. Corporate managers and academics had finally begun to invest in experimentation in radically new solutions. (1984, p. 32)

Consolidating Control of Mass Production

Table 7.2 summarizes major developments in the control of industrial production between the 1890s, roughly the decade of transition between control crisis and revolution and the consolidation of industrial control by the late 1930s. As the brief chronology illustrates, a sustained Control Revolution, centered on new flow technologies, organizational differentiation and specialization, and scientific management, was well under way in industrial production by the time Henry Ford introduced the moving assembly line in 1913.

That the period 1890–1920 marked something approximating the Control Revolution in industrial production has not been lost on business historians. Much the same revolutionary flavor has been captured, for example, by Wickham Skinner, who notes: "Swiftly in the 30-year period from 1890 to 1920 a new management function . . . led to de-

Table 7.2.   Selected developments in the control of production, 1890–1939

Year	Development
1890s	Timekeepers and special clerks, first "staff" employees in many factories, introduced to fill out shop orders, routing slips
1898	Frederick W. Taylor, apostle of what comes to be known as scientific management, begins his time studies
1903	Automotive works (Packard Detroit) arranged so that materials flow from one end to the other with minimum of hauling, carrying Taylor publishes "Shop Management" in journal
Mid-1900s	Railroad's line-and-staff structure adopted for factory production, promoted in engineering journals
1908	In a public demonstration, three Cadillacs are disassembled, the parts mixed but quickly reassembled and the cars driven away—standardization that makes possible the moving assembly line
1909	Ford branch assembly factory built adjacent to Kansas City rail yards, receives knocked-down automobiles on regular freight cars
1911	Taylor publishes *The Principles of Scientific Management*
1913	Ford introduces moving assembly line at its Highland Park factory, reduces time required to make a magneto from 20 to 13 minutes
Mid-1910s	Most industrial processes are still controlled manually using local temperature and pressure gauges, pneumatic valve actuators and off-on switches
1917	Ford River Rouge plant is designed to accommodate production flows, establishes new industrial architecture
Late 1910s	Line-and-staff structure becomes standard for control of mass production in factories
1922	Electric power line commercial carrier placed in operation by Utica (N.Y.) Gas and Electric; transmission lines carry both voices and power, enable distant supervisory control Engineer George S. Radford publishes *The Control of Quality in Manufacturing*, introducing term *quality control*
1924	Continuous sheet steel plant—passing sheets through a series of mills in a tandem train at high speed—begins operations Walter Shewhart, engineer at Western Electric, transforms factory statistics from historical data into production tools useful for replanning and controlling ongoing production

Year	Development
1925	Photoelectric cell, used to count passing objects, publicly demonstrated by Westinghouse
Mid-1920s	General Motors extends control of automobile production to consumer orders, feedback from dealers and demand forecasting
Late 1920s	Control of temperature, pressure, and flow in industrial process plants is still maintained by human operators watching gauges
1930	Shewhart offers the first college course on quality control
1930s	Laboratory analysis develops for quality control in industry
1931	Shewhart publishes *Economic Control of Quality of Manufactured Products*
1939	Elton Mayo's experiments at Western Electric, publicized in book, begin move toward science of human relations in worker control

velopment of the production department and a production manager . . . This was a whole new idea and it was in place by 1920. It added the functions of planning, analysis, operation, improvements, coordination, control, and personnel management . . . This was the first major drastic change in the prior smooth revolution of manufacturing management. It shattered a century-old pattern of technological innovation and investment decisions" (1984, pp. 21–22).

As we might expect, however, this Control Revolution, coming in response to the American Industrial Revolution and the resulting crisis of control, did not occur only, as Skinner would have it, between top corporate management and the factory floor. Control crises followed applications of steam power throughout the material economy, as we saw in Chapter 6, from the crises in rail transportation in the early 1840s to distribution (commission trading and wholesaling) in the 1850s, to production (first in rail mills, then other metal-making and metal-working industries) in the late 1860s, and finally to marketing (first in the continuous-processing industries) in the early 1880s. Similarly, the resolution of these various crises, accumulating in a technological base for the Control Revolution beginning most noticeably in the 1880s and 1890s, might be expected to affect not only industrial production but transportation, distribution, and marketing as well.

Control Revolution in Transportation

Chapter 6 traced the crisis of control in transportation through various phases: concern for safety (after the Great Western crash of 1841), control of freight cars on the interregional trunk lines (early 1850s), monitoring the "foreign" cars of fast-freight and express companies (1860s), and control of vast continental rail networks (1870s). These crises eased before a succession of innovations, from basic organizational technologies like scheduling, multisystem coordination, and centralization of bureaucracy in the 1840s and early 1850s, to Boorstin's "leveling of times and places" through, for example, introduction of refrigerated rail cars and uniform standard time. Control of the speed and volume of rail movements, at first largely a matter of bureaucratic coordination, increasingly came to be built into the railroad infrastructure itself via automatic signals and switches and the standardization of cars and track, all introduced or completed between the late 1860s and early 1880s.

By the adoption of the Interstate Commerce Act of 1887, which established uniform accounting procedures for railroads and imposed control by an Interstate Commerce Commission, the control crisis in the nation's rail system had largely been resolved. Within two decades, however, the rapid development of long-distance military and industrial sea and air transportation would confront communications and control systems engineers with complex new problems. Rather than precipitating anything like the nineteenth-century crisis of control in rail transportation, however, the new problems proved readily tractable, largely as a result of the half-century tradition of work that we have already reviewed on gyroscopes, governors, and related automatic control devices.

Also crucial to control of rapid and long-distance transportation was radio, a wholly new medium of telecommunications that developed almost parallel to the Control Revolution following publication of Maxwell's theory of electromagnetic radiation in 1873. Within the decade and for the remainder of the century, the problem of exploiting radio waves for telecommunications occupied leading scientists throughout the world: Heinrich Hertz and Adolphus Slaby in Germany, Oliver Lodge in England, Edouard Branley in France, Alexander Popov in Russia. Practical results came quickly amid a flurry of innovations: Hertz's apparatus for generating and detecting waves (1886–1889), Branley's coherer receiver (early 1890s), Guglielmo Marconi's wireless telegraphy (1895), John Fleming's vacuum electron tube (1904), Lee de Forest's triode amplifier tube (1906), and Edwin Armstrong's feedback receiver circuit (1913). By 1909, when Marconi received the Nobel Prize in physics (jointly with C. F. Braun) for work in wireless telegraphy, voice broadcasting had become commonplace, with experiments already underway on image transmission (*Scientific American* introduced the term *television* for such transmission in 1907). Broadcast media would prove useful for control not only of transportation but also of consumption, a subject to be taken up in the next chapter.

With the rapid development of radio telephony and automatic control technology by the late 1880s, the fundamental problems of controlling long-distance sea and air transportation had largely been resolved by the mid-1910s, in time for the even more rapid development of military applications during World War I. Thus a revolution begun in nineteenth-century transportation continued unabated into twentieth-century systems. It was grounded in a wide range of generalized information-processing, communication, and control technologies that were used as well for industrial production, distribution, and marketing.

Table 7.3 summarizes the more specific innovations that sustained the Control Revolution in military and industrial transportation through transition to the truly complex systems achieved during World War II. As can be seen from this chronology, the transition involved co-

Table 7.3.   Selected innovations in the control of transportation, 1890–1939

Year	Innovation
1905	Robert Whitehead develops a torpedo with gyroscope which resists deviations in direction through rudder control
1908	Herman Anschutz-Kaempfe patents the Anschutz Gyroscope, the first practical solution to the problem of guiding steel battleships against compass deflections due to turning turrets
1909	Elmer Sperry, Hannibal Ford install gyro-stabilizer in airplane
1911	Master gyrocompass installed on navy ship, successfully tested Sperry gyrocompass includes a complex analog computer to make latitude, course, and speed corrections
1913	Gyroscopic automatic stabilization for aircraft demonstrated Gyro stabilizer installed on a U.S. naval vessel by Sperry
1914	Sperry's son Lawrence wins international competition for a safe plane with biplane fitted with gyroscopic stabilizer

Table 7.3 (*cont.*)

Year	Innovation
1916	Battleship at sea reports and receives orders from Washington, D.C., via radio telephone and regular telephone network in demonstration of possible wartime mobilization Sperry target-bearing and turret-control system includes "battle tracer," analog computer that combines inputs from four electric motors to plot both ship's and target's course and speed
1917	Two-way radio ground control of airplane demonstrated
1918	Hannibal Ford develops a combined bombsight and electrical mechanical computer to determine aircraft's groundspeed and drift, calculate automatically time to release a bomb
1921	Radio beacons for navigation placed in regular service by U.S. Lighthouse Service
1922	Automatic steering Gyro-Pilot installed on Standard Oil ship Radar observations made by the Naval Aircraft Radio Laboratory, which discovers radio equipment can be used to detect ships
1923	A gyropilot designed according to the principles of Nicholas Minorsky's 1922 paper is installed in the battleship *New Mexico*
1929	Automatic pilot, developed by William Green, used successfully on a commercial airliner All-blind airplane flight—guided solely by radio beacon—achieved by Lt. Jimmy Doolittle
1930	Radar detection of airplanes achieved by Naval Aircraft Radio Lab
1931	Automatic pilot enables Wiley Post and Harold Gatty to circumnavigate northern part of earth in eight days, sixteen hours Commercial airliner built with automatic navigation
1932	Gyro-stabilized ship (*Conte di Savoia*, Italian Line) crosses Atlantic All-blind solo flight using standard Air Corps instruments
1933	All-blind cross-country flight completed between College Park, Md., and Newark, N.J., using instrument landing system
1938	Passenger ship equipped with radar (*New York*, Hamburg-American Line) placed in service
1939	U.S. battleship (*New York*) equipped with radar built by the Naval Research Laboratory; RCA awarded contract for six more sets

evolution of technologies of three basic types: information processing (analog computers), communication (radio and radar), and automatic control (gyroscopic compass, stabilizer, and automatic pilot). Also evident in the chronology is the continuing interdependence of innovations for sea and air, and of military and commercial applications. For example, Elmer Sperry designed and installed the first gyrostabilizer for an airplane in 1909 (less than six years after the Wright brothers' first flight), successfully tested another stabilizer in sea trials aboard the USS *Worden*, a 433-ton torpedo-boat destroyer, in 1913, then captured a prize of 50,000 francs offered by the Aero Club of France for a safe plane the following summer. As his son Laurence flew a stabilizer-equipped Curtiss flying boat past the judges' stand, he stood up and waved both arms while his 170-pound mechanic walked six feet out onto the wing (Hughes 1971, pp. 193–200). Sperry developed his gyrocompass in parallel with his gyrostabilizer, according to Hughes (1971, p. 129), "not only in time but also in the way he analyzed prior work, identified inadequacies, and improved upon these in designing a practical device fulfilling an imperative need."

Similar interdependencies can be found throughout the development of sea and air transportation and the technology for its control. Two-way radio control was first demonstrated for a battleship at sea in 1916, between an airplane and ground control—and between two planes— the following year. Radio beacons adopted by the U.S. Lighthouse Service in 1921 were used to guide an all-blind flight eight years later. Gyropilots were first installed on a commercial oil tanker in 1922, a military ship in 1923, a commercial airliner in 1929, and a transatlantic cruise ship in 1932 (Fig. 7.5). The Naval Aircraft Radio Laboratory discovered in 1922 that radio could be used to detect ships at sea; in 1930 it achieved similar "radar" detection of planes in flight. Commercial oceanliners got radar in 1938, U.S. battleships the following year.

Although histories that include precursors to the modern computer largely ignore applications to transportation (Goldstine 1972; Metropolis et al. 1980; Randell 1982; Moreau 1984), our recurrent finding that physical movement, processes, and speed present the most pressing problems of control—and hence information processing—suggests that transportation control technology ought to provide a fertile field for historians of early computing. We have already seen this to be the case for Elmer Sperry.

In 1911, for example, Elmer Sperry invented a mechanical analog computer that embodied a complex mathematical equation—involving a ship's linear speed, the cosine of the angle of its course from true north, and the tangent of latitude—to make automatic corrections to his gyrocompass. Output could either be read from a pointer or me-

Figure 7.5. Automatic control technology made possible commercial aviation in its modern form. On May 15, 1930, less than a year after the first successful use of automatic pilots on commercial airliners, Boeing Air Transport employed on its San Francisco–Chicago route the first airline stewardesses; radar detection of aircraft was first achieved the same year. (Courtesy of Boeing Aircraft.)

chanically coupled directly to the compass (Hughes 1971, pp. 146–147). Five years later Sperry introduced another analog computer, the "battle tracer," in his target-bearing and turret-control system. This computer performed logical operations on inputs from four electric motors to plot both a ship's and its target's course and speed, the tracer moving across a nautical chart to simulate the ship on the sea, a small carriage making pencil marks representing the movements of the target (Hughes 1971, pp. 232–233). Two years later, Hannibal Ford, a Sperry protégé, invented a combined bombsight and electrical mechanical computer which determined the ground speed and drift of an airplane with respect to its target, then automatically calculated and indicated the time to release the bomb (Eames and Eames 1973, p. 63).

To keep these three computers of Sperry and Ford in historical perspective, compare them to what are often heralded as the most important computing innovations of the same period, 1911–1918: Vannevar Bush, considered by some the "father of computing" (Iaciofano 1984, p. 22), patented a device that—pushed over a tract of land—drew a vertical profile showing the elevation of points. Spaniard Leonardo Torres built an end-game chess machine that played to checkmate a rook and king against a human opponent's king. E. G. Fischer and R. A. Harris completed a brass tide predictor, fifteen years in the making, which added thirty-seven components to produce a prediction curve which it also displayed directly on dials (Eames and Eames 1973).

Computer scientists whom I have asked informally to compare these six information-processing problems rate the achievements of Sperry and Ford above those of Bush, Torres, and Fischer and Harris, tentative evidence that the control of transportation—which necessarily involves control of complex movements, processes, and speed—presents a greater challenge for computing than number-crunching per se. Whatever the case, it is likely that other computing devices remain to be unearthed in the history of transportation during the early Control Revolution. Transportation has always been a field with countless and continuing problems of control and hence information processing—a field which historians of computing have not yet begun to explore.

Control Revolution in Mass Distribution

Problems in controlling transportation become, in any material economy, problems of controlling distribution as well. Chapter 6 traced the nineteenth-century crisis of control in distribution through a half-dozen different phases: the nine transshipments of freight between Philadelphia and Chicago (late 1840s) that impeded the new distributional system, the difficulties in keeping track of individual shipments of grain and cotton amid the growing network of warehouses and elevators (early 1850s), the difficulties of traditional mercantile firms in controlling commerce in wheat and corn (late 1850s), similar difficulties among commission merchants (early 1860s) in distributing the first mass-produced consumer goods, the struggle of wholesalers to integrate movements of goods and cash among hundreds of manufacturers and thousands of retailers (1860s), the similar struggle by department stores (late 1860s) to maintain high rates of stock turn, the problems of the new wholesale houses—among the most differentiated organizational structures in the nineteenth century—in integrating a growing number of highly specialized operating units (1870s).

This proliferating chain of crises in the control of distribution progressively eased, as we have seen, before a succession of innovations in information processing and communication. Basic new communication technologies included the telegraph (1840s), a successful transatlantic cable (1866), and the telephone (late 1870s). Both telegraph and telephone as well as a third common carrier—the federal postal system—developed increasingly apart from the great railroad systems. The post office and other transportation systems helped to move a growing number of generalized media of exchange: warehouse receipts and "to arrive" or futures contracts (1850s), postage stamps (1852), through bills of lading (1853), federal paper money (1862), and postal money orders (1864), among others. Distribution of goods, increasingly interregional, national and, by the 1890s, global, came to be controlled by a growing number of new services: regularly scheduled freight services (1830s), express delivery between New York and Boston (1839), freight forwarding (1840s), registered mail (1855), free home delivery (1863), and special delivery (1885).

By the early 1890s, with the Interstate Commerce Commission established in control of the national rail networks and with the railroads themselves establishing car accountant offices to monitor the location and mileage of freight cars on their lines (thereby supplanting the fast-freight and express companies), the crisis in control of America's new continental distribution system had largely been resolved. The next half-century brought consolidation of this Control Revolution in national and increasingly global distribution. The transition from control crisis to Control Revolution in the distributional sectors of the U.S. economy continued smoothly in the 1870s and 1880s with the succession of new information-processing and communication technologies presented in Table 7.4. As this chronology reveals, the innovations included extensions of all of the basic means of control used in the nineteenth century: new communications technologies and common carriers, generalized media of exchange, and distributional services.

Basic new communication technologies that increased control of the international distribution system between the 1890s and World War II included radio (late 1890s), a Pacific cable (1903), transatlantic wire-

Table 7.4. Selected developments in information processing and
communication for control of distribution, 1890–1939

Year	Development
Early 1890s	Car accountant offices—to monitor the location and mileage of cars on railroads—supplant fast-freight and express companies
1891	Coin pay telephones installed widely Travelers' checks devised by American Express Company
1898	Rural free delivery of mail (RFD) systematized by Post Office
1903	Nonexperimental transatlantic radio conversation effected Pacific cable completed from San Francisco to Manila via Honolulu
Mid-1900s	Use of wristwatches, awareness of mechanical time becomes widespread, used to synchronize movement and process
1907	Transatlantic radio message sent on the regular Marconi service
1910	Two-way radio installed in automobile, successfully demonstrated
1913	Post Office inaugurates parcel post service, ending the previous weight limit of four pounds AT&T agrees to allow interconnection to its long-distance network
1914	Post Office begins collections, deliveries in its own automobiles Electric traffic signals, automobile road map introduced
1915	Transcontinental telephone communication demonstrated, commercial service inaugurated Department of Agriculture begins telegraphic "Market News Service" for farm prices and shipments, creates more responsive market
1916	Federal Aid Road Act provides aid for national highway network under constitutional power to establish "post roads"
1917	Precancelled stamps issued
1918	Post Office establishes regular air mail with service between New York City and Washington, D.C.; air mail stamps issued Electric funds transfer system, today known as Fedwire, begins moving money between Federal Reserve and member banks
1920	Pitney Bowes wins first Post Office approval of postage meter

Table 7.4 (*cont.*)

Year	Development
1924	Regular transcontinental air mail established with daily flights (14 stops en route) between New York and San Francisco
1925	Radio facsimile service for transcontinental transmission of photographs begun commercially
1926	Transatlantic radio facsimile service begun commercially; bank check sent across Atlantic by this means
1927	Transatlantic telephone service begun commercially
1929	Air mail service inaugurated to South America from Miami Ship-to-shore radio telephone service begun commercially High-speed telegraph ticker (500 characters per minute) installed
1931	Teletype service begun commercially
1932	Computer pump—indicating quantity of liquid dispensed, price, total amount delivered, cash received—introduced
1936	Amplifiers using electronic negative feedback to eliminate distortion installed as repeaters in coaxial cables
1937	Graduate course in traffic engineering, control and administration established by university (Harvard)
1939	Transatlantic air mail service inaugurated

less telegraph (1907), transcontinental telephone (1915), radio facsimile (early 1920s), transatlantic telephone (1927), high-speed telegraph (1929) and teletype (1931). The Interstate Commerce Act was itself revised in 1910 to cover not only railroads but long-distance telephone service as well. Three years later, in an effort to forestall nationalization, American Telephone and Telegraph reached an agreement with the Attorney General to allow other companies to interconnect to its long-distance network.

In 1908, 2,280 independent telephone companies were not connected to the Bell System, and many homes had to have several instruments to call through different exchanges. By 1921, eight years after the AT&T agreement, the number of companies still unconnected to the Bell System had fallen to 495 (U.S. Bureau of the Census 1975, p. 783). Meanwhile, long-distance calls began to move underground in

1914; by 1936 conversation sped between New York and Philadelphia in a "concentric conducting system," coaxial cable with a conducting outer metal tube encasing but insulated from a central conducting core. This new cable, devised by Lloyd Espenschied and Herman Affel in 1929, proved virtually immune to external electromagnetic noise.

The same New York–Philadelphia line employed as "repeaters" the first application of the Black feedback amplifier. Harold Black of Bell Laboratories, after laboring for years to produce distortion-free amplification in telephone transmissions, finally hit upon the idea of feeding back a small amount of an amplifier's output to input. This method of removing distortion by means of negative electronic feedback, along with Black's mathematical formulation of the new idea (Black 1934), proved to be seminal to the modern theory of automatic control (Mayr 1970, pp. 131–132). Black's feedback amplifier made it possible to carry hundreds of telephone messages simultaneously with great clarity.

The New York–Philadelphia line would in the same year play a pioneering role in yet a third major communication technology. On October 5, 1936, the first intercity television transmission passed from Radio City, New York City, to Philadelphia over the new coaxial cable with Black's new amplifier. In combination, these innovations permitted a large number of telegraph and telephone messages and television images to be sent simultaneously over the same line.

Meanwhile, telephones themselves had become increasingly visible and available to anyone with pocket change. Public telephones, a key tool of commercial travel, appeared after William Gray's 1889 patent for a "coin-controlled apparatus"; the Gray Telephone Pay Station Company began renting them to business establishments in 1891 for 25 percent of the take (Walsh 1950, chap. 9). In that year only 3.7 telephones existed for every thousand Americans; by 1929 it would be 163.1 telephones (U.S. Bureau of the Census 1975, p. 783).

Other technologies that helped to synchronize commercial travel also became commonplace in the first decades of the twentieth century. As a word, *wristwatch* did not appear in English until about 1900. As popular fashion, the devices diffused rapidly through Europe and America after the British army used them to synchronize troop movements in the Boer War, 1899–1902 (Boorstin 1973, p. 362). This development generated the first widespread awareness of clock time (Fig. 7.6). Such rationalization of time helped win public acceptance of Daylight Saving Time—in the name of America's own war effort—in 1917. Radio coordination of motor vehicles also became possible after 1910, when

Chalmers-Detroit successfully tested a two-way system in its auto-mobiles at distances up to three miles. American Telephone and Tele-graph began commercial ship-to-shore radio telephone service on De-cember 8, 1929, when William Rankin, an advertising executive in New York City, called Sir Thomas Lipton, the tea magnate, aboard the liner SS *Leviathan* at sea.

The coevolving networks of transportation and communication served to move still more generalized media of exchange, including travelers' checks (1891), precanceled stamps (1917), and facsimile bank checks (1926). In 1918 the first electric funds transfer system, known today as "Fedwire," eliminated the medium of paper in moving money be-tween the Federal Reserve and member banks; two years later Pitney Bowes eliminated the need for postage stamps when it secured federal approval of metered mail. Even distribution of liquid flows became generalized: in 1932 the Wayne Company of Fort Wayne, Indiana, marketed the first computer pump that automatically calculated, dis-played, and recorded the amount dispensed, its cost, total amount pumped, and payment received or due.

Distribution of all goods became progressively easier during the same period as a result of several new federal postal services: rural free delivery (1898), parcel post (1913), regular air mail (1918), and air service to South America (1929) and to Europe (1939). The telegraphic "Market News Service," inaugurated by the U.S. Department of Ag-riculture in 1915, helped to integrate the distribution of American farm products into a single national system of prices. Here the telegraph and mass media came at last to be applied directly to the problem of the "imperfect" market, the persistent regional variations in prices that had kept the colonial merchants—as we saw in Chapter 5—in "continuous exchange of information about market conditions, prices, and expectations" (Shepherd and Walton 1972, p. 54).

Within a year after Ford introduced the moving assembly line, the U.S. Postal Service began to rely on the mass-produced automobile for collections and deliveries, first in Washington, D.C., but soon on the newer, more sparsely settled RFD routes as well. Two years later

Figure 7.6. Wristwatches became popular in Europe and America after the British army used them to synchronize troop movements in the Boer War, 1899–1902. This widespread rationalization of time helped to win public ac-ceptance of Daylight Saving Time to aid the American war effort in 1917. (Courtesy of the New Jersey Historical Society.)

the federal government began its continuing support of the highway infrastructure with the Federal Aid Road Act, passed by Congress under its constitutional authority to establish "post roads." Growing numbers of vehicles and increasing speed of traffic brought no dramatic crisis but certainly a steadily mounting problem of control: highway fatalities rose from forty-two hundred in 1913 to nearly forty thousand in 1937, an annual total passed only once (1941) until 1962; the average death rate for the period 1933–1937 remains the highest ever recorded in U.S. history. Between 1906 and 1923 the rate for motor vehicles passed that of eight other leading causes of death, including diphtheria, typhoid, and scarlet fever (U.S. Bureau of the Census 1975, pp. 58, 719–720).

As we might expect, the growing problem of highway accidents brought new information and communication technologies for control of traffic: white lines to designate lanes (1911), electric traffic signals and road maps (1914), a national highway route numbering system (1925), police radio control (1933), and permanent state license plates (1937), among many others. In 1936 Pennsylvania State College began the first teacher training course in traffic safety, including both classroom and highway instruction. The following year Harvard University established the first of the nation's many graduate courses in traffic engineering, control, and administration. Even before control of traffic began to assume this highly rationalized modern form, however, the automobile itself would greatly alter control of economic distribution at the retail level.

Market and Retail Control of Distribution

Department stores and mail-order houses began to dominate American retailing by the 1870s, as we have seen, but the early 1920s brought signs of change: large decentralized shopping centers and supermarkets with parking for thousands of automobiles. These new retail institutions reflected an even more fundamental shift in market and retail control of distribution: from continued use of centralized processing and postal communication for mass marketing into the 1910s toward increasing decentralization of outlets and reliance on retailing built around the private automobile after the 1920s. Based on several twentieth-century extensions of solutions first implemented during the nineteenth-century crisis of control in distribution, the revolution in control of marketing and retailing included innovations at all levels of informational activity: preprocessing, organization, programming, processing, and communication.

As in the nineteenth-century struggle to control distribution, many innovations involved preprocessing to facilitate market transactions. Earlier successes at reducing the amount of information necessary to process about commodities—including such innovations as packaging in containers of fixed sizes and weights (1840s), standardized methods of sorting, grading, weighing, and inspecting (early 1850s), fixed prices (1860s), standardized clothing sizes (early 1880s), and periodic presentations via catalog (1880s)—could be extended logically in new types of preprocessing: fully automatic vending machines (Fig. 7.7) that exploited both fixed prices and standardized packages (1897), perfection of the four-color, mail-order catalog (early 1900s), standardization through franchising (c. 1911), preselection like that of the Book-of-the-Month Club (1926), packaging—as opposed to packing—that "sold itself" (late 1920s), and price uniformity enforced through "fair trade" laws (1931).

The functions of retail credit and installment buying, adopted from the neighborhood shopkeeper by large department stores in the late nineteenth century, became increasingly specialized after 1904, when the Fidelity Contract Company formed in Rochester, New York, to purchase installment contracts from retailers. With the spread of the family automobile, installment buying became a major American institution. The Guaranty Securities Company, organized in 1915 in Toledo, Ohio, to finance the installment purchase of Willys-Overland cars, proved so successful that it soon moved to New York to handle twenty-one listed makes. By 1917 forty sizable automobile-sales-finance companies had begun operations; the number exceeded seventeen hundred by 1925. General Motors, noting this success, entered the field in 1919; Ford followed in 1928 with a "Universal Credit Corporation" to help dealers as well as customers (Boorstin 1973, p. 424). Installment credit outstanding for automobiles totaled $304 million in 1919, nearly 43 percent of all consumer goods paper and almost 12 percent of the total U.S. consumer debt. Four years later almost 80 percent of 3.5 million new passenger cars sold on time-payment; by 1929 credit outstanding had reached $1.4 billion, nearly 20 percent of the total consumer debt (U.S. Bureau of the Census 1975, p. 1009).

As indicated by Table 7.5, which summarizes selected developments during the transitional period 1890–1939, the revolution in market and retail control of distribution continued not only in information reduction or preprocessing but also in transaction processing itself. In the nineteenth century such innovations had included differentiation and specialization of bureaucratic control structures (1860s), monitoring of transaction rates via stock turn and other quantitative indicators (late

Figure 7.7. Late nineteenth-century innovations to reduce the amount of information that had to be processed in order to control the distribution of goods included the vending machine. The mechanics of a fully automatic version, introduced in 1897, were extensively explained in Scientific American. *(Courtesy of* Scientific American.*)*

1860s), and the introduction of cash registers to record and control sales (1880s). These innovations in information processing, like the early ones in preprocessing, could be developed logically into new means to control retail distribution. Major innovators included John Hartford, the son of A&P's founder, Charles Walgreen, a Chicago druggist, and Clarence Saunders, a grocer from Memphis, Tennessee.

Saunders's essential idea was to process neither transactions nor commodities as his primary retail function but rather customers them-

Table 7.5. Selected events in the development of market and retail control of distribution, 1890–1939

Year	Development
1894	Sears Roebuck offers wide range of mail-order merchandise in catalog exceeding 500 pages, up from 196 pages previous year
1895	"Cafeteria" restaurant with serving line opens in Chicago
1897	Fully automatic vending machines introduced to dispense gum
1899	Woolworth's Christmas sales approach $500,000; it introduces Christmas bonuses to avert a strike during the crucial period
1900	"Penny" restaurant opens in New York, soon becomes chain of 30
1901	Sears charges fifty cents for its mail-order catalog; preferred customers continue to receive it free
1902	Horn & Hardart opens "Automat" vending restaurant in Philadelphia
1903	Sears establishes its own plant to print its catalog, pioneers use of linotype, four-color printing
1904	Installment finance company organized to purchase installment contracts from retailers
1911	Franchising becomes standard for automotive distribution
1912	U.S. Chamber of Commerce founded in Washington, D.C., by 500 representatives of commercial organizations, trade associations John Hartford, A&P, introduces one-person cash-and-carry economy stores, opens one every three days, doubles chain to a thousand
1913	Drive-through automobile service station appears With inauguration of parcel post service by Post Office, Sears receives five times as many mail orders as in previous year

Table 7.5 (*cont.*)

Year	Development
1916	Charles Walgreen has seven drugstores like general stores with wide range of nonpharmaceutical goods and services, including soda fountains, lunch counters, his own brands of candy, ice cream
	Clarence Saunders opens his first Piggly Wiggly grocery, a new self-service store in which a maze with turnstiles and a checkout require that customers pass by all goods on display
1920s	Drive-in stores, restaurants open in California
1922	Decentralized shopping center (Country Club Plaza) opens, eventually covers 6,000 acres, one-tenth of Kansas City
1923	One of the first supermarkets (Crystal Palace, San Francisco) opens with 68,000 square feet of space, parking for 4,350 cars
1926	Book-of-the-Month Club established with 4,750 members
1928	Richard Franken and Carroll Larrabee publish *Packages That Sell*, pioneering work on the art of packaging and self-service
1931	State (California) passes fair trade law prohibiting resale of trade-marked articles except at price set by producer
1934	Walgreen's opens super-drugstore in Tampa, Fla., presenting merchandise on open display counters rather than in showcases
1936	Robinson-Patman Act, New Deal measure amended to earlier antitrust legislation, culminates most extensive effort to control chain-store price discrimination
1937	A&P begins converting its small economy stores into supermarkets, a word that becomes common during the decade
1939	Drive-in Dairy Queen opens in Moline, Ill.

selves. His first Piggly Wiggly store, opened in 1916, was explicitly designed to process people past merchandise. Turnstiles channeled entering customers into a single aisle, where they could do little else but advance back and forth through a maze of shelves, past all items in stock (packaged, of course, to "sell themselves"), until they reached the exit turnstile, complete with a check-out counter and cash register—and the only employee then at work in the store (Fig. 7.8). Saunders's scheme, what he called a "self-serving" store, can be found to this day in retail establishments throughout the world. It owed its essential processing idea to the buffet-style restaurant (1885), the cafeteria with serving line (1895), and the drive-through auto service station (1913). This latter idea, that retail establishments could be designed to process automobiles, brought drive-in stores and restaurants to California in the 1920s, the drive-in motion picture theater in Camden, New Jersey, in 1933, and the first drive-in Dairy Queen in Moline, Illinois, by 1939.

So successful did Saunders's first grocery become, because its low overhead yielded cheap prices as well as generous profits, that within six years he owned or had franchised more than twelve hundred Piggly Wiggly stores. Unsuccessful stock trading eventually bankrupted him, however, and forced him to forfeit his company to creditors. He soon launched a second successful grocery chain with the unlikely name of "Clarence Saunders, Sole Owner of My Name, Stores, Inc.," only to be ruined by the Great Depression (Boorstin 1975, p. 50). Saunders

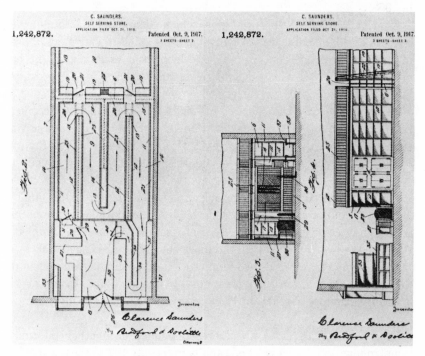

Figure 7.8. The first patent for a means of processing customers and their purchases through a store. Clarence Saunders's floor plan admitted customers through a turnstile (bottom left) and channeled them through a maze of shelves past all goods in stock; the checkout counter (right) was the store's only exit. (Diagrams 2 and 3, U.S. Patent No. 1,242,872 [1917].)

then attempted to mechanize and electrify the transaction processing of the Piggly Wiggly in his "Keedoozle," a word he coined from "Key-Does-All."

In the first Keedoozle store, opened in Memphis in 1937, customers selected goods by inserting a notched rod into keyholes beside sample items displayed behind rows of glass windows, an idea borrowed from the Horn & Hardart Automat Restaurants, established by a Philadelphia baker after 1902 with machinery invented in Sweden, imported from Germany, and materially refined by the American company in the 1930s. To the basic automat idea Saunders added an electric mechanism that automatically recorded customer selections, collected the correspondng items from stock, wrapped them, and released the lot to a conveyor for final bagging upon insertion of the key in a slot following payment. The idea did not prove successful, however, despite Saunders's nearly two decades of efforts to perfect it before his death in 1953.

Charles Walgreen's essential idea, which complemented Saunders's concept of self-service, was to reestablish the small-town general store in the corner drugstore, a place to which local residents might be attracted daily by a wide range of goods and services. Not only drugs but a continually growing variety of personal care and home products were combined with the services provided by newsstand, tobacconist shop, lunch counter, and soda fountain. Walgreen had opened seven such stores in the Chicago area by the time Saunders's first Piggly Wiggly systematized the idea of exposing customers to the widest possible range of goods. Over the next twenty years Walgreen shifted the appeal to his customers from packaging to self-inspection, gradually moving goods from showcases and shelves to open display counters where they could be more readily seen, handled, and even sampled. By 1927 Walgreen had established more than a hundred such stores— more than five hundred in two hundred cities in thirty-seven states at the time of his death in 1939. Most notable was his "super-drugstore," opened in Tampa, Florida, in 1928, a forerunner of the modern shopping mall franchise.

Success for Walgreen's drugstores depended on another informational innovation that developed rapidly in the late 1920s, the packaging that "sold itself." The expression comes from the seminal book, *Packages That Sell*, published by marketing specialists Richard Franken and Carroll Larrabee in 1928. As late as 1920 few household goods came in packages, and unpackaged goods rarely sold under brand names. With the continued rise of trademarks and national brand labels, however, packaged goods increasingly competed—as carefully crafted messages—for shoppers' attention. In 1900, Franken and Larrabee found only 7 percent of advertisements for packaged goods pictured the product; by 1925, 35 percent did.

Their new approach attempted to rationalize the packaging of products, much as Taylor's scientific management had done for their production two decades earlier. Scientific marketing sought to determine, prior to production, which of a wide range of systematically different packages would maximize sales. Subjective reactions to size, shape, material, and color all could be measured, Franken and Larrabee argued, by quantitative techniques adopted from academic psychology. The flat ten-ounce can had proved best for codfish cakes, for example, "because it looked larger than the tall can . . . although both cans had the same cubic capacity."

If goods were to be sold via self-service, transaction processing could be speeded up and yet controlled through the preprocessing of packaging: standardized sizes and weights for standardized prices. If brand-labeled goods were to compete for a shopper's attention from open shelves in the absence of a salesperson, as Saunders and Walgreen intended, then the packages themselves would have to sell—and would need to be scientifically designed toward that end. "The primary function of advertising is to create a desire in the customer's mind to buy a product," Franken and Larrabee wrote of packaging's place in retailing. "This desire, once created, must be carried to the point of sale . . . Even if only a favorable impression has been created, without the desire to buy, the sight of the actual package may turn this impression into active desire on the part of the consumer."

Packaging assumes—in control of consumption—an explicit communication function: to carry the desire to buy, created by advertising, to the point of sale (to rearrange the words of Franken and Larrabee). As Boorstin (1973, pp. 445–446) concludes: "In stores, then, it was more important to display the distinctive brand-named carton than the usable object itself. The 'dummy carton' began to play a leading role in store windows and on counters."

The Movement to Control Retailing

Such radical changes in the control of retail distribution did not go unchecked by higher-level controls. Throughout the period from the 1890s to World War II business and government contrived new organizational structures in attempts to impose bureaucratic control on the nation's commerce. As early as 1870 some forty local chambers of commerce had already formed throughout the United States. In 1903 Congress established the Commerce and Labor Department with its Secretary as the eighth member of President Theodore Roosevelt's cabinet; ten years later Commerce and Labor became separate departments, both with cabinet rank. Congress established the U.S. Commerce Court with six judges in 1910 and the Foreign and Domestic Commerce Bureau, which combined the duties of the Bureau of Manufactures and the Bureau of Statistics, in 1912.

That same year, President William Howard Taft and Secretary of Commerce and Labor Charles Nagel invited about five hundred representatives of trade and commercial associations and individual companies to Washington to found a Chamber of Commerce of the United States. Its headquarters, one of the most imposing buildings in the nation's capital, was completed in 1925. The International Chamber of Commerce, formed in 1920, opened its own headquarters in Paris. Meanwhile, the first Junior Chamber of Commerce had organized in St. Louis by 1915; delegates from twenty-four cities met six years later to organize a national body.

Community groups also organized to oppose the growing systematization of retail distribution and the concentration of its control in a few national companies at the expense of local businesses. Small town and rural retailers had fought the "monopolistic" department stores in the 1860s and 1870s; by the 1880s they mounted a similar effort against the mail-order houses. "What's the use of sending money east when we can buy just as cheap at The Popular Store?" a flyer from that local clothier asked its customers; some merchants publicized the cause by collecting and burning stacks of Montgomery Ward and Sears, Roebuck catalogs in the town square (Boorstin 1975, p. 55) (Fig. 7.9). Because of mail-order competition, small town retailers continually organized to fight the growth of a national distribution system, including rural free delivery in the 1890s and parcel post in the early 1900s. The best organized and most successful opposition arose in the 1920s and 1930s against the most systematic and complete control of market and retail

Figure 7.9. Local resistance to the systematization of retail distribution and the concentration of its control in a few mail-order companies. This flier, printed by a small-town clothing store, helped to promote a public burning of Montgomery Ward and Sears, Roebuck catalogs. (Courtesy of Montgomery Ward & Co.)

distribution: the chain store, a concept (and Americanism) firmly established by the turn of the century.

The first such chain, which proclaimed its ambitions by its name, The Great Atlantic and Pacific Tea Company, included sixty-seven general grocery stores in 1876 and nearly five hundred by 1912. F. W. Woolworth's chain of five-and-tens, begun in 1879, included seven stores in 1886, some sixty by 1901. Walgreen's chain included seven drugstores in 1916, more than a hundred by 1927. For twenty-six different lines of merchandise, the U.S. Federal Trade Commission reported ten chains of two or more stores in 1890, more than a hundred by 1903, more than a thousand by 1922; most sold groceries, clothing, shoes, or drugs (U.S. Bureau of the Census 1975, p. 847). Most rapid expansion of a chain came after 1912, when John Hartford, son of the founder of A&P, introduced the "cash-and-carry" economy store, opening an average of one every three days until 1915, when the chain exceeded a thousand stores. A&P had some fifteen thousand stores by the Great Depression, when it accounted for more than 11 percent of America's food business (Boorstin 1973, p. 110).

Hartford intended his term *cash-and-carry* to communicate its own negation: no credit and no deliveries. As with the installment contract company almost a decade earlier, retailing would continue to differentiate itself from the credit function. Increasingly until the 1930s Hartford applied his idea to small, no-frills, one-person operations much like modern "convenience" stores. Through them, A&P could apply a wide range of controls over distribution: centralized management, regular bulk purchasing, reduction of middlemen, standardization of business practices, product quality and prices, and minimal personal service and credit costs. In all but the last two ways, chain stores duplicated the processing and controlling achievements of an institution from which they otherwise differed radically: the department store, a luxurious palace staffed by hundreds and offering its customers many incidental services, liberal credit, and free deliveries. Like the department store, chain stores could distribute goods at much lower cost than smaller independent operations—at prices that quickly undercut competition.

Not surprisingly, therefore, retailers' opposition to chain stores mounted steadily after 1922, when the National Association of Retail Grocers voted support at its annual convention for special sales taxes on chain store merchandise, zoning laws to limit the number of chain outlets in any neighborhood, and escalating taxes on every store be-

yond the first established by a small owner. Most states eventually
enacted some type of anti-chain-store legislation. In 1931 California
adopted the first "fair trade" law to permit manufacturers to specify
minimum retail prices in order to protect independent stores from
price-cutting competition by chains. After the courts nullified such
laws, Congress passed the Miller-Tydings Act (1937), which exempted
fair trade from antitrust legislation. The previous year Congress had
passed the Robinson-Patman Act, New Deal legislation giving the
Federal Trade Commission new control over chain-store merchandising.

The Rise of Supermarkets and Shopping Centers

Despite such efforts to impose government control on the chain store,
that form of retail distribution continued to flourish. Ironically, Hartford's cash-and-carry, one-person economy stores did not. In 1923 one
of the first supermarkets, San Francisco's Crystal Palace, opened on
a former circus ground in a 68,000-square-foot steel-frame building
with parking for 4,350 cars. Such stores combined the advantages of
both department store and economy chain outlet: like the former, they
offered one-stop shopping and the widest possible selection of goods;
like the latter, they relied on the economies of self-service and cash-
and-carry. They also exploited the same controls over distribution
developed by both older institutions: centralized management, regular
bulk purchasing, reduction of middlemen, standardization of quality
and prices. By 1937, recognizing the advantages of the new form, A&P
had begun systematically to consolidate its four or five one-person
stores in each neighborhood into a single supermarket, thereby reducing the number of its outlets from a high of 15,709 in 1930 to four
thousand after World War II.

The new supermarket buildings *qua* material processors combined
the contributions of Saunders, Walgreen, and Hartford with several
original innovations. Adaptations included Saunders's turnstiles, maze-
like aisles, and check-out counters with cash registers, Walgreen's wide
selection of competing brands displayed on open shelves, and Hartford's cash-and-carry policy and adjacent parking. The most important
new idea proved to be the shopping cart, developed simultaneously
with the supermarket itself in the early 1920s and mass produced in
modern form by the late 1930s. Shopping carts allowed stores to process people rather than transactions—even at high volume and in-

creased speed. At the same time, shopping carts also enabled customers to reduce the number of shopping trips as well as stops, which extended the essential idea of one-stop shopping to time as well as to space but also forced consumers to provide a larger portion of total food storage costs. This processing innovation, combined with the earlier ones of Saunders, Walgreen, and Hartford, enabled supermarkets to sustain unprecedented stock-turn speeds and throughput volumes.

As early as 1937 the Crystal Palace had set sales records of twenty-five *tons* of sugar in an hour, five freightcar-loads of eggs in a month, and an average of nearly a ton of apples per day for an entire year. By the end of World War II various marketing specialists had pegged the lowest limit for a food store's classification as supermarket at an annual volume ranging from one-half to one million dollars. Routinely stocked goods included at least groceries, meat and dairy products, produce, and familiar household items; by the mid-1930s the Crystal Palace also offered liquor, tobacco, jewelry, and drugstores, barber and beauty parlors, and a dry cleaner. Despite this growing variety of goods and services, however, America's independent and chain grocers managed to maintain roughly the same retail trade margins between 1869 and 1947 (Barger 1955).

In the supermarket, well established by the late 1920s, retail distribution approached the ideal of an open processing system dependent on organizational, preprocessing, and related control technologies to sustain rapid throughputs of both products and purchasers for healthy profit at low margin but high volume. This can be clearly seen in the institution's most modern form: check-out cashiers now slide each packaged good past a laser scanner which reads a bar code printed on the label, searches its computer memory for the current price, displays it to the customer, adds it to the total bill, and reduces inventory records by one unit. This processing system, in which a package not only sells itself to the customer but also serves as its own price sticker and inventory check, constitutes one of the most elaborate applications of computer and laser technology to retail distribution. This technology, in the context of the supermarket as processor of people and products, can be seen as an extension of the principle of Clarence Saunders's electrical Keedoozle.

Supermarkets could not have evolved, no more than shopping centers and other modern retail institutions, without the stimulation and control of consumption by national advertisers. As Franken and Larrabee saw in 1928, national advertising created the demand for brand-

name products and packaging reinforced it; all that remained for retailers to do was to attract customers to their stores. Because the new
mass retailers could offer lower prices, they advertised these regularly
against local competition. Local and national advertising, keyed to
modern packaging, promotional, and retailing techniques, would provide the foundation for the Control Revolution's final phrase: bureaucratic control of mass consumption.

8

Revolution in Control of Mass Consumption

Every advertising man knows how the functions of advertising have expanded during the last twenty years . . . That expansion, we are convinced, has only begun—we have taken only the first steps in learning to use and control the power we are working with. The man who would set any limits to the influence of advertising would be presumptuous indeed.

—*Printers' Ink* (1915)

IN 1899 the National Biscuit Company launched the first million-dollar advertising campaign intended to establish a new brand-labeled consumer product, the Uneeda Biscuit. By 1913 National Biscuit ranked seventy-sixth in assets among U.S. industrials. That a baker of soda crackers managed to achieve this lofty position only fourteen years after introducing new advertising for an entirely unprecedented product provides convincing evidence that the crisis in control of consumption had begun to be resolved. It also provided an early model for other companies of how trademarks, consumer packaging, and national advertising could be used to attain bureaucratic control of a market, even for an industry in crisis.

Economic depression in the mid-1890s had demonstrated the shortcomings of the holding company, so named because it had no operating facilities of its own but merely "held" stock in a number of single-unit firms not centrally controlled. To maintain prices high enough to earn even a reasonable return on investment, a holding company had to attract competitors not encumbered by the need for second-order earnings. Price wars often brought ruinous loss of profits; buying out the competition could lead to equally disastrous over-capitalization. Faced with this dilemma, many loose-knit holding companies failed in the mid-1890s, which underscored the need for administrative centralization in horizontal integration. The scramble by business for centralized control generated the American merger movement of 1898–1902, a period that saw 2,643 firms swallowed up by 212 legal consolidations, roughly ten times the rate during the depression years 1894–1897 (Thorelli 1954, pp. 294–303; Nelson 1959, pp. 33–34).

As with many of the other mergers, the National Biscuit Company formed in 1898 as a legal consolidation of three older regional holding companies: American Biscuit (one of the financial failures of the depression), New York Biscuit, and U.S. Baking. By its own admission the company began as "an aggregation of plants" and "thought it necessary, for success, to control or limit competition"; it quickly learned, however, that "either of these courses, if persevered in, must bring disaster" (Chandler 1959, pp. 11–13). Instead, the president of National Biscuit, Adolphus Green, decided to imitate the strategy of Henry Crowell's Quaker Oats: to shift from production in bulk to consumer packaging, with control of consumption by means of trademarks, brand labeling, and national mass advertising, vertical integration backward into purchasing and forward into marketing, and economies of scale resulting from the centralization of production in a small number of mammoth plants (Cahn 1969).

At the time, soda crackers and cookies—like many other foodstuffs intended for household consumption—were sold by retailers loose and in bulk, without brand identification, from often grimy cracker barrels. Green decided to distinguish his own crackers by cutting off their corners, thus making them octagonal—possibly the earliest use of arbitrary physical differentiation to distinguish otherwise ordinary products. Differentiation remains a major alternative to price competition as a marketing strategy (Porter 1980, p. 37); it explains, for example, the recent appearance of toothpaste with stripes. To bypass the cracker barrel, Green's law partner, after some months of experimentation, devised a method of wrapping crackers in a waxpaper-lined cardboard box, watertight packaging that would be promoted in ads featuring a little boy in a raincoat carrying a box of Uneeda biscuits though a storm. The success of this "In-er-Seal" box led *Printers' Ink* to proclaim as early as 1900 that "the package is the very keynote of modern advertising" (Pope 1983, p. 50), and indeed by that year nearly a thousand patents had been issued for folding boxes and related machinery (Franken and Larrabee 1928; Davis 1967).

To advertise his new crackers, Green turned to N. W. Ayer and Son, by then the nation's largest advertising agency with successful campaigns for Montgomery Ward, Procter and Gamble soaps, Hires Root Beer, and Burpee Seeds, among other mass marketing firms (Hower 1949, pp. 115–116). By some accounts Henry N. McKinney, Ayer's leading agent, chose the trade name Uneeda Biscuit (supposedly a more dignified synonym for "cracker") from a list that included Wanta Cracker, Taka Cracker, Hava Cracker, and Usa Cracker (Hower 1949; Cahn 1969); other accounts hold that the name "Uneeda" came from the son of Robert Gair, inventor a decade earlier of the first folding box machine (Boorstin 1973, p. 439).

In any case, Ayer's introduced the product market-by-market in a nationally coordinated campaign that relied heavily on innovative "teaser" ads containing only the word "Uneeda"; billboard, streetcar, and newspaper advertising in each area generated anticipation of the new biscuit's arrival. As a result of what may have been the first advertising budget to reach $1 million for a single year, Uneeda became a household word virtually overnight. Within a year of its introduction the packaged soda cracker, despite being a totally new consumer product, sold at the rate of ten million boxes *per month* (Cahn 1969).

Thus did National Biscuit establish that a trademark, brand labeling, and a massive advertising campaign could be used to generate and control consumer demand and even to restructure an entire industry. In doing so, National Biscuit also saved the institution of the holding company—already evolving into the integrated, multifunctional, and often multinational enterprise that we know today as the modern corporation—from the vicious dilemma of price warring and loss of profits, on the one hand, or buying out competitors and stagnating in overcapitalization on the other.

From Creation of Demand to Its Bureaucratic Control

If advertising can be used to induce brand loyalty in consumers and thereby reduce the elasticity of demand for branded products, then their producers could charge higher prices relative to costs. Although economists do not agree on whether or not advertising has this effect (pro: Comanor and Wilson 1967, 1974; con: Demsetz 1974; Nelson 1975; Galbraith 1967, chap. 18), turn-of-the-century marketing experts did continually urge businessmen to use advertising as a means to control price competition (Pope 1983, p. 72). As brand-name, nationally advertised goods gained in prominence around 1900, retailers themselves began to deemphasize prices in their advertisements (Hower 1943, pp. 269, 274).

Trademark and company reputations, sustained through national advertising, may also enable producers to control entry into a market by competing firms. Although economists do not agree on this possibility either (pro: Bain 1956; con: Schmalensee 1974; Demsetz 1979), turn-of-the-century marketing experts did believe that a national advertising campaign, to the extent that it maintained product differentiation, could constitute a formidable barrier to market entry by competitors. Certainly National Biscuit used the Uneeda campaign as such a barrier to its many regional and private brand competitors, most of whom sought to subvert the strategy through trademark infringement. Even the mighty H. J. Heinz, among the nation's 250 largest industrial companies with fifty-seven separate product lines, could capture only a fraction of the soup market from the much smaller Campbell Company, which enjoyed an established reputation in that product and, because of the resulting volume, sold its products at one-third the price of less-promoted brands. By 1908 *Printers' Ink* estimated that "at least 50 percent of the advertising being done today is for the purpose of creating property in trademarks," a barrier to entry that economist Dorothea Braithwaite (1928) would call a "reputation monopoly" (Pope 1983, pp. 68–69, 88).

The technologies of trademarks, consumer packaging, and mass advertising also provided means by which manufacturers might better control the giant wholesalers and new mass retailers—urban department stores, mail-order houses, and chain stores—rapidly monopolizing the distribution channels for their products. For most manufacturers in the late nineteenth century distribution involved a continual struggle with wholesalers and mass retailers over issues like terms of payment, quantity discounts, advertising allowances, shelf display, and treatment of competitors' products. As long as products remained generic and in bulk, unbranded and unadvertised, wholesalers could control which manufactures reached retailers' shelves; retailers, in turn, could control consumer purchases—usually by concentrating on goods with the highest margins of profit. As one manufacturer who turned to national advertising for increased control described the earlier era: "The manufacturer stood on the merchant's doorstep begging him to buy his product. The merchant then was the King of Commerce, with the manufacturer groveling at his feet" (Pope 1983, p. 78).

Using trademarks and consumer packaging, manufacturers could alter this relationship—the political economy of distribution—in their own favor. Because of the enormous volume of sales generated by national advertising, National Biscuit could integrate forward (Porter and Livesay 1971), buying up the bakery wagons of salesmen who worked on commission, retaining them as salaried employees, and using the fleet to deliver Uneeda crackers directly to grocers. Directed by a network of marketing offices, the salesmen kept to appointed rounds to ensure that crackers reached consumers fresh and unbroken and that grocers worked to sell the product despite its low profit margin (the company held prices down to build volume). In order to use this sales force to maximum efficiency and to meet retailers' demands for a wider product line, National Biscuit expanded within a decade to forty-four items, including the Fig Newton (named for the Boston suburb) and the ZuZu molasses cookie. Its sales fleet also contributed additional free advertising, and not only that which the company painted on the sides of its wagons—several manufacturers offered miniatures of the familiar vehicles, a favorite toy among urban youth in the early twentieth century (Cahn 1969).

Such advertising also served as a substitute for further vertical integration into retailing. Consumer demand alone could force a store to stock an item; continuous monitoring by the manufacturer's salesmen could control its sale. This strategy had been widely discussed before Uneeda: "Once you create a demand by advertising," the journal *Advertising* proclaimed in 1897, "you can compel the trade to handle it on your own terms." After the success of the Uneeda campaign had given National Biscuit control over consumption and hence distribution in precisely this way, *Printers' Ink* restated the strategy even more emphatically: "The manufacturer selling an advertised trademarked article is absolutely independent. The only class to whom he is responsible is the consumer" (Pope 1983, pp. 80, 90).

So potent a weapon was the advertising campaign (the word itself connoted battle) that even the relationship of the agency to its client became transformed by Uneeda's success into one of alliance and conspiracy. Throughout the nineteenth century, when advertising had been viewed as a purely informational service and consisted largely of selling space in publications, agencies routinely served competing clients. Around 1900, however, when Ayer informed National Biscuit that he had accepted the advertising of a rival baker, his prize client—still the agency's largest account—protested strongly on the grounds of conflict of interest. Ayer then quickly adopted the principle—today an industry standard—that an agency ought not to handle competing accounts (Hower 1949, p. 340). Within a decade this principle had become generally accepted (Brown 1912; Pope 1983, p. 163). Thus it might be said that advertising afforded a company greater control not only of competitors, wholesalers, retailers, and consumers, but even—ironically enough—of the agencies that helped them to wage their campaigns.

The Rise of Modern Advertising

As advertising agencies came to be incorporated in the bureaucratic control of mass consumption, the field of advertising began to emerge as a legitimate profession. As we saw in Chapter 6, an increasingly specialized advertising sector had met the crisis in control of consumer demand that followed industrialization through the material economy. The nascent infrastructure for control of consumption included not only the advertising agency itself, after 1841, but a succession of informational innovations for advertisers: indexes of periodicals (1848), a monthly trade journal (1865), book of newspaper circulation (1869), weekly trade paper and sole-agent contract (1875), market surveys for clients (1879), and an industry-wide journal (1888).

As the work of preparing ads moved away from the advertiser and became specialized in the new agencies in the 1890s, formerly freelance copywriters increasingly joined these staffs. More visibly responsible for the content and truthfulness of ads, agencies raised somewhat their ethical standards. At the same time, advertising executives began to found clubs and associations, which not only indicated growing self-awareness as a profession but also represented a modest gesture at self-policing. "It was thought that if they were to sit around the same table once a month there would be a little less throat-cutting and general misbehavior," an early club member later recalled. "It worked out just that way" (Fox 1984, p. 38).

Copywriting became a full-time specialty in the mid-1890s; by the early 1900s the advertising agency in its modern form—with an expanded role for the account executive and specialists in writing, art, and design—had begun to emerge, first at N. W. Ayer and J. Walter Thompson. Increasingly the account executive came to integrate the other specialized functions, often mediating beween the business needs of clients and the egos of the creative staff. Annual advertising expenditures, which had quadrupled to $200 million between 1867 (the earliest year available) and 1880, quadrupled again (to $821 million) by 1904. In 1917, when more than 95 percent of national advertising was handled by agencies, annual expenditures stood at $1.6 billion (U.S. Bureau of the Census 1975, p. 856).

Testimony of advertising's new status came when President Woodrow Wilson created the Creel Committee (named for journalist George Creel) to disseminate information and attempt to shape public opinion during World War I. One division of the committee made posters, another produced motion pictures, another organized professors, novelists, and seventy-five thousand public speakers to explain the war. The Advertising Division, directed by five prominent ad agency executives, placed $1.5 million worth of donated newspaper and magazine space and copy. Most successful of these ads may have been Courtland Smith's "The Greatest Mother in the World," which used a beatific nurse to embody the Red Cross and greatly enhanced the prestige of the organization (chartered by Congress twelve years earlier) then under consideration for a Nobel Peace Prize, which it won later the same year. Artist James Montgomery Flagg's poster of a pointing Uncle Sam declaring "I want YOU for the U.S. Army" proved more enduring—it remains in use today (Fig. 8.1).

If any doubt had remained after the Uneeda campaign about the possibility that advertising might be used to control public opinion or aggregate behavior, the Creel Committee's work served to reaffirm the point. Creel had promised President Wilson "a plain publicity proposition, a vast enterprise in salesmanship, the world's greatest adventure in advertising" (Sullivan 1933, p. 425). After the war Creel concluded, "The work, as a whole, was nothing more than an advertising campaign, and I freely admit that success was won by close imitation of American advertising methods and through the generous and inspirational cooperation of the advertising profession." In 1919 he told advertisers at a national convention: "Before the war, your status was anomalous. Today, by virtue of government recognition as a vital force in American life, you stand recognized as a profession" (Fox 1984, p. 76).

The Creel Committee also served to cement a perceived relationship between advertising and the control through bureaucracy of mass behavior. "In the dim and distant days before the war, reformers were wont to say, 'Oh, if the people would only do so and so!'," *Printers' Ink* editorialized late in 1917. "Under the rigorous prod of Mars, they have learned to say, 'The people *must* do this and that—go out and advertise and make them come through" (Pope 1983, p. 12). Needless

Figure 8.1. An enduring product of the Creel Committee created by President Woodrow Wilson to help shape public opinion during World War I. By distributing millions of copies of this poster, U.S. Army recruiters established the popular symbol of the nation as Uncle Sam (virtually a self-portrait of the artist, James Montgomery Flagg), a testimony to the power mass advertising had achieved by the 1910s. (Courtesy of the United States Army.)

to say, advertising boomed immediately following the war, with expenditures more than doubling to $2.9 billion between 1918 and 1920 (U.S. Bureau of the Census 1975, p. 856).

Table 8.1 summarizes the development of modern advertising throughout the transition period, 1890–1939, between the crisis of control in consumption and the advent of commercial television. As can be seen in the chronology of events in this table, the half-century brought no end to imaginative new ways to stimulate and control the consumption of mass-produced goods. Engineering and advertising, both coming to enjoy new status as professions, produced a continuing stream of innovations in marketing and mass communications still in wide use: billboards standardized to bring interchangeability to outdoor display (1891), trading stamps redeemable for premiums (1891),

Table 8.1. Selected events in the development of advertising and control of consumption, 1890–1939

Year	Development
1890s	Advertising men found clubs, begin modest self-policing
1891	Trading stamps redeemable for premiums introduced Billboards standardized by outdoor advertising agencies
1893	Print patent introduced, first goes to "Heinz" in shape of pickle
Mid-1890s	Copywriting becomes a full-time specialty in advertising agencies; free-lancers flourish
1897	Company (General Electric) creates bureau to publicize itself
1898	Advertising law passed by state (New York) "to prevent misleading and dishonest representations"
1899	Product differentiation, packaging, million-dollar campaign used to establish new consumer product (Uneeda Biscuit) nationally
Early 1900s	Modern advertising agencies emerge with expanded role for account executive, full-time specialists in copywriting, art, design
1902	Mandel Brothers Department Store contracts with *Chicago Tribune* to run full-page ad six days a week throughout the year
1903	Walter Dill Scott, Northwestern University psychologist, publishes *The Theory of Advertising*, first U.S. advertising text

Year	Development
1912	Associated Advertising Clubs of America forms a national committee to promote fair practices
1914	Newspaper rotogravure sections introduced
Mid-1910s	Scientific advertising expands from psychology to other social sciences; economic, business considerations appear in texts
1915	Association of National Advertisers formed
1917	American Association of Advertising Agencies formed President Wilson creates Creel Committee to disseminate war information; produces posters, speeches, motion pictures; Advertising Division places $1.5 million worth of donated copy, space
1919	Illustrated tabloid newspaper (*New York Daily News*) published
1922	Queensboro Realty purchases first advertising time (ten minutes for $100) on radio station (WEAF, New York) Skywriting introduced commercially in America from England
1923	Neon tube advertising signs introduced Daniel Starch publishes textbook for the case method approach to advertising which he introduced at Harvard Business School Edward Bok of *Ladies' Home Journal* establishes set of nine advertising prizes to be awarded by Harvard Business School
1925	Composite photographic layout or "composograph" introduced
1928	Electric sign flasher ("Motogram") installed on four sides of the *New York Times* Building, flashes presidential election returns
1930	Radio advertising course introduced in college (School of Business and Civic Administration, City College of New York)
1932	Advertisers get NBC, CBS networks to allow prices on radio
1937	Animated electric sign—a four-minute cartoon presented using 2,000 light bulbs—displayed on Broadway, New York City
1939	Advertising expenditures for radio exceed those for magazines Commercial television begins on limited basis

patented print forms (1893), newspaper rotogravure sections (1914), commercial radio and skywriting (1922), and neon (1923), electric flashing (1928), and animated (1937) signs (Fig. 8.2).

As Table 8.1 also reveals, advertising became increasingly rationalized during this transitional period of the Control Revolution. In 1897 General Electric became the first U.S. corporation to create its own bureaucratic apparatus to publicize itself and the broader developments in its industry. The following year New York became the first state to enact legislation to "prevent misleading and dishonest representations" in the sale of merchandise; those whose advertising "intended to have the appearance of an advantageous offer, which is untrue or calculated to mislead" would thereafter be guilty of a misdemeanor. In 1903 the first American textbook, *The Theory of Ad-*

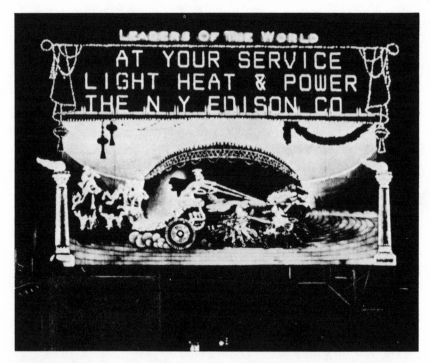

Figure 8.2. Electrical engineering and advertising, both enjoying new status as professions by the turn of the century, combined to produce several innovations in outdoor advertising. In 1910 New York Edison installed this chariot race in colored bulbs atop the Hotel Normandie in Herald Square, New York City; neon tube advertising followed in 1923, electric flashing signs in 1928, animated signs in 1937. (Courtesy of Consolidated Edison Company.)

vertising, appeared under the name of Walter Dill Scott, director of the Psychology Laboratory at Northwestern University (Bartels 1962, pp. 46–48). The following year local ad groups joined in a loose national federation, the Associated Advertising Clubs of America; large advertisers formed the Association of National Advertisers in 1915; and advertising agencies established the American Association of Advertising Agencies two years later (Fox 1984, pp. 68–69).

By this time so-called scientific advertising—the equivalent for mass communication of Taylor's scientific management and the scientific marketing of Franken and Larrabee—had spread from experimental psychology to the other social sciences and to the academic specialization of marketing; economic and other business considerations began to appear in texts (Bartels 1962, pp. 48–49). A well-known text that actually took the title *Scientific Advertising* (1923), written by Lord & Thomas's Claude Hopkins, would be republished in 1952 by psychologist Alfred Politz, who argued in an introduction that "present-day advertising research has a long way to go before it reaches the level of Claude Hopkins's contributions" (Politz 1952); a subsequent edition included an introduction by advertising magnate David Ogilvy, who claims the book had "changed the course of my life" (Ogilvy 1980, 1983, pp. 202–204).

Although Hopkins held that the college-educated ought not to be allowed in his profession, he adhered to experimental methods and played an important role in rationalizing advertising, introducing copy research, test marketing, and sampling by coupon. "Almost any question can be answered, cheaply, quickly and finally, by a test campaign," he wrote (Hopkins 1923). "And that's the way to answer them—not by arguments around the table." This more scientific approach advanced by means of another important book published in 1923, psychologist David Starch's text on the case method approach to teaching advertising principles, which he had introduced at the Harvard Business School (Bartels 1962, p. 54).

In the same year Edward Bok of the *Ladies' Home Journal* established a set of nine annual advertising awards to be administered by the Harvard Business School and selected by a jury of experts. These "Harvard-Bok" awards, as they came to be called, served for seven years to publicize advertising as a profession before their discontinuation in the Great Depression (Fox 1984, p. 106). By that time rationalization of advertising had begun for a new medium: in 1930 the business school at the City College of New York introduced the first graduate course on radio advertising under Frank Arnold, director of development at the National Broadcasting Company.

As this brief chronology suggests, the establishment of advertising as a means to stimulate and control consumption had largely been accomplished by the 1920s. The other half of the strict control loop, the feedback from demand that came to be known as market research, also had developed extensively by the 1920s, as we shall see in a later section of this chapter. Because such control depended on mass communication via mass media, however, advertising—what copywriter John E. Kennedy defined in 1905 as "salesmanship on paper"—would not be fully consolidated in the Control Revolution until the pitchman could himself enter every home in word and image. This began to occur in the 1880s with the daily mass-circulation newspaper and the 1890s with the mass-circulation magazine, both well illustrated with half-tone photographs by the turn of the century. After 1922 the pitchman's voice came into homes as well with the development of yet another new form of mass communication: broadcasting.

The Rise of Modern Mass Media

From early industrialization through the 1880s, as we found in Chapter 6, increasing control of consumption had come through the coevolution of mass media and their messages to attract, hold, and imprint the mass attention. Power-driven printing, the major mass medium before broadcasting, developed rapidly throughout the period of control crisis with a steady progression of innovations: the electric press (1839) and rotary printing (1846), wood pulp and rag paper and the curved stereotype plate (1854), paper-folding machines (1856), the mechanical typesetter (1857), high-speed printing and folding press (1875), half-tone engraving (1880), and linotype (1886). These new technologies reduced costs even as they speeded printing, enabling daily newspaper circulations to stay ahead of population growth in even the largest and fastest growing urban areas.

Stereotyping, the casting of metal in pulp-paper molds to produce printing plates, allowed for print advertisements of more than a single column's width without distracting vertical rules. Despite the great expense of stereotyping machinery, forty-five such systems operated in the United States by 1880. During the 1880s stereotyping itself began to give way to electrotyping, the production of plates by electrically coating with copper or nickel a type mold or an engraving, then pouring molten type metal over this coating (the copper or nickel remains the printing surface). Although stereotyping survived for the print part of newspaper production, electrotyping reduced costs and improved the quality of most other printing, particularly of illustrations. Consequently, photoengraved line illustrations soon competed with half-tone cuts produced by photographing the subject through a wire or glass screen so that the copper printing plate becomes sensitized—and later acid-etched—in a dotted pattern, the larger dots creating darker areas. By the turn of the century various half-tone processes were used by most large U.S. dailies to print photographs regularly (Lee 1937, pp. 97–132).

Half-tone illustrated mass printing brought a new medium of mass advertising: the mass-circulation magazine. Pulp "mail-order magazines," so called because mail-order advertisements engulfed what little popular fiction they contained, had reached circulations of five hundred thousand by the early 1870s, with several million subscribers—mostly rural and poorly educated—in the late 1880s. Polite magazines like *Scribner's* (later *Century*), *Harper's*, and the *Atlantic*, underwritten by large publishing houses and dependent on a highly literate and therefore small readership, could not easily imitate the mail-order success, which depended on a high percentage of advertising pages, low subscription rates (fifty cents a year or less), and staggering promotional budgets (mostly for newspaper advertising) to assure a mass audience. As late as 1887 only four such general circulation magazines claimed as many as 100,000 readers; these four—led by *Century* with 220,000 and *Harper's* with 180,000—had a combined readership of only 600,000 (Mott 1957, p. 8).

In 1891 Frank Munsey, among the first American publishers to appreciate the implications of half-tone reproduction for advertising and mass circulation, launched *Munsey's Magazine*, a middle-brow publication that featured, in the words of Fox (1984, p. 34), "light, topical pieces and many pictures of women in discreet stages of undress." Content alone, however, even pictorial content, could not substitute for low prices and heavy promotion. After two modestly successful years Munsey startled competitors by cutting his magazine's newsstand price to ten cents and reducing annual subscriptions from three dollars to one. When the magazine monopoly refused to handle the new publication, claiming distribution would not be profitable on the percentage of only ten cents, Munsey resorted to newspaper and direct-mail advertising and created his own distribution company to deal directly with news dealers (Britt 1935).

As *Munsey's Magazine's* circulation climbed steadily, other publishers adopted the same formula of half-tone illustration combined with the low prices, high ad volume, and large promotion budgets pioneered earlier by the mail-order magazines. Cyrus Curtis, who had to double the subscription cost of his eight-page *Ladies' Home Journal* to match Munsey's dollar a year, began an expansion campaign—based on $300,000 in credit and notes from N. W. Ayer—that soon included $200,000 a year for advertising. The *Journal's* circulation, 50,000 in 1884, had by 1895 reached 750,000, almost double that of any other magazine (Bok 1923). In that year two sensationalist muckraking magazines, *Cosmopolitan* (established 1886) and *McClure's* (1893), also dropped their newsstand prices to a dime. The readership of *McClure's* expanded so fast that it actually lost money as press runs out-paced funds to be collected on advertising contracted at rates based on earlier, much smaller circulations. Within a decade twenty general magazines would claim circulations of 100,000 or more and a combined readership of 5.5 million (Mott 1957, p. 8).

With the rise of the mass-circulation magazine, a refinement of the essential idea of the mail-order magazine, printing had at last become—some four centuries after Gutenberg—not merely a means of mass production but also a mass *medium*, a new channel for advertising and hence the stimulation and control of mass production itself. Just such a transformation describes the success of E. C. Allen, a book peddler from Augusta, Maine, who in 1869 at age nineteen decided to launch a pulp magazine, the *People's Literary Companion*, solely for the purpose of selling his formula for soap powder by mail order. This single-minded goal encouraged Allen to keep the subscription price low and his own promotional budget high, the double feedback loop by which advertising fosters its own means of mass delivery. By his death in 1891 Allen had become America's leading magazine publisher, a millionaire presiding over a dozen titles with several million readers, a six-story publishing plant, and five hundred employees (Rowell 1906; Pope 1983).

The one element of the modern mass-circulation magazine that Allen did not develop—graphic illustration—awaited further innovation in printing technology. The first half-tone engraving, entitled "Scene in Shantytown, N.Y.," appeared in an American newspaper, the New

York *Daily Graphic*, on March 4, 1880. The following year *Scribner's Monthly* signaled the imminent transformation of American mass media when it adopted a new name: *Century Illustrated Magazine.*

By 1886 the British manufacturer of Pear's Soap had paid twenty-three hundred pounds sterling to artist Sir John Millais for a single painting to use in its advertisements (Presbrey 1929, pp. 98–101). Works of many well-known American artists and illustrators appeared in U.S. periodicals by the 1890s, especially following importation of the color rotogravure press from England after 1904. As late as 1894, only 30 percent of advertisements in four selected publications contained illustrations; the proportion increased steadily to nearly 90 percent by 1919 (Kitson 1921, p. 12). In the same year at least one New York advertising agency could boast an art department of ten people (Pope 1983, p. 140). Thus did the power of visual communication to stimulate and control consumer demand come to be realized—nearly a half-century before commercial television—in the mass-circulation magazine.

New technologies made possible the modern mass-circulation daily newspaper at about the same time. Except for the linotype, developed after 1886, the technological revolution in power mass printing had been essentially completed in 1883, when Joseph Pulitzer took over the New York *World* and transformed it into what most newspaper historians consider America's first modern newspaper. The following half-century, the transitional period of the Control Revolution, brought several printing innovations: the sextuple press (1891), which could print and fold 90,000 four-page newspapers in an hour; the web-fed four-color rotary (1892) and color rotogravure (1904) presses; the automatic plate-casting and finishing machine (1900), which greatly increased the speed of stereotype printing; and the teletypesetter (1932), a paper tape punch and drive for linotype. Innovation also continued outside of the printing plant, including the first journalism school (1908) at the University of Missouri, Columbia, the use of airplanes to cover stories (1920), more portable photography using flashbulbs (1930), radio facsimile transmission of syndicate news photographs direct to newsrooms (1935), microfilming of daily issues (1936), and transmission of a daily newspaper (the St. Louis *Post-Dispatch*) via radio facsimile (1938).

With these technical inventions mass publishers could extend the imagination displayed in the human interest illustration (1870) and illustrated daily newspaper (1873) to other new forms of audience-grabbing content: the comic book (1904), originally compilations of colored cartoons previously published in newspapers; the crossword puzzle (1913), special rotogravure section (1914), illustrated daily tabloid (1919), and composite photographic layout or "composograph" (1925). With the systematization of rural free delivery by the U.S. Post Office in 1898, the local medium of the big city newspaper increasingly integrated entire regions with daily communication (Fig. 8.3). More than a billion periodicals traveled rural mail routes in 1911, nearly two billion by 1929. Daniel Boorstin describes the social impact:

> When a weekly trip to the village post office was the farmer's only way of receiving mail, it was pointless for him to subscribe to a daily newspaper and periodically receive an armful of stale news. Then his needs were best served by the country weeklies. As early as 1902, *Editor and Publisher* noted that "the daily newspapers have never had such a boom in circulation as they have since the free rural delivery was established." Areas with RFD were quickest to subscribe to dailies. Some farmers who never before had a chance to receive a daily ration of fresh news from the city,

Figure 8.3. Rural free delivery in Lafayette, Indiana, c. 1898. With the U.S. Post Office's systematization of RFD in this year, big-city newspapers began to travel rural mail routes (more than a billion copies annually by 1911). By bringing more current news and advertising to isolated farmers, daily newspapers quickly supplanted country weeklies in areas with RFD. (Courtesy of the Smithsonian Institution.)

gorged themselves with two or even three daily papers . . . For the most part, the city dailies which now reached the farmer for the first time brought him the news and advertisements of a wider world. (1973, pp. 135–136)

The growth and decline of the daily newspaper as a means of mass communication in the United States perfectly parallels the transition to the Control Revolution in mass consumption. On the basis of daily circulation per household, the largest decade increase (83 percent) came in 1880–1890, the period when Henry Crowell and most other adopters of continuous-processing technologies first confronted the crisis in control of consumption—the need to create new markets and to stimulate demand. The peak of the newspaper's importance as a mass medium came in the two decades 1910–1930, the transitional period to a sustained Control Revolution, when circulation held steady at an all-time high of 1.3 daily papers per household, up from 0.2 papers in 1850 and 0.9 in 1900 (De Fleur and Ball-Rokeach 1982, pp. 40–41).

Extrapolation from the earliest time series data available suggests that, at the advent of broadcasting in America, total advertising expenditures went 48 percent to newspapers, 18 percent to direct mail, 9 percent to magazines, 3 percent to special business papers, and 2 percent to outdoor display, the remaining 20 percent going to hundreds of other more specialized media like the country weeklies described by Boorstin (U.S. Bureau of the Census 1975, pp. 855–856). Just as advertisers depended primarily on daily newspapers, these same publications depended increasingly on advertising: nearly two-thirds of their revenue came from advertising by 1919, up from about 40 percent in 1879 (Pope 1983, pp. 136–137). In 1902 Mandel Brothers, a Chicago department store, made big news when it contracted with the *Tribune* to run its full-page advertisements six days a week throughout the entire year for a flat annual fee of $100,000 (Boorstin 1973, p. 106). By 1940, however, radio had drawn away 10.3 percent of total U.S. advertising dollars; of these ten percentage points, eight or nine appeared to come at the expense of newspapers (U.S. Bureau of the Census 1975, p. 856).

No better symbol of both the newspaper's peak and decline might be found than the 1920 Presidential election, which pitted Warren G. Harding against James M. Cox, both Ohio newspaper publishers. A few of their readers might well have learned via radio station KDKA in Pittsburgh of Harding's victory, the first news event to be covered by the new broadcast medium (Barnouw 1975, pp. 32–33). Despite continuing innovations in mass publishing, and even though newspapers have always attracted a greater share of advertising than have radio and television combined, after the 1920s bureaucratic control of consumption based on national advertising came increasingly to depend on radio and television broadcasting.

The Rise of Broadcasting

Table 8.2 presents selected events in the development of broadcasting through 1939, the transitional period to a sustained Control Revolution. As the chronology shows, commercial radio broadcasting began in the United States in 1922, commercial television broadcasting in 1939 (on a "limited" basis—meaning stations could invite sponsors to experiment for defrayed costs but could not sell time—until 1941). Radio advertising grew sharply after 1931, as the Great Depression set in and people spent more time at home, but commercials came increasingly to include contests, offers of premiums, and merchandising schemes—shades of Henry Crowell and the 1890s. As Erik Barnouw, a leading historian of broadcasting, describes radio advertising in the early 1930s, "Commercials, which had been brief and diffident in NBC's first days (1926), were becoming long and unrelenting—but successful instruments of merchandising" (1975, p. 73).

Table 8.2. Selected events in the development of broadcast mass media, 1900–1939

Year	Development
1899	Guglielmo Marconi visits America, forms American Marconi, demonstrates radio for U.S. military, precipitates wireless craze
1906	Reginald Fessenden broadcasts Christmas Eve program of song, verse, violin solo, and a speech from Brant Rock, Mass.
1907	De Forest Radio Telephone Company begins broadcasts in New York
1912	News of *Titanic* reaches U.S. via Marconi operator David Sarnoff
1919	General Electric forms Radio Corporation of America (RCA) to take over assets of American Marconi
1920	Sarnoff urges RCA to market a simple "Radio Music Box" for $75 Radio station (KDKA, Pittsburgh) licensed, begins broadcasting
1922	Commercial radio begun, first ten minutes purchased for $100

Year	Development
1924	Network broadcasting extends to Pacific Coast when President Coolidge speech is carried by 23 stations Network sponsored radio program ("The Eveready Hour") carried by stations in three cities
1925	Telecast of an object in motion—a model windmill with turning blades—demonstrated from radio station NOF, Washington, D.C.
1926	National Broadcasting (NBC) organized by GE, RCA, Westinghouse
1927	U.S. Radio Commission created to license broadcasting stations, fix wavelengths and hours of operation Columbia Phonograph Broadcasting System (later CBS) formed Television demonstrated by Commerce Secretary Herbert Hoover in Washington addressing Bell Laboratories group in New York City
1928	Television receivers installed in three homes in Schenectady by General Electric and RCA, regular programs begun Federal Radio Commission issues television license
1929	Color television demonstrated publicly by Bell Laboratories
1930	Mass production of automobile radios begins Home reception of television demonstrated in New York City with half-hour program broadcast 6 miles
1931	Increase in radio advertising, contests, merchandising schemes
1932	"Radio Priest" Coughlin program processes 80,000 fan letters weekly, receives 1.2 million following radio attack on President Hoover Telecast by Democratic National Committee from CBS, New York
1933	Politician (Senator Huey Long) purchases national radio time
1934	Federal Communications Commission (FCC) established to centralize regulation of interstate and foreign commerce by wire and radio
1936	Republicans adapt soap operas to dramatize New Deal horrors, help establish radio as a major tool of national political campaigns
1937	Prerecorded program (description of Hindenburg crash) broadcast coast to coast by NBC's Red and Blue radio networks Mobile television unit—motor van with control apparatus and microwave transmitter—introduced by NBC for outdoor coverage
1938	Radio program broadcast from a tape recording (Millertape) Orson Welles's broadcast "War of the Worlds" causes national panic
1939	Television begins on limited commercial basis, soon on 23 stations

This change might be explained by the rapid diffusion of radio sets. As early as 1916 David Sarnoff—famous as the American Marconi key operator who relayed news of the *Titanic* disaster four years earlier— had urged his company to mass market cheap "Radio Music Boxes," an idea the general manager dismissed as "harebrained" (Barnouw 1975, p. 36). Four years later, after the company had become RCA, Sarnoff presented a merchandising plan predicting sales of 100,000 sets, each priced at $75, the first year. After the immediate success of the first licensed station, KDKA of Pittsburgh (Fig. 8.4), in the same year, even RCA could not keep up with the demand for receivers. Some 60,000 households had them in 1922, 400,000 (1.5 percent) in 1923, 1.25 million (4.7 percent)—compared to Sarnoff's original estimate of one million—in 1924 (U.S. Bureau of the Census 1975, p. 796; Sterling 1984, p. 222).

Economist John Kenneth Galbraith has frequently traced the rise of modern mass media to a new American mass affluence (Galbraith 1956, chap. 8; 1967, chap. 18; 1976, chap. 11). His chain of causation,

Figure 8.4. In 1920 the U.S. Department of Commerce licensed the first radio station, KDKA of Pittsburgh. Broadcasting began with the returns of the presidential election involving Warren G. Harding and James M. Cox, two newspaper editors. Eighteen months later 220 stations crowded America's airwaves. Shown here is a 1921 broadcast of KDKA's pioneering farm program, "The Farmers' Pioneer." (Courtesy of the Westinghouse Broadcasting Company, Inc.)

although not following from the Industrial Revolution with its increases in material processing speed and volume and the resulting need to stimulate and control consumption, might nevertheless be reconciled with such an explanation. According to Galbraith,

a means of mass communication was not necessary when the wants of the masses were anchored primarily in physical need. The masses could not then be persuaded as to their spending—this went for basic foods and shelter. The wants of a well-to-do minority could be managed. But since this minority was generally literate or sought to seem so, it could be reached selectively by newspapers and magazines, the circulation of which was confined to the literate community. With mass affluence, and therewith the possibility of mass management of demand, these media no longer served. Technology, once again, solved the problem that it created. Coincidentally with rising mass incomes came first radio and then television. These, in their capacity to hold effortless interest, their accessibility over the entire cultural spectrum and their independence of any educational qualification, were admirably suited to mass persuasion. Radio and more especially television have, in consequence, become the prime instruments for the management of consumer demand.   (1967, p. 190)

But the early history of broadcasting suggests little justification for Galbraith's emphasis on growing affluence. Even the Great Depression, coming in the first decade of commercial radio, did not much slow the growth of the new mass medium. At the time of the Great Crash of 1929, 10.25 million households (34.6 percent) had radio; 15 million receivers had been produced. Although production continued to fall off for the next three years, distribution of overstock continued: 13.75 million households (45.8 percent) had radio sets in 1930, 16.7 million (55.2 percent) in 1931, 27.5 million (79.9 percent) by the end of the 1930s.

Clearly the Depression did not much dampen the diffusion of radio receivers. The number of U.S. stations—exploding to 556 from 30 in 1923 alone—had already peaked at 681 by 1927 and remained fairly stable near 600 throughout the period 1929–1936, then rose quickly to 851 by 1941. As the regularly scheduled network programming quadrupled from four to sixteen hours per week between 1930 and 1935, average listening time rose nearly forty-five minutes to four hours, forty-eight minutes per day (Sterling 1984, pp. 5, 172, 212, 220, 222). Barnouw assesses the impact of the Depression on radio:

The broadcasting industry, though momentarily jolted by the Depression, had in the long run been helped by it. As theater and film audiences shrank, home audiences grew. Broadcasting had won an almost irrational loyalty among listeners. According to social workers, destitute families that had to give up an icebox or furniture or bedding still clung to radio as to a last link with humanity. Many factors contributed to this. Radio brought into homes President Roosevelt's "Fireside Chats"—an important cohesive force during darkest Depression days. At the same time, troubles over-taking theater and vaudeville were bringing a new surge of talents to radio audiences . . . Meanwhile daytime serials had developed an extraordinary hold over home audiences . . . Many expressed a dire dependence on serials. Thanks to this devotion, many businesses were making a financial comeback through radio sponsorship. If radio was increasingly successful, its tone was also increasingly—and aggressively— commercial.   (1975, pp. 72–73)

Annual advertising expenditures for all media—both broadcast and print—had quadrupled (to $3.43 billion) between 1904 and the Great Crash; they then plummeted four straight years, to the 1933 low of $1.3 billion, and did not regain their 1929 level until 1947 (U.S. Bureau of the Census 1975, p. 856). By sharp contrast, radio advertising—7 percent of all advertising (roughly today's level) by 1935—rose annually thereafter, increasing from $113 million in 1935 (the earliest year available) to $247 million (11 percent) in 1941 and the medium's all-time relative high of 15 percent ($424 million) in 1945. Rapid diffusion of commercial television, introduced in 1939, cut radio's percentage of the total expenditures for advertising in half between 1945 and 1954, although the share for both broadcast media combined stands today relatively steady at 30 percent (Sterling 1984, pp. 83–85).

Broadcasting in Control of Mass Consumption

With the development after 1924 of coast-to-coast broadcasting networks, both radio and television extended the control of consumption pioneered by the mass-circulation publications or at least certain of its essential features: low cost, large audience, ubiquitous advertising. As Sarnoff had envisioned, radios could be mass-produced cheaply: receivers had an average factory value of fifty dollars in 1922, less than thirty-three dollars in 1939—only twenty-seven dollars discounting inflation (Sterling 1984, pp. 212, 214). Like nineteenth-century publishers Allen, Munsey, and Curtis, the radio industry believed in heavy promotion: "The merchandising attention being given to this new means of communication," *Printers' Ink* noted in 1922, "is perfectly astonishing." In contrast to the twenty magazines that had attained circulations of 100,000 by 1905 when two million read Pulitzer's *World*, radio broadcasts in the mid-1920s reached fifty million listeners; by the late 1930s the programming of a single advertising agency received one million fan letters per week (Fox 1984, pp. 152, 160).

Unlike newspapers and magazines, broadcasting could reach people engaged in virtually any kind of waking activity and hence quickly diffused not only through the population but over the hours of the day. Housewives, advertisers originally supposed, had little time for radio; soap operas, introduced in 1929, had become firmly established—along with daytime programming—by 1933. After mass production of car radios began in 1930, a new niche in Americans' daily routines could be filled by broadcasting. One-quarter of all U.S. automobiles had receivers by 1939, one-half by 1951 (when saturation of households exceeded 95 percent). So pervasive did the medium become that listening time for radio did not suffer much even from television: the 1935–1943 peak of four hours, forty-eight minutes for radio has today declined less than one-third, and it has been supplemented by nearly seven hours of television for a total average exposure to broadcasting exceeding ten hours per day (Sterling 1984, pp. 211–223).

As a medium for those who hoped to control mass behavior, radio offered numerous advantages over print media. Like graphics but unlike the printed word, radio could influence illiterates (6 percent of U.S. adults in 1920) and preliterate children, so that Ipana toothpaste, for example, could make its radio pitch for "the one in the red and yellow tube." Unlike newspaper and magazine ads, radio commercials could not be skipped over—they interrupted desired programming and could follow listeners from room to room. Even the penniless and disinterested did not escape as radio gained acceptance in public places and increasingly spilled out of homes and automobiles onto public streets (with the mass marketing of transistor radios after 1954, virtually no escape would be assured). Not only could one listen to radio while engaged in other activities, including reading, one could continue to listen long after becoming too tired to do anything else—so that broadcasting promised (or threatened) to fill every waking moment of the day.

Unlike communication via print, broadcasting could be received by groups—a family in its living room, friends driving to a shopping center, barroom patrons discussing politics. Because it carried the human voice, broadcasting could seem more interpersonal, more intimate, more persuasive than other forms of mass communication. Because it flowed continually, broadcast content could not be scrutinized or saved, a disadvantage compared to print for some advertisers, an advantage for others—and a feature that advertising specialists would learn to exploit. Much as crowds once lined the New York docks, awaiting a ship's arrival from England with the next installment of a Dickens novel, radio audiences now anticipated each new broadcast of their favorite programs, scheduled their lives around them, and remembered these and other unscheduled "media events" throughout their lifetimes. Broadcasting is "the only form of advertising that runs like a train," as one radio specialist put it, "that people wait for, that becomes an event or institution in their lives" (Kendrick 1969, p. 115; Fox 1984, p. 150).

During the reign of the mass print media, as we have seen, advertisers paid rates based on a publication's circulation, well measured after 1914 by the Audit Bureau of Circulations. For radio, by contrast, audience size—measured by A. C. Nielsen after 1935—depended much more on the program (and its time of airing) than on the particular station or network that broadcast it. In other words, the content surrounding a commercial and not its means of delivery proved to be more salient to successful advertising in radio, a fact that led sponsors and their advertising agencies to struggle for greater control over programming. Increasingly they produced the highest-rated shows themselves, leaving the networks to fill less attractive time slots with more educational and cultural—but less popular—programming.

Networks, however, clung to the older print system of charging advertisers according to audience size. While magazine and newspaper editors and publishers might justify higher fees for increased circulation through their efforts to improve their products, radio networks in effect punished a sponsor who produced programming by pegging rates directly according to his degree of success. Advertisers, for their money, enjoyed increased control over communication with prospective customers, derived from largely bypassing the outside editorial control still maintained by the print media. John Kenneth Galbraith makes much the same point about the corporate planning system more generally:

> The planning system has little direct power over channels of written communication . . . The dissenter to the needed beliefs of the planning system has little problem or risk in expressing his dissent. If nothing else,

the fact that most instruments of literary communication—newspapers, magazines, book publishers—must be manned by intellectuals ensures that the goals of the intellectuals will be respected . . . In one area the planning system is uniquely powerful, although less in the propagation of ideas than in general mental conditioning. This is radio and especially television broadcasting . . . These are essential for effective management of demand and thus for industrial planning. The process by which this management is accomplished, the iterated and reiterated emphasis on the real and assumed virtues of goods, is powerful propaganda for the values and goals of the system. It reaches to all cultural levels. In the United States there is no satisfactory noncommercial alternative.   (1967, p. 342)

As radio audiences grew in the 1920s and networks charged higher and higher rates for advertising, sponsors increasingly cut their print budgets to stay on the air. Some newspapers retaliated by refusing to publish radio schedules or advertising. After the Scripps-Howard newspaper chain broke ranks, however, publishers quickly adopted the opposite strategy, increasingly buying into radio and television and actively promoting the new broadcast media. Newspapers owned only 34 stations (5.6 percent) as late as 1929, but 122 (32.7 percent) in 1940. Their percentage of AM stations—although not the number—declined thereafter as publishers' attention shifted to the new FM radio and then to television. Meanwhile, radio passed magazines in advertising revenue in 1939, when newspapers' share fell to 40 percent of the total. Today, magazines and newspapers combined receive only about one-third of advertising expenditures but own about 30 percent of U.S. television stations and about 8 percent of FM and 6 percent of AM radio stations (Sterling 1984, pp. 50–52).

Control over the broadcast mass media became increasingly concentrated via network and group ownership and network affiliation during the same period. In 1929 twelve group owners (those with holdings in more than one market) controlled twenty radio stations (3.3 percent); by 1939 thirty-nine group owners had 109 stations (14.3 percent). Today group ownership stands at about 70 percent for television and 30 percent for radio; the three commercial television networks—ABC, CBS, and NBC—rank among those group owners with the largest combined audiences. In 1929 network affiliation united forty-four stations (6 percent) into two networks (NBC and CBS); the percentage of stations in one of four networks (Mutual appeared in 1934, ABC in 1942) increased through World War II, reaching 50 percent by 1938 and an all-time high of 97 percent in 1947. Today about

70 percent of radio stations are affiliated with one of the same four networks, about 80 percent of television stations with one of three commercial networks (Sterling 1984).

Radio stations did not take long to discover the advantages of networking. On October 7, 1922, hardly a month after the first commercial broadcast, stations WJZ, Newark, New Jersey, and WGY, Schenectady, New York, effected a chain broadcast when they connected to a single field microphone for the World Series baseball game at the Polo Grounds, New York City, with ordinary telegraph lines, which unfortunately did not transmit the highest and lowest frequencies. Less than three months later, stations WEAF, New York, and WNAC, Boston, linked with amplifiers and repeater points for more faithful sound reproduction. On June 21, 1923, the first interregional presidential address, President Harding's World Court speech in St. Louis, was broadcast by station KSD of that city and WEAF in New York.

Control of Public Opinion and Political Behavior

Ironically, it was Calvin Coolidge, inaugurated as President following Harding's sudden death on August 3, 1923, who first used the national radio network to consolidate political control. Despite his nickname of "Silent Cal," Coolidge made numerous radio broadcasts to the nation. The first, his December 6 address to a joint session of Congress, went to radio stations via telephone for simultaneous broadcast in six cities, including Dallas, Kansas City, and Providence.

Two months later, facing nomination and election campaigns, Coolidge made the first overtly "political" broadcast by a President, a Lincoln Day dinner speech to the National Republican Club at the Waldorf Astoria, New York, carried by five stations between Providence and Washington, D.C., and received by an estimated five million listeners. On February 22, 1924, only two weeks after the first coast-to-coast experimental broadcast had been heard by an estimated fifty million people, Coolidge addressed the entire continent from his presidential study using a forty-two-station hookup, the first radio broadcast from the White House. That June fifteen affiliates of the NBC network covered Coolidge's nomination by the Republican Party meeting in Cleveland, the first radio broadcasts from a national political convention (the NBC anchor, Graham McNamee, had announced the two-station World Series hookup in 1922). On October 23, twelve days before the presidential election, Coolidge made the first network broadcast to the West Coast, a forty-five-minute speech dedicating the Chamber of Commerce building in Washington, D.C., which was carried by twenty-three stations, including those in Los Angeles, Portland, and Seattle. The following March 4 twenty-four stations broadcast Coolidge's inauguraton.

President Coolidge emphasized the importance of the new mass media by agreeing to address the annual meeting of the American Association of Advertising Agencies the following year, an occasion that, according to Pope (1983, p. 112), "exemplified the respectability which advertising men had achieved by the 1920s." The President acknowledged before the convention delegates the power of the forces he had used to rally the nation following Harding's death and to assure his own nomination and election. Mass advertising through mass media, Coolidge said, "is the most potent influence in adapting and changing the habits and modes of life, affecting what we eat, what we wear, and the work and play of the whole nation." This new capacity to control mass behavior, the President told the assembled advertisers, "is great power that has been entrusted to your keeping" (Presbrey 1929, p. 625).

Others learned of this power almost by accident. In Detroit Father Charles E. Coughlin, at the Shrine of the Little Flower, had begun radio broadcasts of his sermons (Fig. 8.5). By 1927 this "Radio Priest," as he was called, received four thousand fan letters per week. Five years later he processed eighty thousand letters weekly with the help of ninety-six clerks. After his 1932 radio speech attacking President Herbert Hoover as "the Holy Ghost of the rich, the protective angel of Wall Street," Coughlin received more than 1.2 million letters from listeners. He attempted to build this radio audience into his own political organization, the National Union for Social Justice, and developed it into an American voice for Adolf Hitler before his support faded. As Boorstin (1973, p. 476) describes the new power of broadcasting to mobilize mass behavior, "the privacy of radio reception was an aid to petty would-be dictators, merchants of hate, and demagogues who secured living-room audiences of Americans who might have hesitated to attend one of their public rallies."

Louisiana Senator Huey Long became the first American politician actually to buy radio time to reach a national audience (Williams 1969). In March 1933, after he had introduced his Long Plan for the Redistribution of Wealth, a system to tax the capital of those worth more than $1 million, Long hoped to bypass the press in presenting his plan

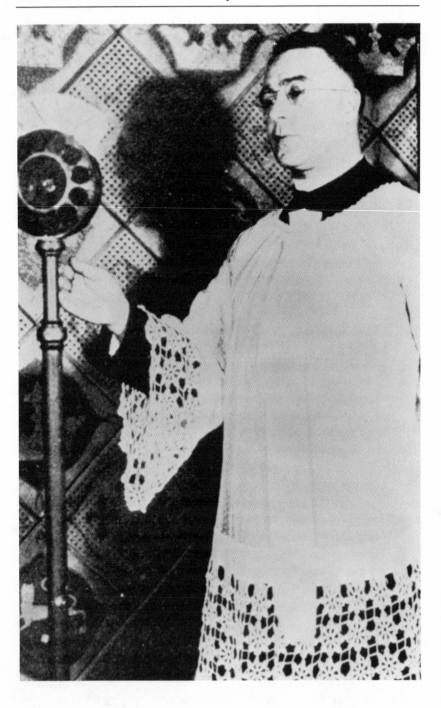

directly to the American public via the NBC radio network, then eighty-eight stations, about 15 percent of those in the country (Sterling 1984, p. 12). In his address Long introduced an innovative use of the radio network to build an audience by tapping into the nation's interpersonal and social networks via its telecommunications network: "Hello friends, this is Huey Long speaking," he began. "And I have some important things to tell you. Before I begin I want you to do me a favor. I am going to talk along for four or five minutes, just to keep things going. While I'm doing it I want you to go to the telephone and call up five of your friends, and tell them Huey is on the air." Such skilled use of the new medium came easily to Long, a celebrity of late-night radio who had dubbed himself "the Kingfish" after a protagonist in the *Amos 'n' Andy* radio series, thereby associating himself with a commercial broadcast image much as presidential aspirants in the 1984 campaign would use the fast-food slogan "Where's the beef?"

By the 1936 presidential campaign broadcasting had established itself as crucial to any national effort to influence public opinion and voting. In a prototype of more recent presidential debates, Michigan Senator Arthur Vandenberg, a leading Republican spokesman, rebutted recorded excerpts of President Roosevelt's speeches. The Republicans also attempted to counter the popular appeal of the president's Fireside Chats by adapting soap operas—called "Liberty at the Crossroads"—to dramatize what they considered horrors of the New Deal. The Democrats, meanwhile, had already moved on to experiment with television: their National Committee made the first political campaign telecast from CBS studios, New York City, on October 11, 1932, three weeks before Roosevelt's first election to the White House, when only a few hundred households had television. Clearly, by the early 1930s politicians had grasped the importance of broadcasting to influence public opinion and to affect mass behavior.

If any conclusive demonstration of this influence had been necessary, it came in 1938. On Halloween eve, Orson Welles's Mercury Theater, a CBS showcase for serious dramatic productions, broadcast an ad-

---

Figure 8.5. Father Charles E. Coughlin broadcasting live from the Shrine of the Little Flower in Detroit, Michigan. By 1927 this "Radio Priest" received four thousand fan letters per week. In 1932 his mail peaked at 1.2 million letters following a single broadcast, an attack on President Herbert Hoover. Coughlin's popularity faded after he attempted to make his political organization, the National Union for Justice, into an American voice for Adolf Hitler.

aptation of H. G. Wells's novel *War of the Worlds* that described a Martian invasion of the United States as reported live by announcers on the scene. More than six million people heard some part of the program and, despite disclaimers of authenticity throughout the broadcast, an estimated one million listeners believed the conquest of earth by Mars to be actually in progress. Before the hour-long program had ended, many telephone exchanges had jammed and hundreds of thousands of people crowded evacuation routes recommended in the dramatization (Koch 1970). Hadley Cantril, a Princeton University social psychologist who conducted 135 postbroadcast interviews to catalog reactions to the program, found general panic across all geographical, educational, and economic groupings (Cantril 1940). "A tidal wave of terror swept the nation," one newspaper reported the following morning.

Government Control of Broadcasting

Radio's growing power to influence public opinion and mass behavior did not go unnoticed by the federal government. Its efforts to control broadcasting began as early as 1912, when a new law required that each transmitter be federally licensed. The requirement went unheeded by many amateur broadcasters, then numbering in the thousands, who delighted—much like today's computer "hackers"—in longdistance mischief, to the extent of allegedly sending fake orders to U.S. Navy ships at sea, purportedly from admirals (Lessing 1956). After the United States declared war on Germany, the navy sealed all amateur radio equipment and seized all commercial shore installations, most of which were American Marconi stations. These served for control of shipping, naval intelligence, and eventually for propaganda, including President Wilson's Fourteen Points, broadcast throughout the world from transoceanic transmitters (Schubert 1928). The official U.S. history of naval communications (U.S. Navy Department 1963) calls this period the "Golden Age."

After the armistice in the fall of 1918 the navy proposed to Congress, in a bill enthusiastically endorsed by the State Department, a naval monopoly of all U.S. broadcasting. Opposition from the amateurs, many of whom had served in war communication, helped to table the bill (Lessing 1956). The navy then pushed for its second choice, a private monopoly. This was established in October 1919 with the Radio Corporation of America (RCA), to which American Marconi, its stations still held by the U.S. government, agreed to transfer all assets. Incorporation papers stipulated that only American citizens could serve as directors or officers and that foreigners could not hold more than 20 percent of the stock; the board included a seat for a federal representative. General Electric replaced British Marconi as parent company; navy admiral W. H. G. Bullard became its government representative. RCA soon had three additional corporate partners, Westinghouse, AT&T, and United Fruit, together representing some two thousand electronics and communications patents (Archer 1938).

Because of this patent monopoly, which seemed to defy the Sherman and Clayton antitrust acts, the federal government—despite its involvement in forming RCA—began to take antimonopoly action. The Federal Trade Commission launched an investigation of radio monopoly in 1923 and the following year issued a formal charge that the partners in RCA had "combined and conspired" to restrain trade and to create a monopoly in the manufacture and sale of radio devices and in domestic and transoceanic broadcasting and telecommunications. After secret negotiations the corporations in 1926 agreed on a response: their stations would be pooled in a new company, National Broadcasting (NBC), to be owned jointly by RCA, GE, and Westinghouse (the latter two still owning large shares of RCA); AT&T would withdraw from active broadcasting by selling its stations or merging them with others, but it would lease under long-term contract the lines to form the NBC network (Archer 1939).

In the same year a U.S. district court ruled, in *United States v. Zenith*, that the Secretary of Commerce had exceeded the licensing authority of the 1912 law, a decision that soon brought anarchy to the airwaves. Congress responded by passing the Radio Act of 1927, which established the Federal Radio Commission to handle licensing. Antimonopoly sentiment found its way into several clauses of the new law, including one repudiating the idea that wavelengths could be owned as permanent property, another specifically barring licenses to parties found guilty of "unfair methods of competition." Within the month RCA voluntarily loosened its patent policy and agreed to license several of its competitors in return for royalties based on sales (Barnouw 1966). The following year, satisfied with this voluntary action, the FTC dropped its monopoly investigation of broadcasting.

The 1929 stock market crash, which exposed many business scandals, brought renewed antitrust zeal throughout the country. In May 1930 the Justice Department filed suit against RCA and its patent partners and demanded termination of the 1919–1921 patent agreements and an end to interlocking ownership and directorates. Late the following year AT&T announced its withdrawal from the earlier cross-licensing agreements, many of which were to expire soon anyway. The following year, wary of the prospect of Franklin Roosevelt gaining the White House and fearing the automatic loss of multimillion-dollar licenses if convicted in the antitrust suit, GE and Westinghouse reluctantly decided to withdraw from RCA. They retained ownership of stations to be managed by NBC, which became a wholly owned subsidiary of RCA (Archer 1939). The signing of the consent decree, less than two weeks after Roosevelt's election, effectively terminated all antitrust actions.

President Roosevelt, who quickly perfected radio broadcasts as a means to control public opinion, especially through his Fireside Chats during the banking crisis of 1933, proposed stronger government control of media: a new Federal Communication Commission (FCC) would consolidate control of broadcasting, replacing the FRC, and control of the telephone system, taken from the Interstate Commerce Commission. After passage of the Communications Act of 1934, the newly formed FCC prepared for intensive study of an emerging medium: television. Scarcely a dozen years after the advent of commercial radio, the nation's infrastructure of electronic mass communication and its control stood largely complete: competing national networks of broadcast stations, linked by cables of the telephone system and financed through advertising, would operate on temporary licenses subject to federal government control (Barnouw 1968) (Fig. 8.6).

The Rise of Market Feedback Technology

As attested by the newspaper photographs of citizens brandishing shotguns and rifles against an impending attack by Martians on Halloween eve in 1938, the broadcast media can serve as an effective means to influence mass behavior. For this influence to constitute *control* in the strong sense discussed in Chapter 1, however, mass communication must be supplemented by a reciprocal flow of information from the mass audience back to the media writers and programmers who seek to attract and hold its attention, to the advertisers who seek to stimulate and control its consumption behavior, and to the politicians who seek to influence its opinions and its vote. Consolidation of the Control Revolution after 1900 brought the steady development of just such market feedback technologies.

Freedom to LISTEN — Freedom to LOOK

As the world grows smaller, the question of international communications and world understanding grows larger. The most important phase of this problem is *Freedom to Listen* and *Freedom to Look*—for all peoples of the world.

Radio, by its very nature, is a medium of mass communication; it is a carrier of intelligence. It delivers ideas with an impact that is powerful... Its essence is freedom—liberty of thought and of speech.

Radio should make a prisoner of no man and it should make no man its slave. No one should

be forced to listen and no one compelled to refrain from listening. Always and everywhere, it should be the prerogative of every listener to turn his receiver off, of his own free will.

The principle of *Freedom to Listen* should be established for all peoples without restriction or fear. This is as important as *Freedom of Speech* and *Freedom of the Press*.

Television is on the way and moving steadily forward. Television fires the imagination, and the day is foreseen when we shall look around the earth from city to city, and nation to nation,

as easily as we now listen to global broadcasts. Therefore, *Freedom to Look* is as important as *Freedom to Listen*, for the combination of these will be the radio of the future.

The "Voice of Peace" must speak around this planet and be heard by all people everywhere, no matter what their race, or creed, or political philosophies. *

David Sarnoff
President and Chairman of the Board,
Radio Corporation of America.

*Excerpts from an address before the United States National Commission for UNESCO.

 RADIO CORPORATION of AMERICA

FREEDOM IS EVERYBODY'S BUSINESS

Figure 8.6. Within a dozen years of the advent of commercial radio, America's infrastructure of electronic mass communication had become essentially what it remains today: a few competing national networks of broadcast stations financed through advertising and operating on temporary licenses subject to federal government control. (Courtesy of the Radio Corporation of America.)

Market feedback, the flow of information from retailers and consumers back to advertisers and others seeking to control mass behavior, can take several major forms: information on sales of advertised
products or of an industry in general; other characteristics of industries
or retail establishments; surveys of mass media audiences or of consumers generally. Technologies for collecting and processing all of
these types of information had appeared by the late 1910s and continued to develop steadily through the 1930s, the period of transition for
the Control Revolution. Most historians of market research (Hower
1949; Bartels 1962; Pope 1983) trace its origins to the "commercial
research" or "trade investigation" that appeared about 1910, roughly
a twenty-year lag behind the rise of mass-circulation media and mass
advertising.

As early as 1879, as we saw in Chapter 6, the N. W. Ayer & Son
advertising agency had a prospective client asking for market research.
A manufacturer of grain threshers, the Nichols-Shepard Company,
insisted that Ayer find out where threshers sold and then compile a
list of newspapers with subscribers in the same areas. In three days,
based on telegrams to state government officials and publishers
throughout the grain belt, Ayer compiled a survey of the thresher
market, which it gave to Nichols-Shepard free in exchange for the
advertising account. Although such market information seems obviously useful to us today, manufacturers and distributors had little
need for it before crisis in the control of consumption made it necessary
and the rise of mass media and mass advertising enabled it to be
exploited. Ralph Hower, Ayer's historian, finds only sporadic and in-
frequent use of market analysis before 1900, when the agency assigned
the task to an undifferentiated "Business-Getting Department" (Hower
1949, pp. 72–77). After studying early advertising more generally,
Pope (1983, p. 141) concludes: "However elementary, Ayer's work
appears more advanced than what other agencies offered at the time.
Market research was slow in getting started."

As can be seen in Table 8.3, however, which summarizes the major
innovations in mass feedback technologies before World War II, market research developed rapidly after 1900, especially in the 1910s and
steadily thereafter, adapting smoothly to the shift in mass communication from print to broadcasting. By 1935 separate professions of
market and survey research had differentiated around a specialized
arsenal of feedback technologies: ad testing (1906), systematic retail
statistics (1910s), questionnaire surveys (1911), coded mailings (1912),

Table 8.3.   Selected innovations in market feedback technologies for control of consumption, 1900–1940

Year	Innovation
1900	Lord & Thomas agency establishes "record of results" department
1906	Lord & Thomas invites clients to test copy in different cities
c. 1910	Market research begins to be widely adopted, formalized
1910s	Statistics on retail operations collected by the Harvard Bureau of Business Research, National Retail Dry Goods Association
1911	Fifty national advertisers support a postcard-questionnaire survey of magazine readership, use it to estimate duplicate circulation Charles Parlin, Curtis Publishing, studies agricultural implement industry, calls 460-page report "commercial research"
1912	Parlin visits all cities of 50,000 population to estimate the business of department stores and dry goods wholesalers J. Walter Thompson agency commissions study of stores by category and state, to be updated every five or six years William Shryer uses concealed code to test mail order ads
1914	Association of American Advertisers organized to verify newspaper circulation figures, establishes Audit Bureau of Circulations
1916	House-to-house market interviewing introduced by *Chicago Tribune* Ad agency (Thompson), corporation (U.S. Rubber) establish their own market research departments Business text published with chapter on qualitative market analysis
1919	Textbook devoted exclusively to market research published
1920	New edition of Thompson retail survey used by 2,300 firms Curtis Publishing conducts saturation survey of Sabetha, Kansas
1921	Percival White develops concept of "measurability of markets"
1923	Radio offer of free autographed photo draws hundreds of letters, convinces advertisers of medium's large potential audience
1926	Curtis Publishing conducts a "Dry Waste Survey" by collecting and analyzing trash of 56 Philadelphia families for four weeks

Table 8.3 (*cont.*)

Year	Innovation
1928	American Tobacco tests appeal of its "Lucky Strike Dance Orchestra" radio show by halting other media ads, finds sharp sales rise
Late 1920s	George Gallup, advertising and journalism professor, begins surveys of newspaper reading habits
1929	U.S. Department of Commerce launches the Census of Distribution
	Archibald Crossley devises a radio audience measurement system based on surveys; results come to be called "Crossley Ratings"
	Survey research textbook published for market researchers
c. 1930	Morris Hansen, U.S. Census Bureau, begins to develop large-scale statistical sampling theory for survey research
1931	Manual published for field work in market research
1933	A. C. Nielsen begins retail-sales index based on druggists' records
1935	Gallup, Elmo Roper begin national surveys using scientific samples
	Nielsen adopts the audimeter to monitor radio audiences
1937	American Marketing Association Committee on Marketing Research Techniques publishes summary of best techniques
1940	R. P. Eastwood publishes *Sales Control by Quantitative Methods* on control of production and sales based on market forecasting

audits of publishers' circulations (1914), specialized market research departments and house-to-house interviewing (1916), research textbooks (1919), saturation (1920) and dry waste (1926) surveys, a census of distribution (1929), sampling theory for large-scale surveys (c. 1930), field manuals (1931), retail sales indices (1933), national opinion surveys, and audimeter monitoring of broadcast audiences (1935).

Compilations of statistics on industries and retailers, like the sporadic efforts of N. W. Ayer in the late nineteenth century, began to be systematized after 1910. Both the Harvard University Bureau of Business Research and the National Retail Dry Goods Association, for example, began to collect and publish data on retail store operations in the 1910s (Bartels 1962, p. 117). Pioneering compilations of industry statistics came from both the Curtis Publishing Company and the J. Walter Thompson advertising agency, both of which established specialized research departments in the 1910s.

When Cyrus Curtis, publisher of the *Ladies' Home Journal* and the *Saturday Evening Post*, purchased the *Country Gentleman*, he hired Charles Parlin, a young Wisconsin schoolteacher, to collect data on the agricultural implement industry, then responsible for most of his new acquisition's advertising. After six months spent interviewing manufacturers, retailers, and farmers themselves, Parlin had compiled a 460-page statistical report on farm implements—who made them, who sold them, who bought them, when, where, how, and why.

Encouraged by advertisers' reception of this report, which revealed many unsuspected facts, Curtis in 1912 sent Parlin on a tour of the hundred largest U.S. cities (those of more than fifty thousand population) to estimate the business of department stores, dry goods wholesalers, and merchant tailors. After traveling thirty-seven thousand miles and conducting 1,121 interviews, Parlin issued a four-volume report, *Department Store Lines*, which distinguished "convenience" purchases (daily, inexpensive, impulsive) from "shopping" purchases (irregular, more expensive, premeditated, postponable). Two years later Parlin completed the five-volume *Automobiles* (1914), a study of the manufacturing and distribution of about a hundred widely sold makes, which he correctly foresaw would decrease in number, and of the components of consumer purchasing decisions. This report increased automotive advertising not only in Curtis publications but in national magazines more generally (Bok 1923; Bartels 1962, pp. 108–109).

In the same year Stanley Resor, head of the J. Walter Thompson agency's branch office in Cincinnati, commissioned a national survey of retail establishments. His immediate motivation came from Red Cross Shoes, a client that Fox (1984, p. 84) describes as "hobbled by a haphazard distribution pattern, in part because no one knew where the appropriate retail outlets were located." Resor, a Yale graduate and student of the Darwinian sociologist William Graham Sumner, drew inspiration for the Thompson agency's solution—the secondary analysis of census statistics—from his favorite text, Henry Thomas Buckle's *History of Civilization in England* (1857), which used this technique to make inferences about the distribution of population and wealth. Resor's study, *Population and Its Distribution* (1912), described consumer populations surrounding all major cities and listed retailers by state and type: hardware, drug, grocery, dry goods, and clothing. Revised and expanded every five or six years, the 1920 edition had 218 pages, sold for $2.50, and found use in 2,300 companies (Fox 1984, pp. 79–94).

By 1929 the U.S. Department of Commerce had launched its own Census of Distribution and thereafter began to collect an increasing number of statistics on wholesale and retail establishments and the goods and services they distribute (U.S. Bureau of the Census 1975, pp. 834–838). In 1933 Arthur C. Nielsen, who would eventually devise the audience ratings for radio and television that still bear his name, devised a novel retail-sales index for druggists based on data from their own records. Buoyed by the demand for his index from drug advertisers, Nielsen developed a "Food Drug Index" and several other retail-sales indices. By 1937, despite the Depression, Nielsen grossed more than $1 million annually by providing feedback on retail sales (Boorstin 1973, p. 154).

Feedback from the Mass Audience

Just as Red Cross Shoes needed market information to control distribution among the rapidly diffusing retail outlets of the 1910s, businesses about this time realized a similar need for information on audiences of the proliferating mass media outlets for their advertising. In 1911, suspecting that advertising dollars might be wasted on overlapping or misdirected mass communication, R. O. Eastman, advertising manager for Kellogg breakfast foods, persuaded fifty other national advertisers to join him in a postcard-questionnaire survey of magazine readership. Results confirmed his suspicions: the rapidly growing number of mass-distribution magazines had unexpectedly complex and overlapping distributions of readers.

Bolstered by the usefulness of such information for targeting advertising and thereby maximizing the audience per dollar spent, Eastman easily persuaded the other companies to support further research. By 1917 he had established his own market research firm. His first client, General Electric, wanted to know what the trademark "Mazda" meant to American consumers. Meanwhile, the publishers themselves, prompted by the knowledge that advertisers had developed independent means to verify precisely circulation claims, began in 1914 to subsidize a new Audit Bureau of Circulation, which they gave authority to examine their own records. Within two decades more than 90 percent of daily U.S. circulation would come under the bureau's audit (Presbrey 1929; Boorstin 1973, p. 150).

By this time, however, advertisers could reach consumers via radio. As we might expect, the idea to use broadcasting in the bureaucratic control of consumption did not come immediately but only after the new medium of mass communication had been sufficiently rationalized. As historian Daniel Boorstin describes the problem, "When David Sarnoff urged his associates to invest in the new wireless 'music box,' they objected that it had no imaginable commercial future, because it depended on 'broadcasting.' That meant, of course, sending out messages 'broadcast' to persons who could not be identified, counted, or located. Since no visible connection was required between broadcaster and receiver, who could tell who was receiving the message? And who would pay for a message sent to nobody in particular?" (1973, p. 154).

The latter question had a quick answer as a result of early and inadvertent market feedback. In January 1923, less than five months after the advent of commercial radio, the cosmetic Mineralava sponsored a broadcast over WEAF, New York, titled "How I Make Up for the Movies," which offered an autographed photo of actress Marion Davies free to listeners. Requests poured in by the hundreds, a surprising response when only sixty thousand households in the entire country had receivers (Sterling 1984, p. 222). The response convinced several advertising agencies to begin advertising products—B. F. Goodrich tires, Eveready batteries, Happiness Candy, Lucky Strike cigarettes—over the same station, which was then charging $10 per minute (Banning 1946; Barnouw 1966).

As the numbers of both advertisers and listeners continued to grow through the 1920s, the cost of commercial radio time also rose. Advertisers, educated by their experience with print media circulations and goaded by newly specialized market research departments (Bartels 1962, p. 109; Pope 1983, p. 142), increasingly demanded to know whom their broadcast messages might reach and the cost relative to other media. But how to know? "Advertising on the airwaves offered the market researcher a puzzling new problem," according to Boorstin (1973, p. 154). "While a publisher could identify his subscribers and could locate his newsstand sales, the radio broadcaster could only guess who might be listening."

More precise measurement of radio audiences came in 1929, when a market researcher, Archibald Crossley, devised a system based on random telephone calls. Introduced publicly the following year, the resulting statistics soon came to be known as the "Crossley Ratings," the equivalent for broadcasting of the print media's Audit Bureau of Circulations. A few years later, sales-index pioneer Arthur C. Nielsen, an engineering graduate of the University of Wisconsin, learned of a device developed at MIT to record the times that a radio receiver had been in operation and the frequencies received. Nielsen decided to install these "audimeters" in a random sample of radio households to record listening times and stations, data that his Chicago office correlated with programs broadcast to determine the relative popularity of shows. Introduced in 1935, these audimeter "Neilsen Ratings" would require seventeen years and $15 million before turning a profit for the company; they became particularly prestigious in television after the early 1950s (Barnouw 1966, 1968). Responding to the challenge of the audimeter, Crossley studies by 1939 had begun to report the number of radio sets—up to 25 percent—left on in empty rooms (Boorstin 1973, p. 155).

Feedback on Mass Consumption

In addition to the sizes of media audiences, advertisers increasingly wanted to measure the effectiveness of the messages that reached them. As early as 1900 Lord & Thomas, a Chicago advertising agency, had established a "Record of Results" department to which it required clients to submit weekly reports of sales of advertised products and of responses to coupon and mail-order solicitations. The agency collated reports from all clients, comparing the results for all advertisements in each publication. By 1906 Lord & Thomas employed eight clerks to process information from six hundred clients in estimating the relative effectiveness of four thousand magazines and newspapers. Thus the agency rationalized control by means of increasingly sophisticated information collection, storage, and processing procedures over the highly ambiguous areas of media reach and ad effectiveness. In the summary of Stephen Fox, who quotes from *Printers' Ink* of April 18, 1906,

> The results were occasionally surprising, showing good returns from small, obscure media and poor yields from large, famous ones. Ad placement, traditionally dependent on a medium's reputation and a sixth sense of the agent's, thus acquired a more rigorous procedure. "We have a positive gauge on mediums and copy," [Morris] Lasker [partner in Lord & Thomas] declared in 1906, "such as is probably to be found nowhere else." When an ad failed, the advertiser normally did not know whether to blame the

copy or the medium. But at L & T, the question was simplified if a dozen different campaigns all bombed in one medium. "We know copy can't be wrong with all," Lasker explained, "so the paper comes under suspicion. If it is really weak, all our advertisers are out within a month, and there is a big aggregate saving." Once under way, the record-of-results department churned out its findings routinely, providing a steady flow of hard data with which to pick media and impress clients.   (1984, p. 61)

The same techniques could be used to test the relative effectiveness of different ads, ad campaigns, and approaches to mass communication more generally. Supported by quantitative evidence, Lord & Thomas after 1906 circulated more than seven thousand copies of a pamphlet, *The Book of Advertising Tests*, which invited new clients to test their former advertisements in one city against the agency's in another— probably the earliest use of comparative market experiments for ad testing. The idea brought in hundreds of letters per week from advertisers (Fox 1984, p. 50).

With the advent of commercial radio in the 1920s, advertisers adopted experimental designs to test the power of the new medium to stimulate consumption. When *The Lucky Strike Dance Orchestra*, produced by Lord & Thomas for American Tobacco, debuted on thirty-nine NBC stations in September 1928, the company stopped advertising Lucky Strike cigarettes in other media. After sales rose 47 percent in November and December, both the show—which would become *Your Hit Parade* and eventually move to television—and the medium became firmly established vehicles for cigarette advertising (Fox 1984, p. 154). During radio's era, 1922–1954 (the latter year being the one in which television first eclipsed the older medium in advertising dollars), cigarette production rose 750 percent while cigars fell 11 percent and pipe tobacco fell 50 percent (U.S. Bureau of the Census 1975, p. 690). Whatever role broadcast advertising might have played in these trends, its widely perceived influence eventually prompted Congress in March 1970 to remove cigarette ads from both radio and television effective the following January.

Just as advertisers sought to measure media audiences and the effectiveness of media messages by the 1910s, they also began to seek feedback from the intended recipients of their messages: individual consumers themselves. William A. Shryer, a Detroit publisher and mail-order advertiser, provided pioneering work on actual consumer behavior. In 1912, eager to maximize returns from his mail-order business, Shryer began to code ("key") his mailings to provide information on which appeals brought which responses from which groups of potential customers (Shryer 1912). Unlike Eastman's postcard-questionnaire survey the previous year, which sought merely to specify audiences for mass-circulation magazines, Shryer sought information on the preferences and behavior of the audience itself.

Contrary to the prevailing but untested theory that advertising had a cumulative effect (Scott 1903, 1908), Shryer found the analog of biological habituation. This major result he called "the law of diminishing returns . . . the real law of advertising" (Shryer 1912, p. 47). The model of cumulative effect, Shryer argued on behalf of his newer theory, presupposed that an advertiser could gradually add to his argument with each potential customer until the balance finally tipped in favor of purchase. This, in turn, presupposed that consumers acted out of reason, while Shryer's experiments with coded mailings suggested that "it would be unprofitable for the advertiser to center his appeals around . . . the exercise of a function so slightly developed" (Shryer 1912, pp. 79, 110).

According to Shryer's analysis, information and reasoning alone did not move people to act. Consumers did not weigh benefits against costs; they responded to stimuli, including veiled suggestions and even mere images. Therefore, Shryer concluded, expressing a view that would characterize advertising practice throughout the 1910s and 1920s (Pope 1983, chap. 6; Fox 1984, chaps. 2, 3): "It is a favorite superstition that because reason is peculiar to the human being it is his prevailing guide to action. Nothing could be much farther from the truth. Man . . . actually . . . is a creature of habits" (Shryer 1912, p. 45).

This view did not imply that market feedback and its scientific analysis would not be useful to control consumption. On the contrary, to the extent that consumers did not act according to reason, their behavior could not be predicted by any advertising "theory" and would require continuous monitoring to detect habituation to messages and other changes in preferences and habit. Just such monitoring of mass populations had begun to develop by the turn of the century in what would become the most widely used of all market feedback technologies: survey research.

The Rise of Survey Research Techology

By the 1890s widespread concern for social reform had created a general interest in social investigation and the collection of social statistical data. As early as 1895 G. Stanley Hall, educational psychologist and founder-president of Clark University, used eight hundred workers to administer sixteen different questionnaires to some eighty thousand schoolchildren (Ross 1972). America's first social science survey of a general population, the Pittsburgh Survey, was completed in 1908 with support from the Russell Sage Foundation; the word *questionnaire* entered popular American discourse from educational psychology about the same time. By 1928 a bibliography of social surveys in the United States listed nearly three thousand titles (Boorstin 1973, pp. 155, 231).

Advertisers and marketers did not take long to adopt similar techniques. By the mid-1910s Stanley Resor of J. Walter Thompson's Cincinnati office, himself a college graduate, had decided that his employees suffered from too much book learning and too little practical experience. Consequently, Resor sent all of his copywriters and artists into the field to ring doorbells, interview housewives, and develop a practical feel for consumer tastes (Fox 1984, p. 84). In 1916 the Chicago *Tribune* began more systematic house-to-house interviewing throughout its metropolitan area to gather consumer information for its advertisers (Boorstin 1973, p. 152).

During the next decade Curtis Publishing greatly advanced consumer research with a series of innovative surveys. In 1920 the company's Research Department conducted a saturation survey of 144 square miles surrounding Sabetha, Kansas, interviewing all but twenty families on brand preferences and buying behavior. Curtis also undertook a "National Pantry Survey," an inventory of cupboard shelves in 3,123 households in eighty-five neighborhoods in sixteen states. In Watertown, New York, Curtis staged an "Every-House-and-Every-Outlet Survey," obtaining detailed information on all but 727 of the city's 28,930 households (97.5 percent). During four weeks in July and August 1926 the company conducted its "Dry Waste Survey": it collected the trash from fifty-six Philadelphia families, fourteen from each income quartile, and quantitatively analyzed the resulting six thousand discarded packages and containers. The findings of such studies not only attracted advertisers to Curtis publications like the *Saturday Evening Post* and *Ladies' Home Journal* but also presumably increased the efficiency and effectiveness of their advertising and distribution—and hence control of consumption (Bok 1923; Bartels 1962; Boorstin 1973, chap. 17).

At the same time George Gallup, a professor of advertising and journalism at Northwestern University, had begun to conduct surveys of newspaper readers to find out what attracted their attention (the picture page led with 85 percent, followed by the most popular comic strip with 70 percent and editorial cartoons with 40–50 percent, a power of graphic images that would be confirmed by television). By 1931 Gallup's interviewers had rung fifteen thousand doorbells to find out which magazine ads people most remembered seeing. They discovered that the appeals advertisers used least frequently actually drew the most attention: sex and vanity, tied for ninth and last place in usage, placed first and second with women; sex scored second with men after quality.

When Gallup published these findings, they quickly became what the journal *Advertising and Selling* described in March 1932 as "probably the most discussed topic of the day" (Fox 1984, p. 138). The following month Raymond Rubicam hired Gallup away from academia to the Young & Rubicam advertising agency, a position he held for sixteen years. Annual billings of the agency, the first to use modern survey techniques to collect and analyze market feedback on a large scale, continued to rise despite the Depression, doubling to $12 million between 1927 and 1935 and rising to $22 million by 1937 and $53 million in 1945, when Young & Rubicam ranked behind only J. Walter Thompson among U.S. agencies (Fox 1984, pp. 139, 332).

Gallup's quantitative and systematized feedback from the mass audience helped to rationalize the stimulation and control of consumption far beyond his own agency. According to Stephen Fox, Gallup "showed that readers preferred short paragraphs, type 'widows' to leave white space at the end of paragraphs, and other type devices—italics, boldface, subheads—to break up the copy and maintain attention. Further, the audience preferred rectangular pictures, not odd shapes, and uncropped photos. Gallup's influence extended outside Y&R, as reflected in the trade's greater use of nudity and sex appeal, a general adoption of comic-strip formats, and the appearance of other readership services" (1984, p. 139).

While Gallup pioneered a more scientific market research, Morris Hansen, a mathematical statistician at the U.S. Census Bureau, worked to extend formal sampling theory to large-scale survey research designs. One implication of Hansen's theory, developed around 1930, was that surveys like the Curtis studies of Sabetha and Watertown and Gallup's magazine study wasted resources—their samples could have been much smaller to yield roughly the same accuracy of results. Diffusion of Hansen's sampling theory and methods in the early 1930s made economically feasible the first national public opinion surveys by

Gallup and Elmo Roper in 1935. Such surveys remain the major mass feedback technology—for politicians as well as other advertisers—in use to this day. In 1971 an article in *Science* magazine ranked the contributions of Hansen and Gallup as two of the forty-two "major advances in social science" between 1900 and 1936 (Deutsch, Platt, and Senghaas 1971).

Professionalization of Market Research

As we might expect, the business of providing market feedback for control of consumption became increasingly specialized and professionalized throughout the transition to the Control Revolution. As we have already seen, market research, the study and application of methodology as distinct from actual marketing, became increasingly specialized and formalized beginning about 1910 (Bartels 1962, p. 106). By 1916 both an advertising agency (J. Walter Thompson) and a major industrial corporation (U.S. Rubber) had specialized market research departments (Bartels 1962, p. 109; Pope 1983, p. 142).

The same year brought a new business text (Shaw 1916) with a chapter devoted exclusively to market research; the field had its own text (Duncan 1919) three years later (Bartels 1962, pp. 111–112). In the early 1920s Percival White, a pioneering theorist in the development of what Robert Bartels (1962) calls "marketing thought," began to develop the concept of "measurability of markets" and its use for prediction and market control; he eventually published the first systematic manual for field workers in market research (White 1931). The first comprehensive survey research text (Reilly 1929) intended for market researchers appeared in 1929; by 1937 a Committee on Marketing Research Techniques, appointed by the American Marketing Association, had published a summary of what it considered to be the best research techniques (Bartels 1962, pp. 114–118).

Three years later the first study of production control based on sales forecasting and methods for controlling sales appeared in print (Eastwood 1940). The title, *Sales Control by Quantitative Methods*, reflects the extension of Alfred Sloan's idea that industrial production and resource allocations can be controlled by market feedback. As the title suggests, a highly rationalized mass market feedback had become firmly established in control of consumption by World War II. As with the control of production and distribution, however, increasingly rationalized control of consumpton would depend on the continued growth of bureaucracy and other more generalized control technologies.

9

Revolution in Generalized Control: Data Processing and Bureaucracy

As every man goes through life, he fills in a number of forms for the records, each containing a number of questions . . . There are thus hundreds of little threads radiating from every man, millions of threads in all. If these threads were suddenly to become visible, the whole sky would look like a spider's web . . . They are not visible, they are not material, but every man is constantly aware of their existence.

—Alexander Solzhenitsyn, *Cancer Ward*

CONTROL of any purposive processing system can be no better than its most generalized and distributed processor of information. As we saw in Chapter 3, bureaucratic organization has served as the generalized means to control all large social systems, tending to develop whenever collective activities need to be coordinated toward some explicit and impersonal goal, that is, to be *controlled*. Chapter 6 described a crisis of control in office technology and bureaucracy in the 1880s, as the growing scope, complexity, and speed of information processing—including inventory, billing, and sales analysis—began to strain the manual handling systems of large business enterprises. This crisis had begun to ease by the 1890s, owing to innovations not only in the processor itself (formal bureaucratic structure) but also in its information creation or gathering (inputs), in its recording or storage (memory), in its formal rules and procedures (programming), and in its processing and communication (both internal and as outputs to its environment).

Innovation in structure developed after the emergence of bureaucracy itself in its modern form by the late 1860s, complete with a dozen or more operating units (in the large wholesale houses) controlled by a hierarchy of salaried managers. Generation of information greatly improved with the modern keyboard typewriter (1873); new inputs came from the stock-ticker (1870), messenger news service (1882), and press clipping service (1884). Information recording and storage capabilities increased with the systematization of office record keeping (early 1870s) and the dictating machine (1885). Programming of management with generalized bureaucratic practices became more extensive after the founding of Wharton (1881) and other university business schools. Information processing improved with the keyboard calculator (1887) and punch-card tabulator (1889); interoffice communication became easier with the desk telephone (1886). Internal bureaucratic control, in the sense of the comptroller or auditor, also developed with the independent accounting firm (1883) and national professional organization of accountants (1886).

Rapid innovation in all of the various components of the bureaucratic controller continued to gather momentum throughout the transitional period of the Control Revolution. As in the earlier response to crisis in control, developments came not only in bureaucratic structure itself but also in the various components of its capability to process information and to control environments both internal and external. Together these developments continued a revolution in the most generalized and distributed control of the material economy, which advanced in only a half-century from the first modern bureaucracies to the advent of computers, a new level of generalized information processing and hence control.

Innovation and Growth in Bureaucracy

Continuing innovation in bureaucratic structure has already been reviewed, up to the War Production Board of World War II, in Chapter 7. As we saw there, the General Electric Company pioneered the centralized, functionally departmentalized organizational structure in the mid-1890s. Within a decade this new information processor, built of the collective cognitive power of hundreds of individual human beings, had been adopted and perfected by several large industrial enterprises, most notably DuPont. By the mid-1920s old and new forms of bureaucracy had been synthesized by General Motors in an organizational structure that had both multiple branches and decentralized control: a general office overseeing autonomous, functionally integrated divisions.

Outside of manufacturing, the modern, functionally departmentalized bureaucratic structure developed first among life insurance companies, in response to marketing control requirements similar to those of General Electric and other mass producers of machinery. For actuarial reasons life insurance companies could not become viable businesses until they had enough customers to spread risks widely and thereby control the rate of return on investments. As with mass production, large volume in life insurance—stimulated and controlled through advertising, direct canvassing by salesmen, and local customer relations offices—also meant lower and hence more competitive unit costs. By the 1890s, confronting these needs and the advantages of generating volume, insurance firms began to open branch offices operated by salaried employees and centrally managed through functionally specialized divisions like sales, operations, and investments (Keller 1963; Buley 1967).

Innovation in bureaucratic structure as an information processor and controller had essentially ended by the mid-1920s. The multidivisional decentralized structure adopted by General Motors, DuPont, and later by U.S. Rubber, General Electric, Standard Oil, and other technologically advanced industries that integrated production with distribution, had emerged by this time. As technological advances permitted control of throughputs and stock turns of increasing volume and speed after World War I, the integrated enterprises adopted new products that exploited by-products of their manufacturing and brought more effective use of their purchasing and marketing organizations. Pioneered by DuPont, which in the 1920s scrambled to sustain managerial structure and facilities greatly expanded during wartime mobilization, diversification of products and markets had by the late 1920s become an explicit strategy of growth for several business enterprises. Alfred Chandler describes the contribution of the new bureaucratic form:

> The multidivisional structure . . . institutionalized the strategy of diversification. In so doing, it helped to systematize the processes of technological innovation in the American economy. The research department in such enterprises tested the commercial viability of new products . . . The executives in the general office, freed from day-to-day operational decisions, determined whether the company's managers could profitably process and distribute these new products . . . If the market was quite different, a new division was formed. By the outbreak of World War II, the diversified industrial enterprises using the divisional organization structure were still few, but they had become the most dynamic form of American business enterprise. (1977, pp. 475–476)

Such distributed control throughout the material economy depended on a rapid increase in the aggregate bulk of bureaucracy, a fact reflected in labor force statistics for the transitional period leading into the Control Revolution. Compared to the total civilian labor force, which grew 28 percent during 1900–1910 (the earliest decade for which statistics are available), managers increased 45 percent and clerical workers 127 percent. During the 1910s, when the labor force grew 13 percent, managers increased 14 percent and clerical workers 70 percent. Over the 1920s, when the labor force grew 15 percent, managers increased 29 percent and clerical workers 28 percent. Even in the 1930s, decade of the Great Depression, when the labor force grew 6 percent and managers only 4 percent, clerical workers increased 15 percent. In 1900 managers and clerks together constituted 8.9 percent of the civilian labor force; by 1940 their portion had nearly doubled to 16.9 percent (U.S. Bureau of the Census 1975, p. 139).

Bureaucratic growth during the period appears even more striking when occupational categories are refined to include only the most generalized office workers (the managers category cited above includes public officials and proprietors; "clerical" includes a wide range of kindred workers). The number of stenographers, typists, and secretaries increased 189 percent during 1900–1910, 103 percent in the 1910s, 40 percent in the 1920s, and 11 percent in the 1930s. Bookkeepers and cashiers increased 93 percent during 1900–1910, 38 percent in the 1910s, 20 percent in the 1920s (they decreased 2 percent over the 1930s). Office machine operators and related clerical workers increased 178 percent during 1900–1910, 102 percent in the 1910s, 30 percent in the 1920s, 31 percent in the 1930s. Accountants and auditors increased 70 percent during 1900–1910, 203 percent in the 1910s, 63 percent in the 1920s, 24 percent in the 1930s. Together, these four categories of office workers more than quadrupled their proportion of the total civilian labor force—from 2.1 to 8.6 percent—between 1900 and 1940 (U.S. Bureau of the Census 1975, pp. 140–141).

Innovation in Office Technology

Growth of bureaucratic control does not depend solely on innovation and growth in the processing structure itself, of course, but also on a spate of innovations in its technology for creating, recording, storing, and processing information, programming its personnel, and communicating with and possibly controlling its internal and external environments. Table 9.1 places selected innovations in these various components of office processing and control during the half-century transition to the sustained Control Revolution in both chronological order and larger bureaucratic context. Innovations that facilitated the production and distribution of information by offices included the mimeograph (1890), multigraph duplicator (1902), and compotype (1925), the portable (1892) and electric (1935) typewriter, and the electric keypunch machine (1923). In addition to these innovations, two new recording and copying devices also improved information storage and retrieval: the Photostat (1910), a photographic machine that copied directly onto sensitized paper, and the Recordak (1927), a check-photographing device that used 16mm motion picture film.

Characteristic of the transition to the Control Revolution in bureaucracy is the history of the Library Bureau of Boston. Founded in 1876 as an offshoot of the American Library Association, the Bureau had discovered an increasingly good business in providing specialized library supplies and equipment not available elsewhere. By 1894, according to the Bureau's *Classified Illustrated Catalog* of that year, it had made an even more startling discovery: "There is hardly a library article on our list that is not also used in offices."

Bolstered by the growing generalization and convergence of information-processing technology across all offices, the Library Bureau began a separate department of Improved Business Methods, joining the ranks of a growing number of "systematizers," what today would be called management consultants, an early application of scientific management to bureaucracy. What had less than twenty years earlier been a professional association for librarians, the first information scientists, now boasted that "among life and fire insurance companies, banks, railways, large manufacturing establishments, and to representative houses in almost every line, it has not only suggested and installed better methods and improved machinery, but it has also effected great savings in expense." Two years later, in March 1896, the Library Bureau became the exclusive agent for Hollerith data-processing equipment in England, France, Germany, and Italy and soon contracted with Travelers' Insurance to compile a year's records using the new system (Austrian 1982, pp. 133–134).

As we might guess, such data-processing technologies flourished in offices during the Control Revolution. Small-scale arithmetic processing continued to increase in speed as a result of the printing adder-

Table 9.1.   Selected events in the development of office technology and bureaucracy for generalized control, 1890–1939

Year	Development
1890	A. B. Dick markets successful "mimeograph," a word he coins
1892	Portable typewriter patented Adding-subtracting machine with printer (Burroughs) introduced
1893	Addressograph machine marketed Four-function calculator introduced
1894	Card punch, sorter adopted by commercial office (Prudential Life)
Mid-1890s	Centralized, functionally departmentalized organizational structure developed by General Electric
1897	Publicity bureau established by corporation (General Electric)
1900	Automatic punch card sorter introduced commercially
1900s	Keyboard calculators replace arithmometers, gain widespread use
1901	United States Steel incorporated with capitalization of $1.4 billion, first corporation to exceed billion dollars in assets
1902	Multigraph duplicator introduced for printing from ribbon or type Plug-board control enhances punch-card tabulation
Mid-1900s	Centralized, functionally departmentalized organizational structure perfected by Du Pont
1906	One of the world's largest bureaucracies, rural free delivery (RFD), largely incorporated within the U.S. postal system
1910	The Photostat, a photographic machine that copies directly onto sensitized paper, introduced by Eastman Kodak
1910s	Printing tabulator of punched cards introduced
1920	Pitney Bowes postage meter approved by U.S. Post Office
1920s	Electric printing calculators marketed
1923	Electric keypunch machine introduced
1925	Compotype, which embosses letters onto aluminum strips for printing, designed and patented (American Multigraph)
Mid-1920s	Multidivisional, decentralized organizational structure with general office and autonomous, integrated divisions adopted by General Motors, other large industrial enterprises
1927	Check photographing machine ("Recordak") using 16mm motion picture film manufactured by Eastman Kodak

Table 9.1 (*cont.*)

Year	Development
1928	Recordak machine installed in bank (Empire Trust, New York City) Multiple-register accumulating calculators introduced Modern 80-column punch card adopted by IBM
1929	Automatic electric stock quotation board—capable of recording one hundred quotations per minute—installed in brokerage firm
Late 1920s	Multiple-register accumulating calculators linked as difference engines to produce data tables
c. 1930	Modern offices contain typewriters, copiers, calculators, dictation equipment, and tapes, communicate with each other using telephone switchboards and multiplexers
1935	Electric typewriter introduced
1939	Book *Management and the Worker* popularizes science of human relations approach to organizational control Ph.D. in accounting conferred (University of Illinois, Urbana)

subtractor (1892), four-function calculator (1893), electric printing calculator (1920s), and multiple-register accumulating calculator (1928). Large-scale data processing, following Herman Hollerith's patenting of the electric punch-card tabulator (1889), greatly increased in speed and volume with the automatic card sorter (1900), switchboard-type card sorter (1902), automatic printing card tabulators (1910s), multiple-register accumulating calculators linked as difference engines to produce data tables (late 1920s), and the modern 80-column punch card (1928).

Necessity at least occasionally proved to be the mother of invention: Prudential Life Insurance, a relatively small company squeezed by the lower unit costs of its larger competitors and in 1891 the first commercial firm to use Hollerith equipment, three years later switched to a mechanical card-perforating machine and sorter devised by one of its actuaries, John Gore, who had previously taught clerks to count cards by listening to the sound they made when riffled under the thumb. By contrast, one of Gore's motorized machines could process fifteen thousand cards per hour, a data-processing system that remained in use at Prudential well into the 1930s—nearly four decades. Because Gore had designed his machines with Prudential's special needs in mind, however, no other company adopted them (Eames and Eames 1973, p. 31; Augarten 1984, p. 78). Prudential apparently grew to appreciate the value of information processing, however. When the Eckert-Mauchly Computer Corporation announced production of the UNIVAC in 1948, it had five contracts: two from the U.S. government, two from market researcher A. C. Nielsen, and one from Prudential (Augarten 1984, p. 161).

Programming of administrators with generalized bureaucratic practices also became more elaborate and widespread after the turn of the century, when university training quickly expanded from the Wharton School's offerings in commercial accounting and law. Both the University of Chicago and the University of California established undergraduate schools of commerce in 1899, New York University and Dartmouth, with its Amos Tuck School of Administration and Finance, in 1900. By the time Harvard opened its Graduate School of Business Administration in 1908, business education had become part of the curriculum on many of the nation's most prestigious campuses.

From the beginning, Harvard's business curriculum included offerings for managers of large multidivisional enterprises, with initial electives on the management of marketing and life insurance firms. By 1914 a mandatory marketing course concentrated not on any specific trade or commodity but on management more generally, including "demand activation, merchandising, pricing" (Copeland 1958, p. 43; Chandler 1977, pp. 466–468), thereby reflecting both the generalization of control and its application to consumption.

Innovation in office communication also flourished during the revolution in generalized bureaucratic control. As can be seen in the chronology in Table 9.1, new developments in communication technology for bureaucracy included the addressograph machine (1893), a postage meter authorized for use by the U.S. Post Office (1920), an automatic electric stock quotation board (1929), and interoffice telephone via private switchboards and multiplexing (c. 1930). Beginning with General Electric in 1897, large businesses increasingly established specialized publicity bureaus to exploit the new technology of communication— mass as well as interoffice—to promote both their industries and themselves.

Meanwhile, internal bureaucratic control continued to develop in the form of the comptroller: twenty thousand accountants and auditors served U.S. business by 1900 (Boorstin 1973, p. 211), although as late as 1929 only 2 percent of the country's accountants calculated by machine (Eames and Eames 1973, p. 91). Such accounting control had attained the status of doctoral work by 1939, when the University of Illinois, Urbana, conferred the nation's first Ph.D. in the subject.

What meager statistics exist for the development of office equipment during 1890–1940 also suggest a rapid growth in generalized information-processing and control technology throughout this transition to a sustained Control Revolution. Although the capital invested in all U.S. manufacturing increased most sharply (131 percent) in constant dollars during the 1880s, capital in the office equipment industry rose most rapidly in the 1890s (194 percent, compared to 67 percent for all manufacturing) and during 1900–1910 (182 percent, versus 81 percent overall); these are more than quadruple the growth rates for office equipment in the decades immediately before or after 1890–1910 (Creamer et al. 1960, table A-8). Similarly, the value of output from all manufacturing increased most sharply (73 percent) during the 1880s, again in constant dollars; output from the office equipment sector rose most rapidly in that decade (191 percent) and during 1900–1910 (171 percent, compared to 49 percent for all manufacturing). These are more than five times the growth rates for office equipment in the decade after 1880–1910 (Creamer et al. 1960, table A-10).

Much the same conclusion about the timing of the Control Revolution in generalized technology can be drawn from the two major types of office equipment, typewriters and office telephones, on which early data are available. Annual output of typewriters more than tripled to nearly one-half million between 1900 and 1921 (the two earliest years for which information is available) and more than doubled again to 1.1 million by 1937. By 1922 more than 13 percent of the nation's public secondary school students had been enrolled in typing classes—almost 17 percent by 1934 and nearly 23 percent in 1949 (the only statistical information available). Between 1920 (the earliest year for which data are available) and 1940 the number of business telephones rose 80 percent to 7.7 million—compared to only a 58 percent increase in residential telephones (U.S. Bureau of the Census 1975, pp. 377, 695, 783).

With the continuing spread of new office technologies, part of the third level of control (as outlined in Chapter 3), in the transition to the Control Revolution, we might expect to find parallel development in the hardware of Level Four: generalized information processing and computing technologies mechanical, electrical, and electronic. Indeed, four separate but interrelated technologies—calculators, punch-card processors, digital and analog computers—all flourished during the critical period 1880–1939, well before the popularly heralded birth of the so-called Computer Age.

Beyond Bureaucracy: Toward a Generalized Hardware of Control

Of the four interrelated lines of generalized information-processing technologies coevolving with bureaucracy in the nineteenth century, all had been well established by 1890. Desk-top calculators, which owe their intellectual origins to the seventeenth-century machinery of Wilhelm Schickard, Blaise Pascal, and Gottfried Leibniz, among others (Flad 1963), had been commercially mass-produced in Europe since about 1820 (Turck 1921; Baxandall 1926). The 1833 design for Charles Babbage's steam-powered Analytical Engine contained the essential components of a digital computer: punch-card input and programming, internal memory ("store"), a central processing unit ("mill"), and output to be printed or set into type (Morrison and Morrison 1961; Hyman 1982). Babbage's son Henry had constructed a working section of the Analytical Engine by 1889 (Babbage 1889).

Essential ideas for the analog computer developed from John Napier's logarithms (1614) and rods (1617), William Oughtred's rectilinear and circular slide rules (c. 1630) and the modern sliding rule (1654), and the calculus of Leibniz (1684) and Sir Isaac Newton (1687). In the 1860s James Thomson, brother of Lord Kelvin, devised the planimeter—an instrument for measuring the area of a plane figure—based on a disc, globe, and cylinder integrator, vital component of analog computers well into the twentieth century. By 1873 Lord Kelvin had adopted his brother's invention to build a tide predictor, probably the first analog computer (Eames and Eames 1973, p. 15). Punched paper cards, developed by Joseph-Marie Jacquard in 1801 to program the patterns woven in power looms (Fig. 6.3) and adopted by Babbage in 1833 for data inputs and programming for his Analytical Engine, had by 1884 been perfected by Herman Hollerith as a medium for electromechanical information processing and tabulation. In 1889 Hollerith received U.S. patent 395,781, entitled "Art of Compiling Statistics," for his electric punch-card tabulator (Austrian 1982, chaps. 2–5).

Thus in 1890, on the threshold of transition from crisis in bureaucratic control to a sustained Control Revolution, four distinct and well-developed types of information-processing and computing hardware could already be used to supplement the third level (bureaucracy) of generalized control. These included mass-produced keyboard adder-subtracters with built-in printers, direct multiplication machines, several large Difference Engines and a part of Babbage's Analytical Engine, several different types of analog computers and Hollerith's electric tabulator, capable of processing an average of ten thousand punch cards per month during normal working hours (Austrian 1982, p. 67). By the turn of the century new additions to this list included mass-produced four-function calculators, a mechanical equation solver capable of finding roots of polynomials up to the ninth degree, and an automatic punch-card sorter (Goldstine 1972; Margerison 1978; Randell 1982).

Table 9.2 illustrates the parallel development of the four technologies—desk-top calculating, digital and analog computing, and punch-card processing—through the transitional period 1880–1939. As the chronologies suggest, innovation in the four fields increased after 1910, especially after widespread applications of electricity in the 1920s. By

Table 9.2.   Selected innovations in more generalized information-processing and computing technologies, 1880–1939

Year	Desk-top calculating	Digital computing	Analog computing	Punch-card processing
1880				
82				
84	Keyboard			Electric
86	add-subtract			tabulator
88	calculator	Part of		
1890	Multiplier	Analytical		
92		Engine		
94	Four-		Equation	
96	function		solver	
98	calculator		80-element	Automatic
1900			harmonic	bin sorter
02			analyzer	Plug-board
04				tabulator
06				
08				
1910			Gyrocompass	
11			computer	
12			Profile	
13			tracer	
14		End-game	80-input tide	
15		chess machine	predictor	

Year	Desk-top calculating	Digital computing	Analog computing	Punch-card processing
16			Battle	
17			tracer	
18				
19				Printing
1920		Electro-		tabulator
21		mechanical		
22		calculator		
23				Electric
24	Electric		Product	keypunch
25	printing		integraph	
26	calculator			
27			Electric	
28	Multiple-	Calculators	network	80-column
29	register	linked as	analyzer	punch
1930	cumulating	difference	Differential	card
31	calculator	engines	analyzer	
32				
33		Mechanical		Punch card
34		programmer		accounting
35			Electrical	machines
36			analog	linked for
37			computer	computing
38			Electronic	
39	Electronic	Bell Labs	analog	
	calculator	Model I	computer	

World War II, the period often cited as the origin of modern computing, the American pioneers of generalized information-processing hardware had already built a half-dozen impressive computing machines: the differential analyzer of MIT engineering professor Vannevar Bush (1930), the first automatic computer general enough to solve a wide variety of problems (Fig. 9.1); the "mechanical programmer" of Columbia University astronomer Wallace Eckert (1933), which linked various IBM punch-card accounting machines to permit generalized and complex computation; Bush's electrical analog computer (1935), more general than his differential analyzer with punched-tape programming; an electronic analog computer (1938) devised at the Foxboro Company; a working prototype of an electronic calculator (1939), under development by John Atanasoff at Iowa State University; and the Bell Laboratories Model I (1939), built by George Stibitz, which AT&T has

Figure 9.1. Analog computers had found a place in both commercial and scientific computing by 1930. Top: an A. C. network analyzer designed in 1929, installed at Westinghouse in 1930, and used to determine electrical load requirements (courtesy of Edwin L. Harder, Pittsburgh). Bottom: Vannevar Bush with his first differential analyzer, built at MIT in 1930 to solve differential equations associated with power surges, failures and blackouts, and a wide range of more general problems (courtesy of the Smithsonian Institution).

claimed to be the first digital computer (an honor that obviously depends largely on the definition of "computer").

Also well developed before World War II were the essential plans for computers that would dominate information-processing hardware into the 1950s. Much as the Control Revolution had fostered radical new conclusions about information processing and decidability from Kurt Gödel, Alonzo Church, Alan Turing, and Emil Post during the 1930s, as discussed in Chapter 2, the same decade brought plans for the embodiment of much the same ideas in mechanical, electrical, and electronic devices. The concepts of information processing, programming, decision, and control and the intellectual stimulation of the relationships among them seemed "in the air" among European and American engineers, mathematicians, and philosophers by the mid-1930s.

Culmination of the Pre–World War II Computer Revolution

In 1934 an engineering student in Berlin, Konrad Zuse, began to design a universal calculating device that anticipated modern computers in several ways, including binary rather than decimal numbers and floating decimal point calculation (the first such applications to a machine), the programming rules of Boolean logic (unknown to Zuse), and the distinctive structure of a concrete open processor of information: punched tape (discarded 35mm movie film) input, a central processing unit, memory, programming, an internal controller, and an output device to display results (the similar structure of Babbage's Analytical Engine was also unknown to Zuse).

Beginning work on a concrete machine in 1936, Zuse had a mechanical prototype in 1938, an electromechanical relay machine in 1939, and by December 1941 the world's first general-purpose, program-controlled calculator in regular operation. Soon two specialized versions that analyzed the wing flutter of Nazi flying bombs displaced a computational office of thirty women at the Henschel Aircraft Company in Berlin. Zuse's fourth general-purpose machine, the Z4, smuggled out of Germany and installed at a technical institute in Zurich in 1950, proved to be what Augarten (1984, p. 97) has called "the only mathematical calculator of any consequence in Continental Europe for several years" (Zuse 1936, 1962, 1980; Schreyer 1939; Speiser 1980).

Meanwhile, the thinking of engineers in the United States had arrived through almost parallel stages at many of the same ideas developed by Zuse. By 1937 an Iowa State University physicist trained in electrical engineering, John Atanasoff, had worked out the idea of a digital calculator quite similar in principle to Zuse's machine. Like Zuse, of whose work he knew nothing, Atanasoff decided to use binary numbers, Boolean logic, and punched paper input and output (cards rather than tape); he, too, separated processing from memory. Unlike Zuse's calculator, and inferior to its design, Atanasoff's machine lacked a central processor and was neither programmable nor automatic—a human operator served almost continuously as its controller. It incorporated several forward-looking advances over Zuse's machine, however: Atanasoff designed his calculator to be electronic, using vacuum tubes rather than electromechanical relays for processing, a bank of capacitors for memory, and an internal electronic timer to synchronize the various operations.

By the end of 1939, supported by a $650 grant from Iowa State, Atanasoff had built the world's first machine to use vacuum tubes to perform arithmetical calculations. Three years later he had operational two rotating drum memories, a major storage device for computers well into the 1950s. As early as 1940, through his personal acquaintance with John Mauchly, Atanasoff had contributed ideas that five years later would be incorporated into the ENIAC, often considered the first modern computer but actually less modern than Atanasoff's calculator in several ways: ENIAC used decimal rather than binary numbers and hence could not exploit Boolean logic; it lacked a general-purpose central-processing unit and only partially distinguished between processing and memory (Atanasoff 1940, 1984; Brainerd 1976; Augarten 1984, pp. 109–131).

In the same year that Atanasoff worked out the design for his electronic calculator, Howard Aiken, a former Westinghouse engineer working as an instructor in applied mathematics at Harvard University, drafted a proposal for an electromechanical calculator, the Mark I, that would quickly become one of the world's most famous machines. Aiken's original proposal argued that a half-dozen scientific fields needed more powerful computing, including purposive monitoring and control of the material economy, what he termed the "science of mathematical economy" (his Harvard colleague, economist Wassily Leontief, had published a formal theory of input-output analysis toward that goal in the same year).

Unlike Zuse and Atanasoff, Aiken not only knew of Babbage's Analytical Engine but included a gloss of the 1833 design in his 1937 proposal, evidence of the intellectual continuity of the Control Revolution over the century past. In many ways Aiken's proposal could even be considered inferior to Babbage's proposed machine: the newer design lacked a differentiated processing structure and any general-purpose central-processing unit (Aiken 1937). After rejection by the Monroe Calculating Machine Company, Aiken's proposal finally won approval in 1939 from IBM, the first entry of what thirty years earlier had been Herman Hollerith's tabulating machine company into a market that thirty years later would make it one of the world's largest corporations, with revenues from information processing as great as the next dozen companies combined (Archbold and Verity 1985, p. 50).

IBM's Mark I, completed in January 1943 at its accounting machine plant in Endicott, New York, at a cost of a half-million dollars, served Harvard University until 1959. Driven by an electric-powered drive-shaft that ran throughout machinery fifty-one feet long but only two feet wide, the "Automatic Sequence Controlled Calculator" contained 760,000 electrical components and five hundred miles of wire. Little more than a giant electromechanical decimal calculator, the Mark I could nevertheless take two 23-digit numbers from paper tape input and within three seconds output their product onto punched cards.

Although it could not execute conditional jumps, an essential feature of programming (as we saw in the algorithm for Maxwell's demon in Chapter 2), the Mark I must be considered America's first program-controlled, general-purpose digital calculator (Oettinger 1962; Bernstein 1963; Ceruzzi 1983). One of its first programmers at Harvard, naval lieutenant Grace Hopper, would within eight years write the first compiler, software to translate a high-level programmer's language directly into binary machine code—an indirect but important legacy of Aiken's 1937 proposal for modern computing (Aiken and Hopper 1946; Augarten 1984, pp. 103–107, 214).

Even before Aiken had received IBM's backing for his proposal, Claude Shannon, then an MIT graduate student employed part-time to tend Vannevar Bush's differential analyzer, had published his master's thesis, "Symbolic Analysis of Relay and Switching Circuits" (Shannon 1938), which applied the propositional calculus of Whitehead and Russell's *Principia Mathematica* (1910–1913) to the design of electrical circuitry. Perhaps the most influential master's thesis ever written, in the words of Augarten (1984, pp. 100–101) it "not only helped transform circuit design from an art into a science, but its underlying message—that information can be treated like any other quantity and be subjected to the manipulation of a machine—had a profound effect on the first generation of computer pioneers."

Shannon's paper also established that programming an electronic digital computer would be a problem not of arithmetic but of logic (Eames and Eames 1973, p. 121; Swartzlander 1976). This same idea, although with implications drawn in the opposite direction, had appeared two years earlier in papers by Alan Turing and Emil Post. Starting not with a machine but with problems of logic, Post (1936) had reduced them to machinelike "primitive acts," Turing (1936) to a computing machine itself.

One of Shannon's examples, published in the *Transactions of the American Institute of Electrical Engineers*, showed how to simplify an "Electric Adder to the Base Two" (Shannon 1938). Working independently, with no knowledge of Shannon's work, Bell Laboratories physicist George Stibitz had already built such an adder on his kitchen table using discarded telephone relays. Encouraged by his boss to develop this "play project" into an aid for solving the complex equations of amplifier and filter design, Stibitz had by October 1939 built, at a cost of $20,000, the Model I calculator using about 450 relays.

Not as sophisticated as Zuse's relay machine, the Model I was permanently wired to solve equations with complex numbers and could not be further programmed. It lacked a general-purpose central-processing unit, a memory, and any clearly defined control unit. Because it input and output via teletype, however, it could be used from anywhere in the telephone system. At the annual meeting of the American Mathematical Society at Dartmouth College in Hanover, New Hampshire, in September 1940, Stibitz installed a few teletypes to demonstrate the Model I in Manhattan—the first use of remote computing via telephone that would come to characterize the "telematic society" (Nora and Minc 1978) thirty years later. By 1946 Stibitz had helped to develop four programmable relay calculators, including the Mark V—a half-million-dollar machine comparable to Zuse's Z4—that introduced the critical feature of conditional jumps in its programming (Stibitz 1940, 1980; Loveday 1977; Ceruzzi 1983).

Judging from this cumulative effort of the late 1930s, then, we might conclude that World War II interrupted work on generalized information-processing and computing technology as much as stimulated it. Consider the momentum of the transition to the Control Revolution in its final years. In 1936 Church, Post, and Turing publish their papers equating decision and computability procedures; Zuse begins building his first universal calculator. In 1937 Shannon publishes his paper equating logic and circuitry; Aiken and Atanasoff work out the designs for their machines; Stibitz builds a binary relay adder. In 1938 the Foxboro Company devises an electronic analog computer; Zuse completes a mechanical prototype of his hardware. In 1939 three seminal calculating machines—Atanasoff's electronic calculator, Zuse's binary relay computer, and the AT&T Model I—are completed; IBM agrees to build Aiken's Mark I.

Even cybernetics, often considered to be a postwar development (Wiener 1948, 1950), was largely anticipated in a paper published in 1940 by a British scientist, W. Ross Ashby. Ashby's paper, according to Miller (1978, p. 487), "stated the fundamental ideas of cybernetics," introducing the term *functional circuit* for negative feedback (Ashby 1940). By the end of 1940 Stibitz had successfully demonstrated telecomputing and Atanasoff had begun conversations with John Mauchly that would help to shape the ENIAC—first postwar forerunner of today's computers.

Information-Processing Hardware and Bureaucratic Control

Not all of the information processors listed in Table 9.2 can be considered general enough to supplement, compete with, or replace bureaucratic structures directly, of course. The harmonic analyzer used to study light waves, Elmer Sperry's gyrocompass computer and battle tracer, Vannevar Bush's profile tracer, Leonardo Torres's end-game chess machine, the tide predictor of E. G. Fischer and R. A. Harris, and the network analyzers used by electric companies to model their power grids were all specialized devices that contributed only indirectly to the development of general-purpose analog and digital computing. Even for innovations like these, however, it is difficult not to conclude that at least a small part of some specialized bureaucratic structures had been supplanted in control. Without Sperry's target-bearing and turret-control system (Hughes 1971, pp. 230–233) would a battleship's plotting room have contained more officers and staff or would it have merely exercised far less gunfire control? Without network analyzers (Hughes 1983, p. 376) would electric companies have employed more engineers and staff or would they have merely exercised less centralized power system control?

In either case, the idea that information-processing and computing hardware might be used to enhance bureaucratic control appears to have emerged only gradually during the transition phase of the Control Revolution. Initial impetus came largely from government bureaucracies. First applications were to statistical compilation and aggregate analysis during the 1870s and 1880s and to larger-scale data processing beginning in the 1890s. Only in the late 1910s and 1920s did bureaucracies begin to realize that the same hardware that processed numerical data might be used to process information more generally and thereby strengthen the control maintained by the entire bureaucratic structure. This might be seen as an early realization of Claude Shannon's more formal demonstration, expressed in his 1938 paper, that information is a commodity that can be processed by machine.

By the turn of this century American government bureaucracies had begun to process information not only for the passive compilation of statistics but also for the active control of individuals: fingerprinting of federal prisoners (1904), collecting tax on personal income (1913), psychologically screening draft inductees (1919), running a national employment service (1933). Innovation in information-processing hardware did not itself cause these changes but it did make them possible.

In 1935, for example, passage of the Social Security Act required the U.S. government to maintain employment records on twenty-six million people, which in turn required the processing of a half-million punch cards per day (Fig. 9.2). The Social Security Administration sustained this level of processing after 1936 by means of 415 card-punches, verifiers, sorters, and collators; the last machine was developed by IBM especially for the task. When the first Social Security benefits came due in 1937, the government's checks passed through the mails in the form of punched cards (Eames and Eames 1973, pp. 108–109), a powerful wedding of the generalized exchange media for data and money.

Such developing use of information-processing technology by government in America and other industrializing countries drew continuing impetus from national censuses. The turning point for the United States occurred in 1879 near the height of the crisis in control of the material economy, when General Francis Amasa Walker, then professor of political economy at Yale University, agreed to direct the 1880 census. Following the recommendations of a special commission, Congress had already removed much of the census work from the hands

Figure 9.2. The Social Security Act of 1935 required that the U.S. government maintain employment records on twenty-six million people, a task that required the processing of a half-million punch cards per day. Left: conservative editorial opinion maintained that the system "depersonalized" individuals by treating them as numbers. Right: when Mrs. Ida Fuller of Ludlow, Vermont, received the nation's first Social Security benefit check in 1937, it arrived in the form of a punched card. (Courtesy of the U.S. Social Security Administration.)

of politically appointed enumerators (Fig. 9.3). Additional statistics would be needed in areas of developing business and social interest, the son of Amasa Walker, a leading economist, argued to Interior Secretary Carl Schurz (Walker 1879). The new superintendent promised to collect such information, far beyond that required by the U.S. Constitution, in 215 different subject areas (compared to five in the previous census), including the first comprehensive census of manufacturers (Munroe 1923; Newton 1968).

Walker's encouragement of invention produced two innovations in data-processing hardware: the Seaton tabulator and the Lanston adding machine. The former, a product of chief clerk of the census Charles Seaton, consisted of a wooden box containing rollers over which a spool of blank paper could be hand-cranked. A slot in the box presented the clerk with a single row of a table in which he might record seven or eight adjacent columns of data before advancing the paper to the next

THE GREAT TRIBULATION.

Figure 9.3. Facing a crisis in control of the U.S. census in the 1870s, Congress removed much of the work from the politically appointed enumerator, depicted as a nosy incompetent in this Saturday Evening Post *cartoon for August 18, 1860. "I jist want to know how many of yez is deaf, dumb, blind, insane, and idiotic," he's shown to be asking, "and how many dollars the old gentleman is worth." (Courtesy of Curtis Publishing Company.)*

row. When a roll had been filled, another clerk would remove it and cut it into regular-sized sheets for counting and consolidation. This was facilitated by the adding machine, developed by a lawyer friend of Seaton, Tolbert Lanston, who would patent a successful monotype machine for casting type in 1887. Lanston's adder allowed entry of numbers as they had been written—left to right—on the Seaton tabulator (Austrian 1982, pp. 9–10).

Along with encouraging these inventions in data-processing hardware, Walker may have made a more important contribution to the Control Revolution by undertaking a census operation that generated so much information that it threatened to overwhelm the processing capabilities of even the most innovative technologies. As described by Eames and Eames (1973, p. 22), "the Eleventh U.S. Census posed a crisis in data processing. Figures from the 1880 census were still being interpreted in 1887. At that rate, especially in view of population increase, the 1890 figures would be obsolete before they could be completely analyzed." Walker, who resigned from the Census Bureau late in 1881 to become president of MIT, returned to Washington every third week to supervise without charge what remained of the tabulation, already well over its appropriated budget (Austrian 1982, p. 11).

From Crisis to Revolution: Hollerith's Contribution

In 1889, facing the prospect of a census that might be superseded by the next one before it could be completely tabulated, the Secretary of the Interior organized a committee of foremost statisticians to investigate faster means of processing data. The committee decided to test three new systems: (1) hand transcription of data, using a different color ink for each characteristic, onto slips of paper that could be sorted by color and counted by hand; (2) transcription onto color-coded cards or "chips," also sorted and counted by hand; and (3) Herman Hollerith's method, which used a keyboard or "pantograph" punch to make holes in predetermined positions in standardized cards, counted individually by means of hand insertion into an electrical circuit-closing press, which had a pin contact for each possible hole location (Hollerith 1889).

Hollerith, a Columbia-trained engineer and veteran of Walker's 1880 census, had first attempted to punch holes in the paper rolls of the Seaton tabulator (the automatic telegraph, developed a decade earlier, also used a paper tape punched with dots and dashes moving over electrical contacts against a metal cylinder). Hollerith eventually abandoned the paper roll for the same reason computer scientists in the 1950s would abandon magnetic tape: serial as opposed to random access. "The trouble was that if, for example, you wanted any statistics regarding Chinamen," Hollerith later wrote, "you would have to run miles of paper to count a few Chinamen" (Austrian 1982, p. 14).

Once again, the railroads had provided bureaucracy its model for information processing and control. During the same period that Hollerith worked to perfect his electrical tabulating system, 1880–1884, he also invented an electropneumatic braking system for railroads, the first patents (1886) to bear his name. As a patent examiner himself at the time, he would have learned a great deal about railroads, since more than half the claims for new inventions that came across his desk were for improved couplers, signaling devices, and other railroad equipment. During field tests of competing braking systems in 1887 Hollerith spent months on the scene and sent notes "taken in the field" to the *Quarterly* of his alma mater, the Columbia School of Mines (Austrian 1982, chap. 4). As his biographer, Geoffrey Austrian, concludes of Hollerith, "his schooling in the pervasive railroad technology of the day had a far greater influence on his development of tabulating machines than the more obvious example of other calculating and adding mechanisms" (1982, p. 36).

Hollerith himself named the railroad as the inspiration for his recording of information about people on punch cards: "I was traveling in the West and I had a ticket with what I think was called a punch photograph . . . The conductor . . . punched out a description of the individual, as light hair, dark eyes, large nose, etc. So you see, I only made a punch photograph of each person" (the punched photograph discouraged vagrants from stealing passengers' tickets and using them as their own [Augarten 1984, p. 73]). Hollerith's first cards, three-and-a-quarter by eight-and-five-eighths inches, closely resembled the railway ticket he claimed as his inspiration; he used a small conductor's punch that fit easily inside his hand to record Baltimore's health data in the first public test of his system. A British journal, *Engineering*, saw in that system the influence of another: "The idea is similar to that exemplified in railway practice in the 'electric slot.' Sometimes a signal is under the control of several signalmen and can only be pulled off by the combined action of them all. Each may pull over his lever, but nothing occurs until all the levers are over when the signal falls."

Austrian finds Hollerith's data-tabulating system—probably the world's first machinery to process information as a material flow—as an even more extensive elaboration of a railroad system:

The railroad itself was a complex system in which many elements had to work together, rather than a single invention. The same may be said of Hollerith's tabulating machines, in which punches, tabulators, and sorters all had to be designed to work in combination with each other as a system, rather than as separate units. Still more important, the concepts of precise

timing, automatic control, and safeguarding against error—important to
the development of railway braking and signaling systems coming into
use—became equally critical considerations in Hollerith's later machines,
in which feeding, sensing, and sorting all had to occur at precise intervals.
Instead of a speeding train of closely coupled cars, imagine a carefully
controlled procession of punched cards speeding through a machine, and
the analogy becomes clearer. Hollerith's automatic control feature—in
which a machine shuts down at the end of one run of cards, records the
totals, and then starts up automatically when another batch comes along—
closely parallels the railroad block signaling system, which prevents a
second train from entering a section of track until it is cleared by the train
ahead of it. Look under the covers of his automatic sorter. It's a railway
freight yard. Only instead of railway cars being switched from a single
track onto separate spurs to make up new trains for further destinations,
you have cards with common designations being sorted into separate
pockets of the machine, arranged for further processing.   (1982, p. 37)

In other words, the concrete open system for processing railroad
cars and trains served Hollerith as the model for processing the information—also as discrete material objects—that would soon serve
in the control of such systems. This fact alone, quite apart from considerations of living systems more generally, suggests that information
processing and communication cannot be understood independently of
the matter, energy, and material processing systems that they control.

By the 1889 census test, Hollerith had already leased his new punch-
card system to health departments in Baltimore, New York City, and
New Jersey and to the Surgeon General's Office in the U.S. War
Department, which after only six months possessed fifty thousand
punched Hollerith cards and found itself largely dependent on the
equipment. The test used 1880 census schedules from 10,491 people in
four districts in St. Louis (there were about forty thousand districts
in the 1890 census). Hollerith recorded the information in less than 73
hours, compared to nearly 111 hours for the chip system and almost
145 hours for the slip system. To tabulate the data took Hollerith less
than six hours, compared to nearly forty-five hours for the chips and
more than fifty-five hours for the slips. Overall, the punch system had
taken only thirteen minutes longer than *half* the time of the next fastest
method of data processing then available. Even at this unprecedented
rate, however, extrapolations from the amount of test data (with no
increase in the number of machines or operators) indicate that Hol-

lerith's system running continuously would have taken nearly fifty-five years to process the entire 1890 census.

The Census Bureau employed ninety-six of Hollerith's tabulating machines, however, and its operators raised average punching rates from five hundred to seven hundred cards per day and daily tabulating rates from ten to eighty thousand persons per day (the one-day record for cards stood at 19,071) (Fig. 9.4). By August 1890 census director Robert Porter had already rebounded nicely from the threatened crisis in the processing of his data, boasting: "With the work force that left this afternoon . . . we could, with these electrical machines, count the entire population of the United States in ten days of seven working hours each . . . could run through the entire population of the earth . . . in less than 200 days, providing places could be found to store the schedules." Hollerith himself alluded to the latter problem: "Consider a stack of schedules of thin paper higher than the Washington Monument . . . Imagine the work required in turning over such a pile of schedules page by page, and recording the number of persons reported on each schedule. This is what was done in *one day* by the population division of the Census Office" (Austrian 1982, pp. 68–69; emphasis in original).

What information-processing systems like Hollerith's would mean for government bureaucracy more generally less than a century later would be spelled out in a 1976 report of the National Commission of Federal Paper Work: "Federal agencies are today churning out forms, reports, and assorted paper work at the rate of over 10 billion sheets per year. That's 4.5 million cubic feet of paper. All of this paper costs the American economy $40 billion per year." In the same year the U.S. Department of Agriculture alone had stored nearly one million cubic feet of records and spent $150 million annually just to print new forms that had helped increase its paper store by 64,000 cubic feet— 36.5 thousand file drawers—or nearly forty four-drawer cabinets per working day (Porat 1977, pp. 144–145). This government bureaucracy alone, not to mention the parallel development of corporate bureaucracy, lends irony to a newspaper article about the punch-card tabulator

Figure 9.4. Herman Hollerith's data-tabulating machinery, probably the world's first mechanical system for processing information as a material flow, reduced the time needed to analyze the U.S. census from nine years in 1880 to less than seven years in 1890. Hollerith's punch-card system also captured the imagination of the new mass public, as evidenced by this cover story in Scientific American for August 30, 1890.

used in the 1890 census, which Hollerith clipped and saved: "The machine is patented," the reporter noted, "but as no one will ever use it but governments, the inventor will not likely get very rich" (Austrian 1982, p. 73).

After the 1889 competition the committee of statisticians had predicted that Hollerith's machines would save some $597,000 over the previous census system that used the Seaton tabulator and Lanston adder. In fact, the new machines saved $5 million and two years time. Unlike the 1880 count, the 1890 census, even though it contained twenty more items (a total of 235), could be analyzed in every possible combination of variables, the most complicated tables produced at no more expense than the simplest ones. Boorstin (1973, p. 172) summarizes the impact of Hollerith's system: "Now it was as easy to tabulate the number of married carpenters 40 to 45 years of age as to tabulate the total number of persons 40 to 45 years of age."

The Census Bureau also concluded that Hollerith's system gave a far better check against error than did hand tabulation (Austrian 1982, pp. 69–70). Compared to the 1880 census, which took nine years and cost $5.8 million, the much more extensive and thorough 1890 count took fewer than seven years. It cost nearly twice as much, however, some $11.5 million (Truesdell 1965). As Augarten (1984, p. 78) surmises, "Hollerith's system apparently possessed hidden costs—the great temptation to use the equipment to the hilt. All those millions of cards, those thousands of watts of electricity, those scores of statisticians, had run up a big bill."

Bureaucracies Worldwide Adopt the Hollerith System

Much the same temptation might have induced bureaucracies continually to seek out new applications for the information-processing technology. As the New York *Post* wrote of census superintendent Porter late in 1890, "Like Alexander, who wept because he had no more worlds to conquer, Porter is said to be sad and tearful because he has no more nation to count" (Austrian 1982, p. 69). For whatever reasons, public and private bureaucracies throughout the world quickly adopted Hollerith's system. They apparently recognized in its processing of information as material flows something akin to their function of organizing their employees into processing, decision, and control structures. Within a year Hollerith machines would be used to tabulate the national censuses of Austria, Canada, and Norway.

Although Hollerith would later assert that he had no interest in selling to businesses until he lost the U.S. census as a customer in 1905, he had attracted a commercial firm—Prudential Life Insurance— even before the first large-scale test of his equipment in 1890. By March 1891 Prudential had two of the systems running in its Newark office (they were replaced in three years by John Gore's more specialized mechanical system). Three years later Hollerith could boast that one hundred million cards had passed through his machines—thirty-five million outside of the U.S. Census Bureau and more than five million in other countries (Austrian 1982, pp. 82–85).

In 1894 a Russian-language book promoting the Hollerith system appeared in St. Petersburg. The following year Hollerith journeyed to Russia—the first of three trips—to persuade the Czar to use Hollerith equipment for the country's first general census, scheduled for 1896. Hollerith eventually sold the Russians five hundred keyboard punches and thirty-five new tabulating machines and loaned them thirty-five used tabulators, which they were to return for use in the 1900 U.S. census. By 1903, laboring under the burden of processing more than three hundred tons of punched cards, a task that would take nine years, the Russians ordered additional Hollerith equipment. As late as the 1930s, after American recognition of the Soviet government, the IBM marketing representative to Russia found the name of Hollerith still revered there, his early tabulators with clock dials preserved as museum pieces (Eames and Eames 1973, p. 27; Austrian 1982, chaps. 10–12) (Fig. 9.5).

Despite Lenin's concern in *The State and Revolution* (1917) to combat the "revisionist" notion that government bureaucracy might be reformed and not "smashed," which reaffirmed Marx's argument in *The Eighteenth Brumaire* (1852, pt. 7), the new Soviet state quickly became a pioneer in the technologies of information processing and control in both bureaucracy and hardware. Soon after the Bolshevik revolution Soviet economists Q. Krassin and G. Grinko, working at a government institute in Moscow, began to develop a theory of central economic planning and control based on continuing data collection, processing, and analysis—a theory that Deutsch et al. (1971, p. 452) rank among the world's nineteen "major advances in social science" during 1900–1920. With the accession of Joseph Stalin after 1924, Soviet thought insisted on the paramount and enduring importance of government bureaucracy. Far from "withering away" under socialism, as Engels had argued in *Anti-Dühring* (1878), the state, according to

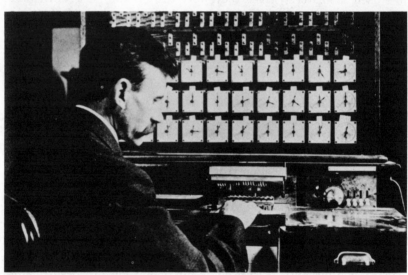

Figure 9.5. The two machines essential to Herman Hollerith's data-processing system. Top: *the card puncher, which included an interchangeable template and dual punch (courtesy of the U.S. Department of the Census).* Bottom: *the electric tabulator with rows of dials, each of which kept a running count of a single category of punched-card inputs (courtesy of the Library of Congress).*

Stalinists, must be reinforced as the controlling influence in the construction of socialism (Miliband 1983).

By the time Stalin had fully consolidated his leadership of the Soviet party and state in 1929, Russia was reported to be the third largest user of data-processing equipment in the world, following only the United States and Germany. The "Gosplan" or Five-Year Plans, an effort to achieve total state regulation of the Soviet economy begun in 1927, "implied a level of control that could hardly have been attempted without machines for the rapid assessment of statistics," according to Eames and Eames (1973, p. 97). "When Russia started renting tabulating equipment, one of the first installations was at the Central Statistical Bureau. Other early users were the Soviet Commissariats of Finance, of Inspection and of Foreign Trade, the Grain Trust, the Soviet Railways, Russian Ford, Russian Buick, the Karkov Tractor plant, and the Tula Armament Works." By 1938 Soviet mathematician L. V. Kantorovich had begun to develop linear programming theory, the solutions of minima-maxima problems that would quickly become central to military and industrial planning and control and remain so today (Bell 1973, pp. 29–30).

Even with its highly centralized state control of the economy, however, Russia continued to lag behind the United States and Germany— where Max Weber (1922) continued to analyze the growth of bureaucratic control—in information-processing technology. As a result of the rapid development of such technology, outlined in the two previous sections, U.S. government bureaucracies also pioneered extensions of control: state (1890) and municipal (1894) employment services, a federal bureau of standards and Union catalog of holdings in some six hundred libraries (1901), state automobile registration (1901), a federal income tax and bureau of labor statistics (1913), a national reserve banking system enacted "to furnish an elastic currency" (1914), and a federal trade commission to regulate commerce (1915).

During World War I the Wilson administration enjoyed unprecedented control of the nation's transportation and communication systems. The War Industries Board, headed by financier Bernard M. Baruch, controlled the production and distribution of virtually all goods and services. Hundreds of subordinate departments, boards, and committees—on everything from seeds, foundry supplies, and automobiles to baby buggies, pocket knives, and candy—collected and processed data for Baruch's board. To sustain such extensive control of the material economy, the government required large amounts of information-processing equipment, so much that Hollerith's former company, then part of the Computing-Tabulating-Recording (C-T-R) that in 1924 would become IBM, suspended all other orders. "The war was doing what the company's salesmen could not hope to do for many years," Hollerith's biographer notes (Austrian 1982, p. 341). "It was making the punched card a daily fact of life for thousands of clerks marshaling the nation's food supply and other resources. And unknown to Hollerith, in faraway Germany some of his machines were being used to chart the journeys of U-boats foraging for Allied shipping."

Meanwhile, the U.S. War Department, one of Hollerith's original customers in 1888, used his equipment to conduct the first large-scale application of psychological testing to control manpower utilization. According to Eames and Eames (1973), "The tests were designed to identify recruits likely to be useless, or even dangerous, in battle . . . The results, coded on cards, were used to fill such specialized personnel needs as 105 scene painters for camouflage work, or 600 chauffeurs who spoke French." At the end of hostilities the Medical Department of the Army turned its Hollerith equipment to a new data set: twenty-three body measurements on each of one hundred thousand enlisted men, which were analyzed to provide the first statistically reliable information for making uniforms for American men (true to stereotype, Texans proved to be the tallest) (Fig. 9.6).

Although C-T-R and other manufacturers of data-processing equipment turned back to their corporate customers after fighting had ended, much of their wartime production found continued use in the federal bureaucracy in new extensions of state control: federal operation of the railroads for twenty-six months (1918–1920), a regular census of distribution and photographing of bank checks (1929), a veterans' administration to centralize federal records on insurance, public health and vocational education (1930), a national income division to monitor the economic depression (1932), a central U.S. statistical office and employment service (1933), a national commission to control security exchanges and over-the-counter markets (1934), a national highway survey and federal punch-card index of fingerprints (1934), the federal administration of social security taxes and benefit payments (1936), a national board to control the "form, style, arrangement and indexing of codifications" of executive agencies (1937) and, by the decade's end, punch-card operations for codebreaking in Washington, D.C., Corregidor, and Pearl Harbor—work that would stimulate the further development of computers during World War II (Goldstine 1972).

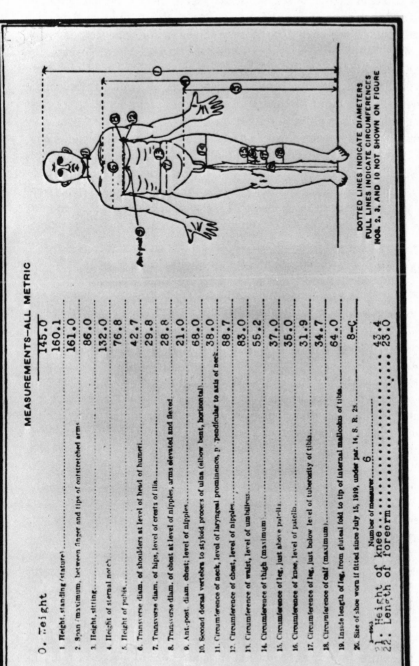

Figure 9.6. *An early application of Hollerith's data-processing system to control the production and distribution of Army uniforms. At the end of World War I the U.S. Army collected twenty-three body measurements on one hundred thousand enlistees, the first statistically reliable information on the shapes and sizes of American men. (U.S. War Department, 1921, p. 61; courtesy of the U.S. Army.)*

Parallel growth in bureaucratic and electromechanical data processing and control could also be found in the private sector of the American economy. As we have already seen in the major beneficiaries of the Library Bureau's "systematizing" and the Soviet Union's Gosplan, the first applications of the new information-processing technologies came in financial institutions, public utilities, railroads, and large manufacturing operations. Life insurance companies ranked among the first companies to see profit in data processing: New York Life, which by the turn of the century had contracted to have its data punched onto Hollerith cards (Austrian 1982, p. 134), adopted about 1903 the nation's first numerical insurance rating system, with values assigned to various factors affecting the insurability of applicants. Telephone companies and other utilities shared a common data-processing problem: the continual recording and billing of large numbers of small amounts.

Data-Processing Equipment in Continuous Control

Aside from the obvious applications to speed up data processing and tabulation, the new hardware soon came to be used—as we might now expect—to control material processing and flows. After Hollerith connected electric adding machines to punch-card fields through his tabulators in 1896, his machinery could be used by railroads for their waybill statistics: weight plus freight, advance and prepaid charges together with codes for what was shipped, who sent and received it, and by what routes (the ultimate extension of the railroads' own punched photographs). This new technology, which in effect enabled the Hollerith system to add as well as to count, greatly reduced the "veritable army of clerks" required in railroad freight departments while at the same time speeding shipping (Eames and Eames 1973, pp. 46–47). After Hollerith devised a tabletop card sorter for the U.S. Agricultural Census in 1901, he rearranged the bins vertically to enable automatic sorting in small railroad offices, even though the new design drew complaints from most other customers as a "backbreaker." Twenty years later, when the railroads no longer dictated commercial design, IBM would make news by "inventing" a horizontal sorter much like Hollerith's 1901 tabletop model (Austrian 1982, pp. 243–244).

Perhaps the most useful applications of the new information-processing hardware for the control of material processing and flows came in the offices of large manufacturing operations and retail establishments. At first punch-card tabulators in these offices merely replaced hand methods of accounting. Within only a few years, however, the new equipment had been applied to two processing tasks too rapid and complex to be performed by bureaucracy alone: the immediate and continuous analysis of sales and costs.

In the nineteenth century, as we saw in Chapter 6, cost accounting had been largely a matter of guesswork at best. After 1900, however, modern factory accounting developed rapidly, pioneered by the same industrial engineers who attempted to make management more "scientific" (Chatfield 1977, chap. 12). Methods to relate overhead costs to the fluctuating flow of materials through a factory proved particularly important. Because the internal statistical data needed to control these flows had already been well defined, their application to cost accounting—given the necessary data-processing capability—proved to be relatively straightforward (Garner 1954, chap. 5).

When Pennsylvania Steel of Philadelphia adopted Hollerith equipment in 1904, for example, most of the information required for cost accounting could already be found—thanks to the work of Frederick Winslow Taylor in the steel industry of the same city—on the time tickets filled out by each worker. By punching data directly from these tickets onto Hollerith cards, various sorting and tabulating operations could supply a running account of the company's business.

Gershom Smith, the newly appointed auditor of Pennsylvania Steel, soon became an outspoken promoter of the new data-processing equipment for control of manufacturing. "Accounting departments," he wrote, "must present the records to management sufficiently promptly to enable those in charge to check excess costs, stimulate production and use the facts before they have become ancient history" (Austrian 1982, p. 201). As a result of such "real-time" control, cost accounting quickly became rationalized enough to merit university study. The first professional organization to include cost accountants as members, in 1915, was a national association of university instructors (Chatfield 1977, p. 153; Chandler 1977, p. 465). When Hollerith stepped down as general manager of his Tabulating Machine Company in 1911, Gershom Smith became his hand-picked successor.

For sales analysis, especially in the offices of the largest retailers, the new information-processing equipment proved similarly invaluable. As early as 1902, Marshall Field, the Chicago dry goods merchant then building one of the world's largest department stores (Twyman 1954), placed an unsolicited order for Hollerith equipment. By the following year the Tabulating Machine Company boasted to other prospective customers about the giant retailer's application of an information-processing system to sales analysis:

> For each invoice a carbon is made and on this is added the class and item numbers of the goods sold. This information and all related information . . . is transferred to punched cards . . . There are now on hand cards for every item sold during the month. In addition to articles by class and item numbers and value, the card also carries such other facts as . . . date, customer, state, route, branch, etc. With a month's sales thus carded, it is clear that to make the business actually done tell its story, it is only necessary to group by any one fact or combination of facts. (Austrian 1982, pp. 204–205)

In similar ways the new technologies of information processing and control came to be applied to a wide range of bureaucratic functions: purchasing records, inventory, overhead allocations, payroll analysis, shipping costs, sales projections, market forecasting. The larger the enterprise and the more diversified the throughputs it controlled, the more crucial the new technologies would become.

At first these technologies appeared to "de-skill" the office work force. Manual bookkeeping "would need the personal attention of someone of marked ability," the British journal *Engineering* observed in 1902. "But when the data are punched on cards, the job can be put in the hands of a girl (sic)." By 1920, however, because of the increasingly complex tasks to which the new machinery had been applied along with the coevolution of continually more complex equipment, IBM had established a school to train its own executives in the use of its products. After General Electric enrolled five of its administrators in the school in 1934, IBM officially began a Customer Executive School offering short intensive courses in how to operate IBM equipment (Eames and Eames 1973, pp. 48, 90).

By 1928 the five largest manufacturers of information-processing equipment—Remington Rand, National Cash Register, Burroughs, IBM, and Underwood Elliott Fisher—had profits exceeding $32 million on revenues of almost $180 million. In 1939, despite the Depression, their profits totaled $18.6 million on nearly $177 million in revenues. By 1984 the top five companies—IBM, Digital Equipment, Burroughs, Control Data, and NCR (three of the five from the 1928 and 1939 lists)—had a combined income of nearly $7.7 billion on revenues exceeding $66 billion, part of total worldwide information-processing revenues exceeding $130 billion (Augarten 1984, pp. 183–184; Archbold and Verity 1985, pp. 37, 50–51).

At least in rough outline, then, the shape of the modern information-processing industry appears to have been well established—in corporate leadership, growth rate, and profit margins—before World War II. Just as the relative size of the information work force doubled between 1880 and 1900 and again by 1940, it would double yet a third time and become one-half of the total labor force by the 1980s (Table 1.2). An increasing number of those information workers serve the data- and word-processing and computing technology of modern bureaucracies, the most recent development along with microprocessors and genetic reprogramming of the continuing revolution in generalized control technology.

10

Conclusions: Control as Engine of the Information Society

The great scientific revolution is still to come. It will ensue when men systematically use scientific procedures for the control of human relationships and the direction of the social effects of our vast technological machinery . . . The story of the achievement of science in physical control is evidence of the possibility of control in social affairs.

—John Dewey, *Philosophy and Civilization* (1931)

ONLY SINCE World War II have the industrial economies of the United States, Canada, Western Europe, and Japan appeared to give way to information societies, so named because the bulk of their labor force engages in informational activities and the wealth thus generated comes increasingly from informational goods and services. Although all human societies have depended on hunting and gathering, agriculture, or the processing of matter and energy to sustain themselves, such material processing, it would seem, has begun to be eclipsed in relative importance by the processing of information.

How did this come to be—and why? Despite scores of technical and popular books and articles documenting the advent of the Information Society no one seems to have even raised, much less answered, these crucial questions. Among the many things that human beings value, how did information, embracing both goods and services, come to dominate the world's largest and most advanced economies? Material culture has also been crucial throughout human history, and yet capital did not begin to displace land as an economic base until the Industrial Revolution. To what comparable technological and economic "revolution" might we attribute the similar displacement of the industrial capital base by information and information-processing goods and services, or the overshadowing of the Industrial by the Information Society?

The answer, as we have seen, is the Control Revolution, a complex of rapid changes in the technological and economic arrangements by which information is collected, stored, processed, and communicated and through which formal or programmed decisions can effect societal control. From its origins in the last decades of the nineteenth century the Control Revolution has continued unabated to this day and in fact has accelerated recently with the development of microprocessing technologies. In terms of the magnitude and pervasiveness of its impact upon society, intellectual and cultural no less than material, the Control Revolution appears to be as important to the history of this century as the Industrial Revolution was to the last. Just as the Industrial Revolution marked an historical discontinuity in the ability to harness energy, the Control Revolution marks a similarly dramatic leap in our ability to exploit information.

Why did the Control Revolution begin in America at mid-nineteenth century, closely following the Industrial Revolution? Such questions of timing become easier to answer if we consider, as we did in Chapter 5, that national economies constitute concrete open processing systems engaged in the continuous extraction, reorganization, and distribution of environmental inputs to final consumption. Until the last century these functions, even in the largest and most developed national economies, still were carried on at a human pace, with processing speeds enhanced only slightly by draft animals and wind and water power and with system control increased correspondingly by modest bureaucratic structures. So long as the energy used to process and move material throughputs did not much exceed that of human labor, individual workers in the system could provide the information processing required for its control.

Once energy consumption, processing and transportation speeds, and the information requirements for control are seen to be interrelated, the Industrial Revolution takes on new meaning. By far its greatest impact from this perspective was to speed up society's entire material processing system, thereby precipitating a crisis of control, a period in which innovations in information-processing and communication technologies lagged behind those of energy and its application to manufacturing and transportation.

Crisis and Revolution

Table 10.1 summarizes the crisis in control that moved progressively through the American economy of the nineteenth century, from trans-

Table 10.1.   Selected crises in the control of transportation, production, distribution, and consumption, 1840–1889

Year	Crisis
1841	Western Railroad collision kills two, injures seventeen; Massachusetts Legislature investigation
1849	Freight must be processed through nine transshipments between Philadelphia and Chicago, impeding distributional networks
1851–54	Erie Railroad, first trunk line connecting East and West, begins operations in "utmost confusion," misplaces cars for months
1850s	With the growing network of grain elevators and warehouses, and the mounting demand for mass storage and shipment, transporters have increasing difficulty keeping track of individual shipments of grain and cotton
1850s & 1860s	Mercantile firms are increasingly unable to control the growing commerce in wheat, corn, and cotton Commission merchants are increasingly unable to handle the distribution of mass-produced consumer goods
1860s	With the advent of fast-freight and express companies, railroads experience difficulty monitoring the location and mileage of "foreign" cars on their lines Wholesalers scramble to integrate movement of goods and cash among hundreds of manufacturers and thousands of retailers Petroleum producers adopting continuous-processing technologies increase output three to six times while halving unit costs, confront need to stimulate consumption, differentiate products, build brand loyalty
Late 1860s	Rail mills adopting Bessemer process struggle to control increased speeds of steel production Large wholesalers and retailers like department stores confront need to maintain high rates of stock turn
1870s	Railroad companies (except the Pennsylvania) delay building large systems because they lack means to control them Producers of basic materials—iron, copper, zinc, glass—struggle to maintain competitively fast throughputs within their plants Large wholesale houses, among the most differentiated organizational structures in the nineteenth century, find need to integrate a growing number of highly specialized operating units

Year	Crisis
1882	Henry Crowell, adopting continuous-processing technology to oatmeal, produces twice national consumption, confronts need to create new markets
1880s	Metalworking industries—from castings and screws to sewing machines, typewriters, and electric motors—struggle to process materials at the volume and speed of the metal producers
	Producers of flour, soap, cigarettes, matches, canned foods, and film adopt continuous-processing technologies, confront needs to create new markets and to stimulate and control consumption
	Growing scope, complexity and speed of information processing—inventory, billing, sales analysis—needed to run large business begins to strain capacity of manual handling systems

portation (railroads) to distribution (commission trading and whole-saling), then to production (rail mills, other metal-making and metalworking industries), and finally to marketing (continuous-processing industries). As we have seen, what began as a crisis of safety on the railroads in the early 1840s hit distribution in the 1850s, production in the late 1860s, and marketing and the control of consumption in the early 1880s.

As the crisis of control spread through the material economy, it inspired a continuing stream of innovations in control technology. These innovations, effected by transporters, producers, distributors, and marketers alike, reached something of a climax by the 1880s. With the rapid increase in bureaucratic control and a spate of innovations in industrial organization, telecommunications, and the mass media, the technological and economic response to the crisis—the Control Revolution—had begun to remake societies throughout the world by the beginning of this century.

Table 10.2 presents a selective summary of the more important innovations in information technology that constituted the nineteenth-century Control Revolution, at least in the United States, and its growth through the 1930s toward an Information Society. This list of innovations reveals a steady development of organizational, information-processing, and communication technology over at least the

Table 10.2. Selected innovations in the control of production, distribution, and consumption and in more generalized control, 1830–1939

Year	Production	Distribution	Consumption	Generalized
1830		Scheduled		
32		freight line		
34			Penny press	
36				
38	Machine-tool	Telegraph		
1840	factory	Through-freight	Daguerreotype	
42	American System	forwarding	Advertising	Large-scale
44	of manufacture		agency	formal
46		Packaging	Hoe press	organization
48	Standardized	Commodity	Newspaper	
1850	wire gauge	exchanges	association	
52	Commissioned	Postage stamp	Wood pulp, rag	Hierarchical
54	industrial	Through bill	paper	process con-
56	consultants	Registered mail	Iteration copy	trol system
58		Railroad scale	Typesetter	Formal line-
1860	Continuous-	Futures	Display type	and-staff
61	processing		Advertising	control
62	technology	Paper money	of Christmas	
63		Fixed prices		
64		Postal money		
65		order	Premiums	
66		Transatlantic	for coupons	Modern bureau-
67		cable		cracies with
68	Bessemer	Traveling	Newspaper cir-	multiple de-
69	processing	salesmen	culation book	partments
1870	Continuous		Trademark law	
71	processing		Human-interest	
72	of materials	Mail-order	advertising	
73	Shop-order		Illustrated	Typewriter
74	accounting	Large chain	daily paper	with QWERTY
75	Plant design	of stores	Advertising	keyboard
76	to speed	Telephone	weekly	
77	processing			
78		Telephone	Full-page	
79		switchboard,	advertising	
1880		exchanges		
81			New trademark	Business school
82			law	Dow Jones news
83		Uniform stan-	Mass daily	Accounting firm
84	Rate-fixing	dard time	Newspaper	Bonding company
85	department,	Special de-	syndicate	
86	cost control	livery mail	Linotype	Desktop
				telephone

Year	Production	Distribution	Consumption	Generalized
87	Time recording			
88			Ad journal	Punch-card
89		Car accountant	National pub-	tabulator
1890	Staff time-	offices	licity stunt	Mimeograph
91	keepers for	Pay telephone,	Standardized	Multiplier
92	routing	travelers'	billboards	
93		checks	Print patent	Addressograph
94			Full-time	Four-function
95		Cafeteria	copywriters	calculator
96				Centralized,
97		Vending machine	Corporate pub-	departmental
98	Time studies	Rural free	licity bureau	corporate
99		delivery	Million-dollar	organization
1900			ad campaign	Automatic card
01			Modern adver-	sorter
02		Automat	tising agency	Plug-board
03	Auto plant	Pacific cable	Advertising	tabulator
04	designed for		textbook	
05	processing	Wristwatches		
06	Factory control		Advertising	
07	by line-and-	Transatlantic	copy testing	
08	staff	radio		
09	Auto branch	Gyrocompass		
1910	assembly	Two-way auto	Formal market	Photostat
11	Scientific	radio	research	
12	management	Franchising	Mail-order ad	
13	Moving auto	Parcel post	testing	
14	assembly	Aircraft gyro-	Circulation	
15	Unattended	stabilizer	audit bureau	
16	substations	Self-service	Household	
17	River Rouge	store	market inter-	
18	processing	Air mail,	viewing	
19	architecture	Fedwire	Market	Printing
1920		Metered mail	research	tabulator
21		Drive-ins	textbook	Postage meter
22	Distant control	Shopping center	Commercial	
23	of electrical	Supermarket	radio	Electric key-
24	transmission	Transcontinen-	National net-	punch
25	Demand feed-	tal air mail,	work radio	Decentralized
26	back control	facsimile	Dry waste	corporate
27	Pneumatic	Transatlantic	survey	organization
28	proportional	telephone	Flasher sign	Multiple-
29	controller	Aircraft auto-	Radio ratings	register
		matic pilot		cumulating
				calculator

Table 10.2. *(cont.)*

Year	Production	Distribution	Consumption	Generalized
1930	Quality control		Automobile	
31	course, text	Teletype	radio	
32		service		
33	PID controller		Retail-sales	Machines linked
34	Pneumatic		index	for computing
35	transmitter		National polls	Electric
36	Lab analysis	Modern coaxial	Audimeter	typewriter
37	for quality	cable	ratings	
38	control	Radar	Animated sign	
39	Human relations	Transatlantic	Commercial	Electronic
	textbook	air mail	television	calculator

decades of the 1850s through 1880s, a period that lags industrialization by perhaps ten to twenty years. Remarkable, in light of the Clark-Bell sequence discussed in Chapter 5, is the sharp periodization of the listing. Among the three economic sectors, virtually all of the major innovation in control through the 1860s can be found in distribution; much of that in the 1870s and later comes in production or consumption. Similarly, most of the important listings for distribution come before 1870, nearly all of those for production and consumption come after this date (major innovations in generalized control appear more sporadically throughout the period).

No less remarkable is the similar periodization in the development of information-processing, communication, and control technologies. Each of the major sectors of the economy tended to exploit a particular area of information technology: transportation concentrated on the development of bureaucratic organization, production on the organization of material processing, including preprocessing, division of labor, and functional specialization; distribution concentrated on telecommunications, marketing on mass media. These relationships, combined with those for the three economic sectors discussed above, account for the patterns in nineteenth-century control technology evident in Table 10.2.

Most bureaucratic innovation arose in response to the crisis of control in the railroads; by the late 1860s the large wholesale houses had fully exploited this form of control. Innovation in telecommunications (the telegraph, postal reforms, and the telephone) followed the movement of the crisis of control to distribution. Innovation in organizational technology and preprocessing (the shop-order system of accounts, routing slips, rate-fixing departments, cost control, uniform accounting procedures, factory timekeepers, and specialized factory clerks) followed the movement of the control crisis into the production sector in the 1870s. Most innovations in mass control (full-page newspaper advertising, a trademark law, print patents, corporate publicity departments, consumer packaging, and million-dollar advertising campaigns) came after the late 1870s with the advent of continuous-processing machinery and the resulting crisis in control of consumption. Along with these innovations came virtually all of the basic mass communications technologies still in use a century later: photography, rotary-power printing, motion pictures, the wireless, magnetic tape recording, and radio.

Despite such rapid changes in mass media and telecommunications technologies, the Control Revolution also represented a restoration—although with increasing centralization—of the economic and political control lost at more local levels during the Industrial Revolution. Before this time, control of government and markets had depended on personal relationships and face-to-face interactions; by the 1890s, as we saw in Part III, control began to be reestablished by means of bureaucratic organization, the new infrastructures of transportation and telecommunications, and system-wide communication via the new mass media.

If the Control Revolution was essentially a response to the Industrial Revolution, however, why does it show no sign of abating more than a century later? As we saw in Chapter 7, three forces seem to sustain its development. First, energy utilization, processing speeds, and control technologies have continued to coevolve in a positive spiral, advances in any one factor causing—or at least enabling—improvements in the other two. Second, additional energy has increased not only the speed of material processing and transportation but their volume and predictability as well. This, in turn, has further increased both the demand for control and the returns on new applications of information technology. Increases in the volume of production, for example, have brought additional advantages to increased consumption, which manufacturers have sought to control using the information technologies of market research and mass advertising. Similarly, the increased reliability of production and distribution flows has increased the economic returns on informational activities like planning, scheduling, and forecasting. Third, information processing and flows need themselves to be controlled, so that informational technologies must continue to be applied at higher and higher layers of control—certainly an ironic twist to the Control Revolution.

Information in Control

Given that a revolution in control did begin in response to a crisis generated by the Industrial Revolution, why have the technologies of information processing, preprocessing, programming, and communication played such a major part in the Control Revolution? In short, why the new centrality of information?

No study of technological innovation or economic history alone can possibly hope to answer this question, I argued in Part I, no more than the history of organic evolution can explain the importance of information to all living things. In both cases the reasons why information plays a crucial role will not be found in historical particulars but rather in the nature of all living systems—ultimately in the relationship between information and control. Life itself implies purposive activity and hence control, as we found in Chapter 2, in national economies no less than in individual organisms. Control, in turn, depends on information and activities involving information: information processing, programming, decision, and communication.

Inseparable from control are the twin activities of information processing and reciprocal communication. Information processing is essential to all purposive activity, which is by definition goal directed and must therefore involve the continual comparison of current states to future goals. Two-way interaction between controller and controlled must also occur to communicate influence from the former to the latter and to communicate back (as feedback) the results of this action.

Each new technological innovation extends the processes that sustain human social life, thereby increasing the need for control and for improved control technology. Thus, technology appears autonomously to beget technology and, as argued in Part II, innovations in matter and energy processing create the need for further innovation in information processing and communication. Because technological innovation is increasingly a collective, cumulative effort whose results must be taught and diffused, it also generates an increased need for technologies of information storage and retrieval.

Foremost among the technological solutions to the crisis of control—in that it served most other control technologies—was the rapid growth in the late nineteenth century of formal bureaucracy and rationalization. The latter includes what computer scientists now call *preprocessing*, a complement to the control exercised by bureaucracy through information processing, increasingly using computers and microprocessors. Perhaps most pervasive of all rationalization is the increasing tendency to regulate interpersonal relationships in terms of a formal set of impersonal, quantifiable, and objective criteria, changes that greatly facilitate control by both government and business. The complex social systems that arose with the growth of capitalism and improved transportation and communication would have overwhelmed any information-processing system that operated on a case-by-case basis or by the particularistic considerations of family and kin that characterized preindustrial societies.

Another explanation for the increasing importance of information in modern economies is suggested by the purposive nature of living systems. All economic activity is by definition purposive, after all, and requires control to maintain its various processes to achieve its goals. Because control depends on information and informational activities, these will enter the market, as both goods and services, in direct relationship to an economy's demand for control. But if control is in fact crucial to *all* living systems, why has the economic demand for control—in the form of informational goods and services—increased so sharply, thereby precipitating the rise of the Information Society? Economic activity might indeed depend on control, and control on information, but why do these relationships seem relatively so much more important now than a century ago?

The Information Society

The Information Society has not resulted from recent changes, as we have seen, but rather from increases in the speed of material processing and of flows through the material economy that began more than a century ago. Similarly, microprocessing and computing technology, contrary to currently fashionable opinion, do not represent a new force only recently unleashed on an unprepared society but merely the most recent installment in the continuing development of the Control Revolution. This explains why so many of the components of computer control have been anticipated, both by visionaries like Charles Babbage and by practical innovators like Daniel McCallum, since the first signs of a control crisis in the early nineteenth century.

The progress of industrialization into the nineteenth century, with the resulting crisis of control, the technological and economic response that constituted the Control Revolution, and the continuing development of the Information Society, including the telematic stage just now emerging—together these factors account for virtually all of the societal changes noted by contemporary observers as listed in Table 1.1. These include the rise of a new information class (Djilas 1957; Gouldner 1979), a meritocracy of information workers (Young 1958), postcapitalist society (Dahrendorf 1959), a global village based on new mass media and telecommunications (McLuhan 1964), the new industrial state of increasing corporate control (Galbraith 1967), a scientific-technological revolution (Richta 1967; Daglish 1972; Prague Academy 1973), a technetronic era (Brzezinski 1970), postindustrial society (Touraine 1971; Bell 1973), an information economy (Porat 1977), and the micro millennium (Evans 1979).

The various transformations these observers identify may now be seen to be subsumed by major implications of the Control Revolution: the growing importance of information technology; the parallel growth of an information economy and its control by business and the state; the organizational basis of this control and its implications for social structure, whether Young's meritocracy or Djilas's new social class; the centrality of information processing and communication, as in McLuhan's global village; the information basis of Bell's postindustrial society; and indeed the growing importance of information and knowledge throughout modern culture. In short, particular attention to the material aspects of information processing, communication, and control promises to make possible a synthesis of a large proportion of this literature on contemporary social change.

Despite the Control Revolution's importance for understanding contemporary society, however, especially the continuing impact of computers and microprocessors, the most useful lesson relates to our understanding of social life more generally. The rise of the Information Society itself, more than even the parallel development of formal information theory, has exposed the centrality of information processing, communication, and control to all aspects of human society and social behavior. It is to these fundamental informational concepts, I believe, that we social scientists may hope to reduce our proliferating but still largely unsystematic knowledge of social structure and process.

References

Index

References

For works of primarily historical interest, generally those published before 1960, citation is to the year of *first* publication except for ancient texts. When the text used was not this edition, page numbers and other references are to the edition (including year of publication) listed after the publisher. In the citation "Kant 1788," for example, the year refers to the first German publication; references are to the English-language edition published by Bobbs-Merrill in 1956. Citations to works in languages other than English are to translations whenever available.

Abbot, Waldo, and Richard L. Rider. 1957. *Handbook of Broadcasting: The Fundamentals of Radio and Television*, 4th ed. New York: McGraw-Hill.

Abelson, Philip H. 1983a. "Biotechnology: An Overview." *Science* 219 (February 11): 611–613.

———— 1983b. "New Biotechnology Companies." *Science* 219 (February 11): 609.

Ackermann, Wilhelm. 1924. "Begründung des 'tertium non datur' mittels der Hilbertschen Theorie der Widerspruchsfreiheit." *Mathematische Annalen* 93: 1–36.

Adams, Charles F., Jr. 1868. "The Railroad System." Pp. 333–429 in *Chapters of Erie, and Other Essays*, by Charles F. Adams, Jr., and Henry Adams. Boston: James R. Osgood, 1871.

Adams, Henry. 1918. "The Dynamo and the Virgin (1900)." Chap. 25, pp. 379–390 in *The Education of Henry Adams: An Autobiography*. Boston: Houghton Mifflin Sentry, 1961.

Adler, Julius. 1976. "The Sensing of Chemicals by Bacteria." *Scientific American* 234(4): 40–47.

Adler, Mortimer J., ed. 1952. *The Great Ideas: A Syntopicon of Great Books of the Western World*, vol. 2. Chicago: Encyclopedia Britannica.

Aiken, Howard H. 1937. "Proposed Automatic Calculating Machine." Pp. 195–201 in *Origins of Digital Computers: Selected Papers*, ed. Brian Randell. New York: Springer-Verlag, 3rd ed., 1982.

Aiken, Howard H., and Grace M. Hopper. 1946. "The Automatic Sequence Controlled Calculator." Pp. 203–222 in *Origins of Digital Computers: Selected Papers*, ed. Brian Randell. New York: Springer-Verlag, 3rd ed., 1982.

Alberts, Robert C. 1973. *The Good Provider: H. J. Heinz and His Fifty-seven Varieties*. Boston: Houghton Mifflin.

Albion, Robert Greenhalgh. 1938. *Square-Riggers on Schedule: The New York Sailing Packets to England, France, and the Cotton Ports.* Princeton, N.J.: Princeton University Press.

——— 1939. *The Rise of New York Port, 1815–1860.* New York: Scribner, 1970.

——— 1941. "Early Nineteenth-Century Shipowning—A Chapter in Business Enterprise." *Journal of Economic History* 1(1):1–11.

Alford, B. W. E. 1973. *W. D. and H. O. Wills and the Development of the U.K. Tobacco Industry, 1786–1965.* London: Methuen.

Althusser, Louis. 1965. *For Marx,* trans. Ben Brewster. New York: Random House Vintage, 1969.

——— 1971. *Lenin and Philosophy, and Other Essays,* trans. Ben Brewster. New York: Monthly Review Press.

Amihud, Yakov, ed. 1976. *Bidding and Auctioning for Procurement and Allocation.* Proceedings of a conference at the Center for Applied Economics, New York University. New York: New York University Press.

Archbold, Pamela, and John Verity. 1985. "A Global Industry: The *Datamation* 100." *Datamation* 31(11): 36–182.

Archer, Gleason L. 1938. *History of Radio.* New York: American Historical Society.

——— 1939. *Big Business and Radio.* New York: American Historical Company.

Ardrey, Robert. 1970. *The Social Contract: A Personal Inquiry into the Evolutionary Source of Order and Disorder.* New York: Atheneum.

Aristotle. 1885. *Politica,* trans. Benjamin Jowett. Oxford: Oxford University Press, Clarendon.

——— 1912. *De Generatione Animalium,* trans. Arthur Platt. Pp. 715–789 in *The Works of Aristotle,* vol. 5, ed. J. A. Smith and W. D. Ross. Oxford: Oxford University Press, Clarendon.

——— 1931. *De Anima,* trans. J. A. Smith. Pp. 402–435 in *The Works of Aristotle,* vol. 3, ed. W. D. Ross. Oxford: Oxford University Press, Clarendon.

Arnold, Horace L., and Fay L. Faurote. 1915. *Ford Methods and the Ford Shops.* New York: Arno, 1972.

Aron, Raymond. 1961. *18 Lectures on Industrial Society,* trans. M. K. Bottomore. London: Weidenfeld and Nicolson, 1967.

——— 1966. *The Industrial Society: Three Essays on Ideology and Development.* New York: Simon and Schuster, Clarion, 1967.

Arrow, Kenneth J. 1951a. "Alternative Approaches to the Theory of Choice in Risk-Taking Situations." *Econometrica* 19(4): 404–437.

——— 1951b. *Social Choice and Individual Values.* New York: Wiley.

——— 1962. "The Economic Implications of Learning by Doing." *Review of Economic Studies* 29(3): 155–173.

——— 1964. "Control in Large Organizations." *Management Science* 10(3): 397–408.

Arthur, Henry B. 1976. "The Structure and Uses of Auctions." Pp. 187–202 in *Bidding and Auctioning for Procurement and Allocation,* ed. Yakov

Amihud. Proceedings of a conference at the Center for Applied Economics, New York University. New York; New York University Press.

Ashby, W. Ross. 1940. "Adaptiveness and Equilibrium." *Journal of Mental Science* 86: 478–483.

Ashe, Thomas. 1808. *Travels in America, Performed in 1806*. Newburyport, Mass.: William Sawyer.

Atanasoff, John V. 1940. "Computing Machine for the Solution of Large Systems of Linear Algebraic Equations." Pp. 315–335 in *Origins of Digital Computers: Selected Papers*, ed. Brian Randell. New York: Springer-Verlag, 3rd ed., 1982.

———— 1984. "Advent of Electronic Digital Computing." *Annals of the History of Computers* 6(3): 229–282.

Atherton, Lewis E. 1949. *The Southern Country Store, 1800–1860*. Baton Rouge: Louisiana State University Press.

Augarten, Stan. 1984. *Bit by Bit: An Illustrated History of Computers*. New York: Ticknor and Fields.

Austrian, Geoffrey D. 1982. *Herman Hollerith: Forgotten Giant of Information Processing*. New York: Columbia University Press.

Averitt, Robert T. 1968. *The Dual Economy: The Economics of American Industry Structure*. New York: Norton.

Axelrod, Robert. 1980a. "Effective Choice in the Prisoner's Dilemma." *Journal of Conflict Resolution* 24(1): 3–25.

———— 1980b. "More Effective Choice in the Prisoner's Dilemma." *Journal of Conflict Resolution* 24(3): 379–403.

———— 1984. *The Evolution of Cooperation*. New York: Basic.

Axelrod, Robert, and William D. Hamilton. 1981. "The Evolution of Cooperation." *Science* 211 (March 27): 1390–96.

Babbage, Charles. 1832. *On the Economy of Machinery and Manufactures*. London: Charles Knight, 3rd ed., 1833.

Babbage, Henry Prevost, ed. 1889. *Babbage's Calculating Engines, Being a Collection of Papers Relating to Them, Their History and Construction*. Charles Babbage Institute Reprint for the History of Computing Series, vol. 2. Cambridge, Mass.: MIT Press, 1984.

Bain, Joe S. 1956. *Barriers to New Competition: Their Character and Consequences in Manufacturing Industries*. Cambridge, Mass.: Harvard University Press.

Banning, William P. 1946. *Commercial Broadcasting Pioneer: The WEAF Experiment, 1922–1926*. Cambridge, Mass.: Harvard University Press.

Barach, Arnold B. 1971. *Famous American Trademarks*. Washington: Public Affairs Press.

Barbour, Violet. 1930. "Dutch and English Merchant Shipping in the Seventeenth Century." *Economic History Review* 2(2): 261–290.

Barger, Harold. 1955. *Distribution's Place in the American Economy Since 1869*. National Bureau of Economic Research no. 58, General Series. Princeton, N.J.: Princeton University Press.

Barnouw, Erik. 1966. *A History of Broadcasting in the United States*, vol. 1, *A Tower of Babel, to 1933*. New York: Oxford University Press.

———— 1968. *A History of Broadcasting in the United States*, vol. 2, *The Golden Web, 1933–1953*. New York: Oxford University Press.

———— 1975. *Tube of Plenty: Evolution of American Television*. London: Oxford University Press.

Bartels, Robert. 1962. *The Development of Marketing Thought*. Homewood, Ill.: Richard D, Irwin.

Barthes, Roland. 1957. *Mythologies*, trans. Annette Lavers. St. Albans, England: Paladin, 1973.

———— 1968. *Elements of Semiology*, trans. Annette Lavers and Colin Smith. New York: Hill and Wang.

———— 1977. *Image, Music, Text*, trans. Stephen Heath. London: Fontana.

Baughman, James P. 1968. *Charles Morgan and the Development of Southern Transportation*. Nashville, Tenn.: Vanderbilt University Press.

Baxandall, David. 1926. *Calculating Machines and Instruments*. London: Science Museum, rev. ed., 1975.

Beadle, George W. 1964. "Nature and Man's Mind." Commencement Address, May 31, Kansas State College. Unpublished.

Beadle, George W., and Mary Beadle. 1966. *The Language of Life*. Garden City, N.Y.: Doubleday.

Beck, Melinda. 1984. "Can We Keep the Skies Safe?" *Newsweek* 103(5): 24–31.

Beckman, Theodore N., Nathanael H. Engle, and Robert D. Buzzell. 1937. *Wholesaling: Principles and Practice*. New York: Ronald Press, 3rd ed., 1959.

Beer, Samuel H. 1969. *British Politics in the Collectivist Age*, rev. ed. New York: Random House, Vintage.

Bell, Daniel. 1960. *The End of Ideology: On the Exhaustion of Political Ideas in the Fifties*. New York: Free Press, rev. ed. 1965.

———— 1973. *The Coming of Post-Industrial Society: A Venture in Social Forecasting*. New York: Basic Books.

———— 1976. "Foreword: 1976." Pp. ix–xxii in *The Coming of Post-Industrial Society: A Venture in Social Forecasting*, pap. ed. New York: Basic Colophon.

———— 1979. "The Social Framework of the Information Society." Pp. 163–211 in *The Computer Age: A Twenty-Year View*, ed. Michael L. Dertouzos and Joel Moses. Cambridge, Mass.: MIT Press.

———— 1980. "Introduction." Pp. vii–xvi in Simon Nora and Alain Minc, *The Computerization of Society: A Report to the President of France*. Cambridge, Mass.: MIT Press.

Bendix, Reinhard. 1960. *Max Weber: An Intellectual Portrait*. Garden City, N.Y.: Doubleday.

Bentham, Jeremy. 1789. *Introduction to the Principles of Morals and Legislation*. New York: Macmillan, Hafner, 1948.

Berg, Howard C. 1975. "How Bacteria Swim." *Scientific American* 233(2): 36–44.

Berg, Howard C., and Robert A. Anderson. 1973. "Bacteria Swim by Rotating their Flagellar Filaments." *Nature* 245 (October 19): 380–382.

Berger, Brigitte, and Peter L. Berger. 1976. *Sociology: A Biographical Approach*. New York: Basic.

Bergson, Henri. 1907. *Creative Evolution*, trans. A. Mitchell. New York: Random House, Modern Library, 1944.

Berkeley, Edmund Callis. 1962. *The Computer Revolution*. Garden City, N.Y.: Doubleday.

Berlin, Brent, and Paul Kay. 1969. *Basic Color Terms: Their Universality and Evolution*. Berkeley: University of California Press.

Berman, Paul J., and Anthony G. Oettinger. 1975. *The Medium and the Telephone: The Politics of Information Resources*, Working Paper 75–8 (December 15). Cambridge, Mass.: Harvard University Program on Information Technologies and Public Policy.

Bernstein, Jeremy. 1963. *The Analytical Engine: Computers, Past, Present and Future*. New York: Random House.

Bezold, Clement, ed. 1978. *Anticipatory Democracy: People in the Politics of the Future*. New York: Random House, Vintage.

Bierstedt, Robert S. 1963. *The Social Order: An Introduction to Sociology*, 2nd ed. New York: McGraw-Hill.

——— 1975. "Comment on Lenski's Evolutionary Perspective." Pp. 154–158 in *Approaches to the Study of Social Structure*, ed. Peter M. Blau. New York: Free Press.

Bitting, A.W. 1916. *Processing and Process Devices*. Research Laboratory, National Canners Association, Bulletin no. 6 (December). Washington: National Capital Press.

Black, Harold S. 1934. "Stabilized Feedback Amplifiers." *Bell System Technical Journal* 13(1): 1–18.

Blanqui, Jérôme Adolphe. 1837. *History of Political Economy in Europe*, trans. Emily J. Leonard. New York: G. P. Putnam's Sons, 1880.

Blumer, Herbert. 1937. "Social Psychology." Pp. 144–198 in *Man and Society*, ed. Emerson P. Schmidt. New York: Prentice-Hall.

——— 1969. *Symbolic Interactionism: Perspective and Method*. Englewood Cliffs, N.J.: Prentice-Hall.

Boalt, Gunnar, Robert Erikson, Harry Gluck, and Herman Lantz. 1971. *The European Orders of Chivalry*. Stockholm: P. A. Norstedt.

Bohannan, Paul. 1955. "Some Principles of Exchange and Investment among the Tiv." *American Anthropologist* 57(1): 60–70.

Bok, Edward W. 1923. *A Man from Maine*. New York: Scribner's.

Bonner, John Tyler. 1980. *The Evolution of Culture in Animals*. Princeton, N.J.: Princeton University Press.

Boorman, Scott A., and Paul R. Levitt. 1980. *The Genetics of Altruism*. New York: Academic Press.

Boorstin, Daniel J. 1958. *The Americans: The Colonial Experience*. New York: Random House, Vintage.

——— 1973. *The Americans: The Democratic Experience*. New York: Random House, Vintage.

——— 1975. *Portraits from "The Americans: The Democratic Experience."* New York: Random House.

———— 1978. *The Republic of Technology: Reflections on Our Future Community.* New York: Harper and Row.

Boulding, Kenneth E. 1953. *The Organizational Revolution: A Study in the Ethics of Economic Organization.* New York: Harper.

———— 1964. *The Meaning of the Twentieth Century: The Great Transition.* New York: Harper and Row.

Bradbury, Jack W. 1977. "Social Organization and Communication." Pp. 1–72 in *Biology of Bats*, vol. 3, ed. William A. Wimsatt. New York: Academic Press.

Brainerd, John G. 1976. "Genesis of the ENIAC." *Technology and Culture* 17(3): 482–488.

Braithwaite, Dorothea. 1928. "The Economic Effects of Advertisement." *Economic Journal* 38(1): 16–37.

Branson, H. R. 1953. "Information Theory and the Structure of Proteins." Pp. 84–104 in *Essays on the Use of Information Theory in Biology*, ed. Henry Quastler. Urbana: University of Illinois Press.

Breed, Warren. 1971. *The Self-Guiding Society.* New York: Free Press.

Bridge, James Howard. 1903. *The Inside History of the Carnegie Steel Company: A Romance of Millions.* New York: Aldine.

Briggs, Asa. 1977. "The Pleasure Telephone: A Chapter in the Prehistory of the Media." Pp. 40–65 in *The Social Impact of the Telephone*, ed. Ithiel de Sola Pool. Cambridge, Mass.: MIT Press.

Brinton, C. Crane. 1933. *English Political Thought in the Nineteenth Century.* London: Benn.

Britt, George. 1935. *Forty Years—Forty Millions: The Career of Frank A. Munsey.* New York: Farrar and Rinehart.

Brough, James. 1963. *Auction!* Indianapolis, Ind.: Bobbs-Merrill.

Brown, Lawrence A. 1965. *Models for Spatial Diffusion Research—A Review.* Technical Report no. 3, ONR Spatial Diffusion Study. Evanston, Ill.: Department of Geography, Northwestern University.

Brown, Lester R. 1972. *World Without Borders.* New York: Random House.

Brown, Roger. 1973. *A First Language: The Early Stages.* Cambridge, Mass.: Harvard University Press.

Brown, Seth. 1912. *Advertising Agency Relations.* Chicago: Seth Brown.

Bruchey, Stuart Weems. 1956. *Robert Oliver, Merchant of Baltimore, 1783–1819.* Johns Hopkins University Studies in Historical and Political Science, series 74, no. 1. Baltimore: Johns Hopkins Press.

———— comp. and ed. 1967. *Cotton and the Growth of the American Economy, 1790–1860: Sources and Readings.* New York: Harcourt, Brace and World.

Brzezinski, Zbigniew. 1970. *Between Two Ages: America's Role in the Technetronic Era.* New York: Viking Press.

Buck, Norman Sydney. 1925. *The Development of the Organisation of Anglo-American Trade, 1800–1850.* New Haven, Conn.: Yale University Press.

Buckle, Henry Thomas. 1857. *History of Civilization in England.* New York: Ungar, 1964.

Buley, Roscoe Carlyle. 1967. *The Equitable Life Assurance Society of the United States, 1859–1964.* New York: Appleton-Century-Crofts.

Burchfield, R. W., ed. 1972. *A Supplement to the Oxford English Dictionary*, vol. 1. Oxford: Oxford University Press, Clarendon.

Burck, Gilbert. 1964. "Knowledge: The Biggest Growth Industry of Them All." *Fortune* (November): 128–131 ff.

Burnham, David. 1983. *The Rise of the Computer State*. New York: Random House.

Butler, Samuel. 1872. *Erewhon*. New York: Penguin, 1970.

Buttrick, John. 1952. "The Inside Contract System." *Journal of Economic History* 12(3): 205–221.

Cahn, William. 1969. *Out of the Cracker Barrel: The Nabisco Story from Animal Crackers to Zuzus*. New York: Simon and Schuster.

Cairns-Smith, A. G. 1971. *The Life Puzzle*. Edinburgh: Oliver and Boyd.

―――― 1977. "Synthetic Life for Industry." Pp. 405–410 in *The Encyclopedia of Ignorance*, ed. Ronald Duncan and Miranda Weston-Smith. New York: Pocket Books.

Cantril, Hadley. 1940. *The Invasion from Mars: A Study in the Psychology of Panic*. Princeton, N.J.: Princeton University Press.

Carlyle, Thomas. 1850. *Latter-Day Pamphlets*. New York: Charles Scribner's Sons, 1898.

Carroll, John B. 1964. "Words, Meanings and Concepts." *Harvard Educational Review* 34(2): 178–202.

Cassady, Ralph, Jr. 1967. *Auctions and Auctioneering*. Berkeley: University of California Press.

Catterall, Ralph C. H. 1903. *The Second Bank of the United States*. Decennial Publications of the University of Chicago, 2nd series, vol. 2. Chicago: University of Chicago Press.

Cavalli-Sforza, L. L., and M. W. Feldman. 1981. *Cultural Transmission and Evolution: A Quantitative Approach*. Princeton, N.J.: Princeton University Press.

Cavalli-Sforza, L. L., M. W. Feldman, K. H. Chen, and S. M. Dornbusch. 1982. "Theory and Observation in Cultural Transmission." *Science* 218 (October 1): 19–27.

Ceruzzi, Paul E. 1983. *Reckoners: The Prehistory of the Digital Computer, from Relays to the Stored Program Concept, 1935–1945*. Westport, Conn.: Greenwood.

Chandler, Alfred D., Jr. 1956. *Henry Varnum Poor: Business Editor, Analyst, and Reformer*. Cambridge, Mass.: Harvard University Press.

―――― 1959. "The Beginnings of 'Big Business' in American Industry." *Business History Review* 33(1): 1–31.

―――― 1962. *Strategy and Structure: Chapters in the History of the Industrial Enterprise*. Cambridge, Mass.: MIT Press.

―――― 1965a. "The Railroads: Pioneers in Modern Corporate Management." *Business History Review* 39(1): 16–40.

―――― comp. and ed. 1965b. *The Railroads, the Nation's First Big Business: Sources and Readings*. New York: Harcourt, Brace and World.

―――― 1967. "The Large Industrial Corporation and the Making of the Modern American Economy." Pp. 71–101 in *Institutions in Modern America:*

Innovation in Structure and Process, ed. Stephen E. Ambrose. Baltimore: Johns Hopkins Press.

—— 1972. "Anthracite Coal and the Beginnings of the Industrial Revolution in the United States." *Business History Review* 46(2): 141–181.

—— 1977. *The Visible Hand: The Managerial Revolution in American Business.* Cambridge, Mass.: Belknap Press of Harvard University Press.

Chase, Ivan D. 1980. "Cooperative and Noncooperative Behavior in Animals." *American Naturalist* 115(6): 827–857.

Chatfield, Michael. 1977. *A History of Accounting Thought.* Rev. ed. Melbourne, Fla.: Krieger.

Chomsky, Noam. 1965. *Aspects of the Theory of Syntax.* Cambridge, Mass.: MIT Press.

—— 1968. *Language and Mind.* New York: Harcourt, Brace and World.

Church, Alonzo. 1936a. "A Note on the Entscheidungsproblem." *Journal of Symbolic Logic* 1: 40–41; 101–102.

—— 1936b. "An Unsolvable Problem of Elementary Number Theory." *American Journal of Mathematics* 58: 345–363.

Clark, Colin. 1940. *The Conditions of Economic Progress.* London: Macmillan, 3rd ed., 1957.

Clark, John G. 1966. *The Grain Trade in the Old Northwest.* Urbana: University of Illinois Press.

—— 1970. *New Orleans, 1718–1812: An Economic History.* Baton Rouge: Louisiana State University Press.

Clark, Thomas D. 1944. *Pills, Petticoats and Plows: The Southern Country Store.* Indianapolis, Ind.: Bobbs-Merrill.

Clark, Victor S. 1916. *History of Manufactures in the United States, 1607–1860.* Washington: Carnegie Institution.

Coase, Richard H. 1937. "The Nature of the Firm." *Economica* 4(16): 386–405.

Cochran, Thomas C. 1948. *The Pabst Brewing Company: The History of an American Business.* New York: New York University Press.

Cohen, Ira. 1971. "The Auction System in the Port of New York, 1817–1837." *Business History Review* 45(4): 488–510.

Cole, Arthur Harrison. 1926. *The American Wool Manufacture*, vol. 1. Cambridge, Mass.: Harvard University Press.

Cole, H. S. D., Christopher Freeman, Marie Jahoda, and K. L. R. Pavitt, eds. 1973. *Models of Doom: A Critique of the Limits to Growth.* New York: Universe Books.

Coleman, James S., Elihu Katz, and Herbert Menzel. 1957. "The Diffusion of an Innovation among Physicians." *Sociometry* 20: 253–270.

Collins, James H. 1924. *The Story of Canned Foods.* New York: E. P. Dutton.

Collins, Randall. 1979. *The Credential Society: An Historical Sociology of Education and Stratification.* New York: Academic.

Comanor, William S., and Thomas A. Wilson. 1967. "Advertising Market Structure and Performance." *Review of Economics and Statistics* 49(4): 423–440.

———— 1974. *Advertising and Market Power*. Cambridge, Mass.: Harvard University Press.

Comaroff, John L. 1978. "Rules and Rulers: Political Processes in a Tswana Chiefdom." *Man* 13(1): 1–20.

Comaroff, John L., and Simon Roberts. 1981. *Rules and Processes: The Cultural Logic of Dispute in an African Context*. Chicago: University of Chicago Press.

Comte, Auguste. 1852. *System of Positive Polity*, vol. 2, *Social Statics, or the Abstract Theory of Human Order*. London: Longmans, Green, 1875.

Copeland, Melvin T. 1958. *And Mark an Era: The Story of the Harvard Business School*. Boston: Little, Brown.

Copley, Frank Barkley. 1923. *Frederick W. Taylor, Father of Scientific Management*, 2 vols. New York: Harper.

Corina, Maurice. 1975. *Trust in Tobacco: The Anglo-American Struggle for Power*. New York: St. Martin's.

Creamer, Daniel, Sergei P. Dobrovolsky, and Israel Borenstein. 1960. *Capital in Manufacturing and Mining: Its Formation and Financing*. National Bureau of Economic Research Study in Capital Formation and Financing, no. 6. Princeton, N.J.: Princeton University Press.

Crick, Francis H. C. 1966. "The Genetic Code—Yesterday, Today, and Tomorrow." *Cold Spring Harbor Symposia on Quantitative Biology* 31: 3–9.

Crick, Malcolm. 1976. *Explorations in Language and Meaning: Towards a Semantic Anthropology*. New York: Wiley Halsted.

Crozier, Michel. 1973. *The Stalled Society*. New York: Viking Press.

Cuff, Robert. 1984. "Mobilizing American Production for World War II." Unpublished.

Cummings, Richard Osborn. 1941. *The American and His Food*. New York: Arno Press and the New York Times, 2nd rev. ed., 1970.

Curtis, Helena. 1975. *Biology*. 2nd ed. New York: Worth.

Cutting, James E., and Burton S. Rosner. 1974. "Categories and Boundaries in Speech and Music." *Perception and Psychophysics* 16(3): 564–570.

Daglish, Robert, ed. 1972. *The Scientific and Technological Revolution: Social Effects and Prospects*. Moscow: Progress Publishers.

Dahrendorf, Ralf. 1959. *Class and Class Conflict in an Industrial Society*. Stanford, Calif.: Stanford University Press.

———— 1964. "Recent Changes in the Class Structure of European Societies." Pp. 291–336 in *A New Europe?* ed. Stephen R. Graubard. Boston: Houghton Mifflin.

Dale, Ernest. 1956. "Contributions to Administration by Alfred P. Sloan, Jr. and GM." *Administrative Science Quarterly* 1(1): 30–62.

Dance, Frank E. X., and Carl E. Larson. 1976. *The Functions of Human Communication: A Theoretical Approach*. New York: Holt, Rinehart, and Winston.

Danhof, Clarence H. 1969. *Change in Agriculture: The Northern United States, 1820–1870*. Cambridge, Mass.: Harvard University Press.

Darwin, Charles R. 1859. *On the Origin of Species*. Facsimile of 1st ed. Cambridge, Mass.: Harvard University Press, 1975.

—— 1871. *The Descent of Man and Selection in Relation to Sex*. New York: Appleton, 2nd ed., 1913.

David, Paul A. 1970. "Learning by Doing and Tariff Protection: A Reconsideration of the Case of the Ante-Bellum United States Cotton Textile Industry." *Journal of Economic History* 30(3): 521–601.

Davis, Alec. 1967. *Package and Print: The Development of Container and Label Design*. New York: Clarkson N. Potter.

Davis, Charles S. 1939. *The Cotton Kingdom in Alabama*. Montgomery: Alabama State Department of Archives and History.

Davis, Kingsley. 1949. *Human Society*. New York: Macmillan.

Davis, Martin, ed. 1965. *The Undecidable: Basic Papers on Undecidable Propositions, Unsolvable Problems, and Computable Functions*. Hewlett, N.Y.: Raven Press.

Davis, Martin, and Reuben Hersh. 1973. "Hilbert's 10th Problem." *Scientific American* 229(5): 84–91.

Dawkins, Richard. 1976. *The Selfish Gene*. New York: Oxford University Press.

—— 1980. "Good Strategy or Evolutionarily Stable Strategy?" Pp. 331–367 in *Sociobiology: Beyond Nature/Nurture? Reports, Definitions and Debate*, ed. George W. Barlow and James Silverberg. American Association for the Advancement of Science Selected Symposia Series no. 35. Boulder, Col.: Westview.

—— 1982. *The Extended Phenotype: The Gene as the Unit of Selection*. San Francisco: W. H. Freeman.

Deese, Sara L. 1957. "Trade: World Statistics." Pp. 348–353 in *Encyclopaedia Britannica: A New Survey of Universal Knowledge*, vol. 22. Chicago: Encyclopedia Britannica.

de Finetti, Bruno. 1937. "Foresight: Its Logical Laws, Its Subjective Sources," trans. Henry E. Kyburg, Jr. Pp. 93–158 in *Studies in Subjective Probabilities*, ed. Henry E. Kyburg, Jr., and Howard E. Smokler. New York: Wiley, 1964.

De Fleur, Melvin L., and Sandra J. Ball-Rokeach. 1982. *Theories of Mass Communication*, 4th ed. New York: Longman.

Demsetz, Harold. 1974. "Two Systems of Belief about Monopoly." Pp. 164–184 in *Industrial Concentration: The New Learning*, ed. Harvey J. Goldschmid, H. Michael Mann, and J. Fred Weston. Boston: Little, Brown.

—— 1979. "Accounting for Advertising as a Barrier to Entry." *Journal of Business* 52(3): 345–360.

de Roover, Florence Edler. 1941. "Partnership Accounts in Twelfth Century Genoa." *Bulletin of the Business Historical Society* 15(6): 87–92.

de Roover, Raymond. 1953. "The Commercial Revolution of the Thirteenth Century." Pp. 80–85 in *Enterprise and Secular Change: Readings in Economic History*, ed. Frederic C. Lane and Jelle C. Riemersma. Homewood, Ill.: Irwin.

—— 1963. "The Organization of Trade." Pp. 42–118 in *The Cambridge Eco-*

nomic History of Europe, vol. 3, *Economic Organization and Policies in the Middle Ages*, ed. M. M. Postan, E. E. Rich, and Edward Miller. Cambridge: Cambridge University Press.

Dertouzos, Michael L. 1979. "Individualized Automation." Chap. 3, pp. 38–55 in *The Computer Age: A Twenty-Year View*, ed. Michael L. Dertouzos and Joel Moses. Cambridge, Mass.: MIT Press.

Dertouzos, Michael L., and Joel Moses, eds. 1979. *The Computer Age: A Twenty-Year View*. Cambridge, Mass.: MIT Press.

Deutsch, Karl W., John Platt, and Dieter Senghaas. 1971. "Conditions Favoring Major Advances in Social Science." *Science* 171 (February 5): 450–459.

Dewey, John. 1931. *Philosophy and Civilization*. New York: Minton, Balch.

Dickinson, H. W., and Rhys Jenkins. 1927. *James Watt and the Steam Engine*. Oxford: Oxford University Press.

Dizard, Wilson P., Jr. 1982. *The Coming Information Age: An Overview of Technology, Economics, and Politics*. New York: Longman.

Djilas, Milovan. 1957. *The New Class: An Analysis of the Communist System*. New York: Praeger.

Dordick, Herbert S., Helen G. Bradley, and Burt Nanus. 1981. *The Emerging Network Marketplace*. Norwood, N.J.: Ablex.

Douglas, Mary, and Baron Isherwood. 1979. *The World of Goods: Towards an Anthropology of Consumption*. New York: Norton.

Driesch, Hans A. E. 1908. *The Science and Philosophy of the Organism*, 2 vols. Gifford Lectures, University of Aberdeen, 1907–1908. London: A. & C. Black, 1929.

Drucker, Peter F. 1959. *Landmarks of Tomorrow*. New York: Harper and Row.

——— 1969. *The Age of Discontinuity*. New York: Harper and Row.

Dubbey, J. M. 1978. *The Mathematical Work of Charles Babbage*. Cambridge: Cambridge University Press.

Duncan, Carson S. 1919. *Commercial Research: An Outline of Working Principles*. New York: Macmillan.

Durden, Robert F. 1975. *The Dukes of Durham, 1865–1929*. Durham, N.C.: Duke University Press.

Durkheim, Emile. 1893. *The Division of Labor in Society*, trans. George Simpson. New York: Free Press, 1933.

——— 1915. *The Elementary Forms of the Religious Life*, trans. Joseph Ward Swain. New York: Free Press, 1965.

——— 1928. *Socialism and Saint-Simon (Le socialisme)*, ed. Alvin W. Gouldner, trans. Charlotte Sattler. Yellow Springs, Ohio: Antioch Press, 1958.

Eames, Charles, and Ray Eames. 1973. *A Computer Perspective*. Cambridge, Mass.: Harvard University Press.

Eastwood, Robert Parker. 1940. *Sales Control by Quantitative Methods*. New York: Columbia University Press.

Eavenson, Howard N. 1942. *The First Century and a Quarter of American Coal Industry*. Baltimore: Waverly Press.

Eco, Umberto. 1976. *A Theory of Semiotics*. Bloomington: Indiana University Press.

Eddington, Arthur S. 1928. *The Nature of the Physical World*, Gifford Lectures, 1927. Cambridge: Cambridge University Press.

Eichner, Alfred S. 1976. *The Megacorp and Oligopoly: The Micro Foundations of Macro Dynamics*. Cambridge: Cambridge University Press.

Eigen, Manfred, and Ruthild Winkler. 1981. *Laws of the Game: How the Principles of Nature Govern Chance*, trans. Robert and Rita Kimber. New York: Harper and Row, Colophon.

Eisenberg, John F., N. A. Muckenhirn, and R. Rudran. 1972. "The Relation between Ecology and Social Structure in Primates." *Science* 176 (May 26): 863–874.

Eisenstadt, Shmuel N., ed. 1972. *Post-Traditional Societies*. New York: Norton.

Ekeh, Peter P. 1974. *Social Exchange Theory: The Two Traditions*. Cambridge, Mass.: Harvard University Press.

Ellul, Jacques. 1964. *The Technological Society*, trans. John Wilkinson. New York: Knopf.

Elster, Jon. 1979. *Ulysses and the Sirens: Studies in Rationality and Irrationality*. Cambridge: Cambridge University Press.

Elton, Charles S. 1927. *Animal Ecology*. New York: Macmillan.

Emerson, Harrington. 1911. *Efficiency as a Basis for Operation and Wages*. New York: Engineering Magazine.

——— 1913. *The Twelve Principles of Efficiency*. New York: Engineering Magazine.

Emlen, Stephen T. 1975. "The Stellar-Orientation System of a Migratory Bird." *Scientific American* 233(2): 102–111.

Engels, Friedrich. 1878. *Anti-Dühring*. Moscow: Progress, 1975.

Etzioni, Amitai. 1968. *The Active Society: A Theory of Societal and Political Processes*. New York: Free Press.

Evans, Christopher. 1979. *The Micro Millennium*. New York: Washington Square/Pocket Books.

Evans, Lawrence B. 1977. "Impact of the Electronics Revolution on Industrial Process Control." *Science* 195 (March 18): 1146–1151.

Fabricant, Solomon. 1949. "The Changing Industrial Distribution of Gainful Workers: Some Comments on the American Decennial Statistics for 1820–1940." *Studies in Income and Wealth*, vol. 11. New York: National Bureau of Economic Research.

Fagen, Robert M. 1980. "When Doves Conspire: Evolution of Nondamaging Fighting Tactics in a Nonrandom-Encounter Animal Conflict Model." *American Naturalist* 115(6): 858–869.

Feuer, Lewis S. 1969. *Marx and the Intellectuals: A Set of Post-Ideological Essays*. Garden City, N.Y.: Anchor Books.

Fisher, R.A. 1930. *The Genetical Theory of Natural Selection*. Oxford: Oxford University Press, Clarendon.

Fishlow, Albert. 1965. *American Railroads and the Transformation of the Ante-Bellum Economy*. Cambridge, Mass.: Harvard University Press.

Flad, Jean-Paul. 1963. "Les Trois Premières Machines à Calculer: Schickard

(1623), Pascal (1642), Leibniz (1673)." *La Conference au Palais de la Découverte*, no. D93. Paris: Université de Paris.

Flint, James. 1822. *Letters from America*, ed. Reuben Gold Thwaites. Cleveland, Ohio: Arthur H. Clark, 1904.

Fogel, Robert William, and Stanley L. Engerman. 1971a. "A Model for the Explanation of Industrial Expansion During the Nineteenth Century: With Application to the American Iron Industry." Pp. 148–162 in *The Reinterpretation of American Economic History*, ed. Robert William Fogel and Stanley L. Engerman. New York: Harper and Row.

—— eds. 1971b. *The Reinterpretation of American Economic History*. New York: Harper and Row.

Ford, Henry. 1923. *My Life and Work*. Garden City, N.Y.: Doubleday, Page.

Forester, Tom, ed. 1980. *The Microelectronics Revolution*. Cambridge, Mass.: MIT Press.

Foucault, Michel. 1966. *The Order of Things: An Archaeology of the Human Sciences*. New York: Vintage, 1970.

Fox, Stephen. 1984. *The Mirror Makers: A History of American Advertising and Its Creators*. New York: Random House.

Franken, Richard B., and Carroll B. Larrabee. 1928. *Packages That Sell*. New York: Harper.

Frantz, Joe B. 1951. *Gail Borden: Dairyman to a Nation*. Norman: University of Oklahoma Press.

Frazer, James George. 1918. *Folk-Lore in the Old Testament: Studies in Comparative Religion, Legend and Law*, vol. 2. New York: Macmillan, 1923.

Frege, Gottlob. 1879. "Begriffsschrift: A Formula Language, Modeled upon that of Arithmentic, for Pure Thought," trans. Stefan Bauer-Mengelberg. Pp. 1–82 in *From Frege to Gödel: A Source Book in Mathematical Logic, 1879–1931*, ed. Jean van Heijenoort. Cambridge, Mass.: Harvard University Press, 1967.

Fuller, Thomas. 1642. *The Holy State and the Profane State*, 2 vols., ed. Maximilian Graff Walten. New York: Columbia University Press, 1938.

Furlong, Lawrence. 1796. *The American Coast Pilot*. Newburyport, Mass.: Blunt and March.

Galbraith, John Kenneth. 1956. *American Capitalism: The Concept of Countervailing Power*. Rev. ed. Boston: Houghton Mifflin.

—— 1967. *The New Industrial State*. Boston: Houghton Mifflin, 3rd rev. ed., 1978.

—— 1976. *The Affluent Society*, 3rd ed., rev. Boston: Houghton Mifflin.

Gallie, Duncan. 1978. *In Search of the New Working Class*. Cambridge: Cambridge University Press.

Gamow, George. 1954. "Possible Relation between Deoxyribonucleic Acid and Protein Structure." *Nature* 173: 318.

Garner, S. Paul. 1954. *Evolution of Cost Accounting to 1925*. University: University of Alabama Press, reprint ed., 1976.

Gartner, Alan, and Frank Riessman. 1974. *The Service Society and the Consumer Vanguard*. New York: Harper and Row.

Gerard, Ralph W. 1960. "Becoming: The Residue of Change." Pp. 255–267 in

Evolution after Darwin, University of Chicago Centennial, vol. 2, *Evolution of Man: Man, Culture, and Society*, ed. Sol Tax. Chicago: University of Chicago Press.

Gerth, Hans H., and C. Wright Mills, eds. 1946. *From Max Weber: Essays in Sociology*. New York: Oxford University Press.

Gibb, George Sweet. 1950. *The Saco-Lowell Shops: Textile Machinery Building in New England, 1813–1949*. Cambridge, Mass.: Harvard University Press.

Giddens, Anthony. 1979. *Central Problems in Social Theory: Action, Structure, and Contradiction in Social Analysis*. London: Macmillan.

Giedion, Siegfried. 1948. *Mechanization Takes Command: A Contribution to Anonymous History*. New York: Norton, 1969.

Gierke, Otto von. 1880. *The Development of Political Theory*, trans. Bernard Freyd. New York: Norton, 1939.

——— 1881. *Political Theories of the Middle Age*, trans. Frederic William Maitland. Cambridge: Cambridge University Press, 1900.

——— 1913. *Natural Law and the Theory of Society, 1500 to 1800*, trans. Ernest Barker. Cambridge: Cambridge University Press, 1934.

Ginsberg, Morris. 1965. *On Justice in Society*. London: Heinemann.

Gintis, Herbert. 1970. "The New Working Class and Revolutionary Youth." *Continuum* 8(1, 2): 151–152.

Gluckman, Max. 1967. *The Judicial Process among the Barotse of Northern Rhodesia*. 2nd ed. Manchester, England: Manchester University Press.

Gödel, Kurt. 1931. "On Formally Undecidable Propositions of the *Principia Mathematica* and Related Systems, I," trans. Elliott Mendelson. Pp. 5–38 in *The Undecidable: Basic Papers on Undecidable Propositions, Unsolvable Problems, and Computable Functions*, ed. Martin Davis. Hewlett, N.Y.: Raven Press, 1965.

Goffman, Erving. 1961. *Asylums: Essays on the Social Situation of Mental Patients and Other Inmates*. Garden City, N.Y.: Doubleday Anchor.

——— 1967. *Interaction Ritual: Essays on Face-to-Face Behavior*. Garden City, N.Y.: Doubleday Anchor.

——— 1971. *Relations in Public: Microstudies of the Public Order*. New York: Basic.

Goldstine, Herman H. 1972. *The Computer from Pascal to von Neumann*. Princeton, N.J.: Princeton University Press.

Goodrich, Carter. 1960. *Government Promotion of American Canals and Railroads, 1800–1890*. New York: Columbia University Press.

Gorz, André. 1968. *Strategy for Labor*. Boston: Beacon Press.

Gouldner, Alvin W. 1979. *The Future of Intellectuals and the Rise of the New Class*. New York: Seabury Press, Continuum.

Granovetter, Mark S. 1973. "The Strength of Weak Ties." *American Journal of Sociology* 78(6): 1360–80.

Gras, Norman S. B. 1939. *Business and Capitalism: An Introduction to Business History*. New York: Crofts.

——— 1953. "Capitalism—Concepts and History." Pp. 66–79 in *Enterprise and Secular Change: Readings in Economic History*, ed. Frederic C. Lane and Jelle C. Riemersma. Homewood, Ill.: Irwin.

Gras, Norman S. B., and Henrietta M. Larson. 1939. *Casebook in American Business History*. New York: Crofts.

Gray, James. 1954. *Business without Boundary: The Story of General Mills*. Minneapolis: University of Minnesota Press.

Gray, Lewis Cecil. 1941. *History of Agriculture in the Southern United States to 1860*, 2 vols. Carnegie Institution of Washington, no. 430. New York: Peter Smith.

Greene, Asa. 1834. *The Perils of Pearl Street, Including a Taste of the Dangers of Wall Street, by a Late Merchant*. New York: Betts & Anstice and Peter Hill.

Grelling, Kurt. 1939. "Zur Logik der Sollsatze." *Unity of Science Forum* (January): 44–47.

Griliches, Zvi. 1957. "Hybrid Corn: An Exploration in the Economics of Technical Change." *Econometrica* 25: 501–522.

Grinker, Roy R., ed. 1967. *Toward a Unified Theory of Human Behavior: An Introduction to General Systems Theory*, 2nd ed. New York: Basic.

Gurevich, Michael. 1961 *The Social Structure of Acquaintanceship Networks*. Doctoral dissertation, Massachusetts Institute of Technology.

Hägerstrand, Torsten. 1952. *The Propagation of Innovation Waves*. Lund Studies in Geography, Series B, Human Geography, no. 4. Lund, Sweden: Department of Geography, Royal University of Lund.

——— 1953. *Innovation Diffusion as a Spatial Process*, trans. Allan Pred. Chicago: University of Chicago Press, 1967.

Halacy, Dan S. 1970. *Charles Babbage: Father of the Computer*. New York: Crowell-Collier.

Haldane, J.B.S. 1932. *The Causes of Evolution*. Ithaca, N.Y.: Cornell University Press, 1966.

Halmos, Paul. 1970. *The Personal Society*. London: Constable.

Hamilton, William D. 1964. "The Genetical Theory of Social Behaviour, I and II." *Journal of Theoretical Biology* 7(1): 1–16; 17–52.

——— 1967. "Extraordinary Sex Ratios." *Science* 156: 477–488.

——— 1975. "Innate Social Aptitudes in Man: An Approach from Evolutionary Genetics." Pp. 133–155 in *Biosocial Anthropology*, ed. Robin Fox. New York: Wiley.

Hammond, Matthew B. 1897. *The Cotton Industry: An Essay in American Economic History*. Pt. 1, The Cotton Culture and the Cotton Trade. American Economic Association (December). New York: Macmillan.

Harrington, Virginia D. 1935. *The New York Merchant on the Eve of the Revolution*. New York: Columbia University Press.

Hart, Herbert L. A. 1961. *The Concept of Law*. Oxford: Oxford University Press, Clarendon.

Hartley, R. V. L. 1928. "Transmission of Information." *Bell System Technical Journal* 7(3): 535–563.

Hartman, William D., and Henry M. Reiswig. 1973. "The Individuality of

Sponges." Pp. 567–584 in *Animal Colonies: Development and Function Through Time*, ed. Richard S. Boardman, Alan H. Cheetham, and William A. Oliver, Jr. Stroudsburg, Pa.: Dowden, Hutchinson, and Ross.

Hatt, Paul K., and Nelson N. Foote. 1953. "Social Mobility and Economic Advancement." *American Economic Review* 43(2): 364–378.

Hawkes, Nigel. 1971. *The Computer Revolution*. New York: Dutton.

Heaton, Herbert. 1948. *Economic History of Europe*. Rev. ed. New York: Harper.

Helvey, T. C. 1971. *The Age of Information: An Interdisciplinary Survey of Cybernetics*. Englewood Cliffs, N.J.: Educational Technology Publications.

Herbrand, Jacques. 1930. "Investigations in Proof Theory: The Properties of True Propositions," trans. Burton Dreben and Jean van Heijenoort. Pp. 524–581 in *From Frege to Gödel: A Source Book in Mathematical Logic, 1879–1931*, ed. Jean van Heijenoort. Cambridge, Mass.: Harvard University Press, 1967.

———— 1931. "On the Consistency of Arithmetic," trans. Jean van Heijenoort. Pp. 618–628 in *From Frege to Gödel: A Source Book in Mathematical Logic, 1879–1931*, ed. Jean van Heijenoort. Cambridge, Mass.: Harvard University Press, 1967.

Herodotus. 1910. *The History of Herodotus*, vol. 1, trans. George Rawlinson, ed. E. H. Blakeney. New York: Dutton, 1940.

Hicks, John R. 1939. *Value and Capital: An Inquiry into Some Fundamental Principles of Economic Theory*. Oxford: Oxford University Press, Clarendon, 2nd ed., 1946.

Hilbert, David. 1899. *Foundations of Geometry*, trans. Leo Unger. La Salle, Ill.: Open Court, 2nd ed., 1971.

———— 1900. "Mathematical Problems: Lecture Delivered before the International Congress of Mathematicians at Paris in 1900," trans. Mary Winston Newson. *Bulletin of the American Mathematical Society* 8 (1902): 437–479.

Hiltz, Starr Roxanne, and Murray Turoff. 1978. *The Network Nation: Human Communication via Computer*. Reading, Mass.: Addison-Wesley.

Hobbes, Thomas. 1651. *Leviathan: On the Matter, Form, and Power of a Commonwealth Ecclesiastical and Civil*, ed. Michael Oakeshott. New York: Macmillan, 1962.

Hofstadter, Douglas R. 1979. *Gödel, Escher, Bach: An Eternal Golden Braid*. New York: Random House, Vintage.

———— 1982. "Metamagical Themas: Is the Genetic Code an Arbitrary One, or Would Another Code Work as Well?" *Scientific American* 246(3): 18–29.

Holland, Donald R. 1974. "Volney B. Palmer: The Nation's First Advertising Agency Man." *Pennsylvania Magazine of History and Biography* 98(3): 353–381.

Hollerith, Herman. 1889. "An Electric Tabulating System." Pp. 133–143 in *Origins of Digital Computers: Selected Papers*, ed. Brian Randell. New York: Springer-Verlag, 3rd ed., 1982.

Homans, George Caspar. 1961. *Social Behavior: Its Elementary Forms*. New York: Harcourt Brace Jovanovich, rev. ed., 1974.

Hooker, Barbara I. 1968. "Birds." Pp. 311–337 in *Animal Communication: Techniques of Study and Results of Research*, ed. Thomas A. Sebeok. Bloomington: Indiana University Press.

Hopkins, Claude C. 1923. *Scientific Advertising*. New York: Chelsea House, 1980.

Hornung, Clarence P. 1959. *Wheels across America*. New York: A. S. Barnes.

Hower, Ralph M. 1943. *The History of Macy's of New York, 1858–1919: Chapters in the Evolution of the Department Store*. Cambridge, Mass.: Harvard University Press.

——— 1949. *The History of an Advertising Agency: N. W. Ayer and Son at Work, 1869–1949*. Rev. ed. Cambridge, Mass.: Harvard University Press.

Hoyle, Fred. 1964. *Man in the Universe*. New York: Columbia University Press.

Huebner, S. S. 1911. "The Functions of Produce Exchanges." *Annals of the American Academy of Political and Social Science* 38(2): 319–353.

Hughes, Thomas Parke. 1971. *Elmer Sperry: Inventor and Engineer*. Baltimore: Johns Hopkins University Press.

——— 1983. *Networks of Power: Electrification in Western Society, 1880–1930*. Baltimore: Johns Hopkins University Press.

Hunter, Louis C. 1949. *Steamboats on the Western Rivers: An Economic and Technological History*. Cambridge, Mass.: Harvard University Press.

——— 1951. "The Heavy Industries Before 1860." Pp. 172–189 in *The Growth of the American Economy*, ed. Harold F. Williamson, 2nd ed. Englewood Cliffs, N.J.: Prentice-Hall.

Huxley, Aldous. 1932. *Brave New World*. New York: Harper and Row, 1979.

Huxley, Julian. 1923. "Courtship Activities in the Red-Throated Diver *(Colymbus stellatus Pontopp.)*: Together with a Discussion of the Evolution of Courtship in Birds." *Journal of the Linnean Society of London, Zoology* 53: 253–292.

——— 1960. "The Openbill's Open Bill: A Teleonomic Enquiry." *Zoologische Jahrbücher Abteilung für Anatomie und Ontogenie der Tiere* 88: 9–30.

——— 1966. "A Discussion on Ritualization of Behaviour in Animals and Man: Introduction." *Philosophical Transactions of the Royal Society of Britain* 251: 249–271.

Hyman, Anthony. 1982. *Charles Babbage: Pioneer of the Computer*. Princeton, N.J.: Princeton University Press.

Iaciofano, Carol. 1984. "Computer Time Line." Pp. 20–34 in *Digital Deli: The Comprehensive, User-Lovable Menu of Computer Lore, Culture, Lifestyles and Fancy*, ed. Steve Ditlea. New York: Workman.

Ionescu, Ghita, ed. 1976. *The Political Thought of Saint-Simon*. Oxford: Oxford University Press.

Jakobson, Roman. 1960. "Closing Statement: Linguistics and Poetics." Pp. 350–377 in *Style in Language*, ed. Thomas A. Sebeok. Cambridge, Mass.: MIT Press and Wiley.

——— 1972. "Verbal Communication." Pp. 38–44 in *Communication*, ed. *Scientific American* Magazine. San Francisco: W. H. Freeman.

Jenkins, Clive, and Barrie Sherman. 1979. *The Collapse of Work*. London: Eyre Methuen.

Jenkins, John Wilber. 1927. *James B. Duke, Master Builder: The Story of Tobacco, Development of Southern and Canadian Water-Power and the Creation of a University*. New York: George H. Doran.

Jeremy, David J. 1973. "Innovation in American Textile Technology during the Early 19th Century." *Technology and Culture* 14(1): 40–76.

Jones, Fred Mitchell. 1937. "Middlemen in the Domestic Trade of the United States, 1800–1860." *Illinois Studies in the Social Sciences* 21(3): 1–81.

Judson, Horace Freeland. 1979. *The Eighth Day of Creation: Makers of the Revolution in Biology*. New York: Simon and Schuster, Touchstone.

Kahn, Herman. 1970. *Forces for Change in the Final Third of the Twentieth Century*. Croton-on-Hudson, N.Y.: Hudson Institute.

Kant, Immanuel. 1788. *Critique of Practical Reason*, trans. Lewis W. Beck. Indianapolis, Ind.: Bobbs-Merrill, 1956.

——— 1790. *Critique of Judgement*, trans. J. C. Meredith. Oxford: Oxford University Press, 1952.

Katz, Elihu. 1957. "The Two-Step Flow of Communication: An Up-to-Date Report on an Hypothesis." *Public Opinion Quarterly* 21: 61–78.

Keeton, William T., and Carol Hardy McFadden. 1983. *Elements of Biological Science*. 3rd ed. New York: Norton.

Keller, Morton. 1963. *The Life Insurance Enterprise, 1885–1910: A Study in the Limits of Corporate Power*. Cambridge, Mass.: Belknap Press of Harvard University Press.

Kelley, Etna M. 1954. *The Business Founding Date Directory*. Scarsdale, N.Y.: Morgan and Morgan.

Kendon, Adam, and Andrew Ferber. 1973. "A Description of Some Human Greetings." Pp. 591–668 in *Comparative Ecology and Behaviour of Primates*, ed. Richard P. Michael and John H. Crook. Proceedings of a conference held at the Zoological Society, London, November 1971. New York: Academic.

Kendrick, Alexander. 1969. *Prime Time: The Life of Edward R. Murrow*. Boston: Little, Brown.

Keynes, John Maynard. 1930. *A Treatise on Money*, vol. 1, *The Pure Theory of Money*. New York: Harcourt, Brace.

Killick, John R. 1974. "Bolton Ogden & Co.: A Case Study in Anglo-American Trade, 1790–1850." *Business History Review* 48(4): 501–519.

Kimball, George E., and Philip M. Morse. 1951. *Methods of Operations Research*. New York: Wiley.

Kirkland, Edward Chase. 1948. *Men, Cities and Transportation: A Study in New England History, 1820–1900*, 2 vols. Cambridge, Mass.: Harvard University Press.

——— 1961. *Industry Comes of Age: Business, Labor, and Public Policy, 1860–1897*. *Economic History of the United States*, vol. 6. New York: Holt, Rinehart and Winston.

Kitson, Harry Dexter. 1921. "Minor Studies in the Psychology of Advertising: From the Psychological Laboratory of Indiana University." *Journal of Applied Psychology* 5(1): 5–13.

Kleene, Stephen C. 1936. "General Recursive Functions of Natural Numbers." *Mathematische Annalen* 112: 727–742.

Koch, Howard. 1970. *The Panic Broadcast: Portrait of an Event*. Boston: Little, Brown.

Konishi, Masakazu, and Fernando Nottebohm. 1969. "Experimental Studies in the Ontogeny of Avian Vocalizations." Pp. 29–48 in *Bird Vocalizations: Their Relations to Current Problems in Biology and Psychology*, ed. Robert A. Hinde. Cambridge: Cambridge University Press.

Korte, Charles, and Stanley Milgram. 1970. "Acquaintance Networks between Racial Groups: Application of the Small World Method." *Journal of Personality and Social Psychology* 15(2): 101–108.

Kranzberg, Melvin, and Carroll W. Pursell, Jr., eds. 1967. *Technology in Western Civilization*, 2 vols. New York: Oxford University Press.

Krige, J. D. 1939. "Some Aspects of Lovhedu Judicial Arrangements." *Bantu Studies* 13: 113–129.

Kroeber, Alfred L., and Clyde Kluckhohn. 1952. *Culture: A Critical Review of Concepts and Definitions*, Papers of the Peabody Museum of American Archaeology and Ethnology, vol. 47, no. 1. Cambridge, Mass.: Peabody Museum, Harvard University.

Krooss, Herman E., and Martin R. Blyn. 1971. *A History of Financial Intermediaries*. New York: Random House.

Kuhlmann, Charles B. 1951. "Processing Agricultural Products after 1860." Pp. 432–453 in *The Growth of the American Economy*, 2nd ed., ed. Harold F. Williamson. Englewood Cliffs, N.J.: Prentice-Hall.

Kuhn, Thomas S. 1957. *The Copernican Revolution: Planetary Astronomy in the Development of Western Thought*. Cambridge, Mass.: Harvard University Press.

———— 1962. *The Structure of Scientific Revolutions*. Chicago: University of Chicago Press, 2nd ed. enl., 1970.

Kummer, Hans. 1971. *Primate Societies: Group Techniques of Ecological Adaptation*. Chicago: Aldine-Atherton.

Lacan, Jacques. 1966. *Ecrits: A Selection*, trans. Alan Sheridan. New York: Norton.

Lambert, Isaac E. 1941. *The Public Accepts: Stories behind Famous Trade-Marks, Names and Slogans*. Albuquerque: University of New Mexico Press.

Lamberton, Donald M., ed. 1974. *The Information Revolution*. Annals of the American Academy of Political and Social Science, vol. 412. Philadelphia: American Academy of Political and Social Science.

Lane, Frederic C. 1944. *Andrea Barbarigo, Merchant of Venice, 1418–1449*. Johns Hopkins University Studies in Historical and Political Science, series 62, no. 1. Baltimore: Johns Hopkins University Press.

———— 1953. "Family Partnerships and Joint Ventures in the Venetian Republic." Pp. 86–101 in *Enterprise and Secular Change: Readings in Eco-

nomic History, ed. Frederic C. Lane and Jelle C. Riemersma. Homewood, Ill.: Irwin.

Large, Peter. 1980. *The Micro Revolution*. London: Fontana.

—— 1984. *The Micro Revolution Revisited*. Totowa, N.J.: Rowman and Allanheld.

Laurie, Peter. 1981. *The Micro Revolution: Living with Computers*. New York: Universe Books.

Lebergott, Stanley. 1964. *Manpower in Economic Growth: The American Record since 1800*. New York: McGraw-Hill.

—— 1966. "United States Transport Advance and Externalities." *Journal of Economic History* 26(4): 437–461.

Lee, Alfred McClung. 1937. *The Daily Newspaper in America: The Evolution of a Social Instrument*. New York: Macmillan.

Lee, Guy A. 1937. "The Historical Significance of the Chicago Grain Elevator System." *Agricultural History* 11(1): 16–32.

Lehninger, Albert L. 1975. *Biochemistry: The Molecular Basis of Cell Structure and Function*. 2nd ed. New York: Worth.

Leland, Mrs. Wilfred C., and Minnie D. Millbrook. 1966. *Master of Precision: Henry M. Leland*. Detroit: Wayne State University Press.

Lenin, Vladimir I. 1917. *The State and Revolution: Marxist Teaching on the State and the Tasks of the Proletariat in the Revolution*. Peking: Foreign Languages Press, 1965.

Lenneberg, Eric H. 1967. *Biological Foundations of Language*. New York: Wiley.

Leontief, Wassily W. 1941. *The Structure of the American Economy, 1919–1929: An Empirical Application of Equilibrium Analysis*. Cambridge, Mass.: Harvard University Press.

Lessing, Lawrence P. 1956. *Man of High Fidelity: Edwin Howard Armstrong, A Biography*. Philadelphia: Lippincott.

Levins, Richard. 1970. "Extinction." Pp. 75–107 in *Some Mathematical Questions in Biology*, vol. 1, ed. Murray Gerstenhaber. *Lectures on Mathematics in the Life Sciences*. Providence, R.I.: American Mathematical Society.

Lévi-Strauss, Claude. 1949. *The Elementary Structures of Kinship*, trans. and ed. Rodney Needham. Boston: Beacon, rev. ed., 1969.

—— 1958. *Structural Anthropology*, trans. Claire Jacobson and Brooke Grundfest Schoepf. New York: Basic, 1963.

—— 1962. *The Savage Mind*. Chicago: University of Chicago Press, 1966.

Lewin, Kurt. 1931a. "The Conflict between Aristotelian and Galileian Modes of Thought in Contemporary Psychology," trans. Donald K. Adams. *Journal of Genetic Psychology* 5: 141–177.

—— 1931b. "Environmental Forces in Child Behavior and Development," trans. Donald K. Adams. Pp. 94–127 in *A Handbook of Child Psychology*, ed. Carl Murchison. Worcester, Mass.: Clark University Press.

—— 1931c. "The Psychological Situations of Reward and Punishment." Pp. 114–170 in *A Dynamic Theory of Personality: Selected Papers*, trans. Donald K. Adams and Karl E. Zener. New York: McGraw-Hill, 1935.

Lewis, Russell. 1973. *The New Service Society*. London: Longman.

Lichtheim, George. 1963. *The New Europe: Today and Tomorrow*. New York: Praeger.

Lindsay, Peter H., and Donald A. Norman. 1977. *Human Information Processing: An Introduction to Psychology*. 2nd ed. New York: Academic.

Littauer, Sebastian B. 1950. "The Development of Statistical Quality Control in the United States." *American Statistician* 4(5): 14–20.

Litterer, Joseph A. 1963. "Systematic Management: Design for Organizational Recoupling in American Manufacturing Firms." *Business History Review* 37(4): 369–391.

Livesay, Harold C. 1975. *Andrew Carnegie and the Rise of Big Business*, ed. Oscar Handlin. Boston: Little, Brown.

Locke, John. 1690. *Of Civil Government: Two Treatises*. London: J. M. Dent, 1924.

Lorenz, Konrad Z. 1963. *On Aggression*, trans. Marjorie Kerr Wilson. New York: Harcourt, Brace, and World, 1966.

Loveday, Evelyn. 1977. "George Stibitz and the Bell Labs Relay Computers." *Datamation* 23(9): 80–85.

Lovejoy, Arthur O. 1936. *The Great Chain of Being: A Study of the History of an Idea*. William James Lectures, Harvard University, 1933. Cambridge, Mass.: Harvard University Press.

Lovelock, J. E. 1979. *Gaia: A New Look at Life on Earth*. Oxford: Oxford University Press.

Luce, R. Duncan. 1959. *Individual Choice Behavior: A Theoretical Analysis*. New York: Wiley.

Luce, R. Duncan, and Patrick Suppes. 1965. "Preference, Utility, and Subjective Probability." Pp. 249–410 in *Handbook of Mathematical Psychology*, vol. 3, ed. R. Duncan Luce, Robert R. Bush, and Eugene Galanter. New York: Wiley.

Luhmann, Niklas. 1984. "Modes of Communication and Society." Unpublished speech. Fakultät für Soziologie, Universität Bielefeld, F.R.G.

Lumsden, Charles J., and Edward O. Wilson. 1981. *Genes, Mind, and Culture: The Coevolutionary Process*. Cambridge, Mass.: Harvard University Press.

Lyons, John. 1970. *Noam Chomsky*. New York: Viking.

Mabey, Richard. 1970. *Food Connexions*. London: Penguin Education.

MacArthur, Robert H. 1965. "Ecological Consequences of Natural Selection." Pp. 388–397 in *Theoretical and Mathematical Biology*, ed. Talbot Howe Waterman and Harold J. Morowitz. New York: Blaisdell.

McCafferty, E. D. 1923. *Henry J. Heinz: A Biography*. New York: Bartlett Orr.

Machlup, Fritz. 1962. *The Production and Distribution of Knowledge in the United States*. Princeton, N.J.: Princeton University Press.

———— 1980. *Knowledge: Its Creation, Distribution, and Economic Significance*, vol. 1. Princeton, N.J.: Princeton University Press.

MacIver, Robert M. 1937. *Society: A Textbook of Sociology*. New York: Farrar and Rinehart.

McKie, James W. 1959. *Tin Cans and Tin Plate: A Study of Competition in*

Two Related Markets. Cambridge, Mass.: Harvard University Press.

McLaughlin, John F. 1980. *Mapping the Information Business.* Cambridge, Mass.: Program on Information Resources Policy, Harvard University.

McLennan, Hugh. 1970. *Synaptic Transmission,* 2nd ed. Philadelphia: Saunders.

McLuhan, Marshall. 1964. *Understanding Media: The Extensions of Man.* New York: McGraw-Hill.

MacNab, Robert M., and Daniel E. Koshland, Jr. 1972. "The Gradient-Sensing Mechanism in Bacterial Chemotaxis." *Proceedings of the National Academy of Sciences* 69(9): 2509–2512.

McNally, Robert. 1974. *Biology: An Uncommon Introduction.* San Francisco: Canfield.

Madison, James H. 1974. "The Evolution of Commercial Credit Reporting Agencies in Nineteenth-Century America." *Business History Review* 48(2): 164–186.

Maine, Henry J. S. 1861. *Ancient Law: Its Connection with the Early History of Society and Its Relation to Modern Ideas.* Boston: Beacon, 1963.

Malinowski, Bronislaw. 1922. *Argonauts of the Western Pacific: An Account of Native Enterprise and Adventure in the Archipelagoes of Melanesian New Guinea.* New York: Dutton, 1961.

—— 1926. *Crime and Custom in Savage Society.* New York: Harcourt, Brace.

—— 1934. "Introduction." Pp. 17–72 in H. Ian Hogbin, *Law and Order in Polynesia: A Study of Primitive Legal Institutions.* New York: Harcourt, Brace.

—— 1945. *The Dynamics of Culture Change: An Inquiry into Race Relations in Africa,* ed. Phyllis M. Kaberry. New Haven, Conn.: Yale University Press.

—— 1948. *Magic, Science and Religion, and Other Essays.* Garden City, N.Y.: Doubleday Anchor, 1954.

Mallet, Serge. 1963. *La Nouvelle Classe Ouvrière.* Paris: Editions du Seuil.

Mally, Ernst. 1926. *Grundgesetze des Sollens: Elemente der Logik des Willens.* Graz, Austria: Leuschner and Lubensky.

Mansfield, Edwin. 1961. "Technical Change and the Rate of Imitation." *Econometrica* 29 (October): 741–766.

—— 1963. "Intrafirm Rates of Diffusion of an Innovation." *Review of Economics and Statistics* 45 (November): 348–359.

Marcuse, Herbert. 1964. *One-Dimensional Man: Studies in the Ideology of Advanced Industrial Society.* Boston: Beacon Press.

Margerison, Tom A. 1978. "Computers." Chap. 48, pp. 1150–1203 in *A History of Technology,* vol. 7, *The Twentieth Century, c. 1900 to c. 1950,* pt. 2, ed. Trevor I. Williams. Oxford: Oxford University Press, Clarendon.

Marler, Peter. 1975. "On the Origin of Speech from Animal Sounds." Pp. 11–37 in *The Role of Speech in Language,* ed. James F. Kavanagh and James E. Cutting. Cambridge, Mass.: MIT Press.

Marquette, Arthur F. 1967. *Brands, Trademarks, and Good Will: The Story of the Quaker Oats Company.* New York: McGraw-Hill.

Marris, Robin. 1964. *The Economic Theory of Managerial Capitalism.* New York: Free Press.

Marschak, Jacob. 1968. "Economics of Inquiring, Communicating, and Deciding." *American Economic Review* 58(2): 1–8.

Martin, James. 1978. *The Wired Society.* Englewood Cliffs, N.J.: Prentice-Hall.

────── 1981. *The Telematic Society: A Challenge for Tomorrow.* Englewood Cliffs, N.J.: Prentice-Hall.

Martin, James, and David Butler. 1981. *Viewdata and the Information Society.* Englewood Cliffs, N.J.: Prentice-Hall.

Martin, James, and Adrian R. D. Norman. 1970. *The Computerized Society.* Englewood Cliffs, N.J.: Prentice-Hall.

Martin, Samuel E. 1964. Review of *Universals of Language*, ed. Joseph H. Greenberg. *Harvard Educational Review* 34(2): 353–355.

Marx, Karl. 1843. "Critique of Hegel's Philosophy of the State." Pp. 151–202 in *Writings of the Young Marx on Philosophy and Society*, trans. and ed. Lloyd D. Easton and Kurt H. Guddat. Garden City, N.Y.: Doubleday Anchor, 1967.

────── 1852. *The Eighteenth Brumaire of Louis Bonaparte.* New York: International Publishers, 1963.

Mattingly, Ignatius G., Alvin M. Liberman, Ann K. Syrdal and Terry Halwes. 1971. "Discrimination in Speech and Nonspeech Modes." *Cognitive Psychology* 2(2): 131–157.

Maxwell, James Clerk. 1865. "A Dynamical Theory of the Electromagnetic Field." *Philosophical Transactions of the Royal Society of London* 155: 459–513.

────── 1868. "On Governors." *Proceedings of the Royal Society of London* 16: 270.

────── 1871. *Theory of Heat.* Westport, Conn.: Greenwood, 3rd ed., 1970.

May, Earl Chapin. 1938. *The Canning Clan: A Pageant of Pioneering Americans.* New York: Macmillan.

Maynard Smith, John. 1972. "Game Theory and the Evolution of Fighting." Pp. 8–28 in *On Evolution.* Edinburgh: Edinburgh University Press.

────── 1974. "The Theory of Games and the Evolution of Animal Conflicts." *Journal of Theoretical Biology* 47(1): 209–221.

Maynard Smith, John, and G. R. Price. 1973. "The Logic of Animal Conflict." *Nature* 246 (November 2): 15–18.

Mayo, Elton. 1945. *The Social Problem of an Industrial Civilization.* Boston: Division of Research, Graduate School of Business Administration, Harvard University.

Mayr, Ernst. 1961. "Cause and Effect in Biology." *Science* 134: 1501–1506.

────── 1974a. "Behavior Programs and Evolutionary Strategies." *American Scientist* 62(6): 650–659.

────── 1974b. "Teleological and Teleonomic: A New Analysis." *Boston Studies in the Philosophy of Science* 14: 91–117.

────── 1976. *Evolution and the Diversity of Life: Selected Essays.* Cambridge, Mass.: Belknap Press of Harvard University Press.

Mayr, Otto. 1970. *The Origins of Feedback Control*. Cambridge, Mass.: MIT Press.

—— 1976. "Maxwell and the Origins of Cybernetics." Pp. 168–188 in *Philosophers and Machines*, ed. Otto Mayr. New York: Neale Watson.

Mead, George Herbert. 1934. *Mind, Self, and Society: From the Standpoint of a Social Behaviorist*, ed. Charles W. Morris. Chicago: University of Chicago Press.

Mead, Margaret. 1970. *Culture and Commitment: A Study of the Generation Gap*. New York: Doubleday, Natural History Press.

Meadows, Donella H., Dennis L. Meadows, Jorgen Randers, and William W. Behrens III. 1972. *Limits to Growth: A Report for the Club of Rome's Project on the Predicament of Mankind*. New York: Universe Books.

Menger, Karl. 1934. *Morality, Decision, and Social Organization: Toward a Logic of Ethics*, trans. Eric van der Schalie. Boston: Reidel, 1974.

—— 1939. "A Logic of the Doubtful: On Optative and Imperative Logic." Pp. 53–64 in *Reports of a Mathematical Colloquium*, vol. 2. Notre Dame University. Bloomington: Indiana University Press.

Menzel, Randolf, J. Erber, and T. Masuhr. 1974. "Learning and Memory in the Honeybee." Pp. 195–217 in *Experimental Analysis of Insect Behaviour*, ed. L. Barton Browne. New York: Springer-Verlag.

Merriman, Roger Bigelow. 1925. *The Rise of the Spanish Empire in the Old World and in the New*, vol. 3, *The Emperor*. New York: Macmillan.

Merton, Robert K. 1968. *Social Theory and Social Structure*. Enl. ed. New York: Free Press.

Metcalfe, Capt. Henry. 1885. *The Cost of Manufactures and the Administration of Workshops, Public and Private*. New York: Wiley, 3rd ed., 1907.

Metropolis, Nicholas C., Jack Howlett, and Gian-Carlo Rota, eds. 1980. *A History of Computing in the Twentieth Century: A Collection of Essays*. International Research Conference on the History of Computing, Los Alamos Scientific Laboratory, 1976. New York: Academic.

Meynaud, Jean. 1968. *Technocracy*, trans. Paul Barnes. London: Faber and Faber.

Michael, Donald N. 1968. *The Unprepared Society: Planning for a Precarious Future*. New York: Harper and Row, Colophon.

Mikell, I. Jenkins. 1923. *Rumbling of the Chariot Wheels*. Columbia, S.C.: The State Company.

Milgram, Stanley. 1967. "The Small-World Problem." *Psychology Today* 1(1): 60–67.

Miliband, Ralph. 1983. "The State." Pp. 464–468 in *A Dictionary of Marxist Thought*, ed. Tom Bottomore. Cambridge, Mass.: Harvard University Press.

Mill, John Stuart. 1848. *Principles of Political Economy, with Some of their Applications to Social Philosophy*, 2 vols. Boston: Little, Brown.

—— 1859. *On Liberty*, ed. David Spitz. New York: Norton, 1975.

Miller, George A. 1956a. "The Magical Number Seven, Plus or Minus Two: Some Limits on Our Capacity for Processing Information." *Psychological Review* 63(2): 81–97.

—— 1956b. "Psychology's Block of Marble." *Contemporary Psychology* 1(8): 252–253.

Miller, George A., Eugene Galanter, and Karl H. Pribram. 1960. *Plans and the Structure of Behavior*. New York: Holt, Rinehart and Winston.

Miller, James Grier. 1978. *Living Systems*. New York: McGraw-Hill.

Minorsky, Nicholas. 1922. "Directional Stability of Automatically Steered Bodies." *Journal of the American Society of Naval Engineers* 34(2).

Moore, Sally Falk. 1969. "Comparative Studies: Introduction." Pp. 337–348 in *Law in Culture and Society*, ed. Laura Nader. Chicago: Aldine.

Moore, Wilbert E. 1951. *Industrial Relations and the Social Order*, rev. ed. New York: Macmillan.

Moreau, René. 1984. *The Computer Comes of Age: The People, the Hardware, and the Software*, trans. Jack Howlett. Cambridge, Mass.: MIT Press.

Morgenstern, Oskar. 1976. Foreword to *Bidding and Auctioning for Procurement and Allocation*, ed. Yakov Amihud. Proceedings of a conference at the Center for Applied Economics, New York University. New York: New York University Press.

Morrison, Philip, and Emily Morrison, eds. 1961. *Charles Babbage and His Calculating Engines: Selected Writings by Charles Babbage and Others*. New York: Dover.

Mott, Edward Harold. 1901. *Between the Ocean and the Lakes: The Story of Erie*. New York: John S. Collins.

Mott, Frank Luther. 1957. *A History of American Magazines*, vol. 4, *1865–1885*. New York: Appleton.

Moynihan, Martin H. 1976. *The New World Primates: Adaptive Radiation and the Evolution of Social Behavior, Languages, and Intelligence*. Princeton, N.J.: Princeton University Press.

Munroe, James Phinney. 1923. *A Life of Francis Amasa Walker*. New York: Holt.

Murray, Sir James A. H., ed. 1933. *The Oxford English Dictionary*, 13 vols. Oxford: Oxford University Press, Clarendon.

Nadel, S. F. 1953. "Social Control and Self-Regulation." *Social Forces* 31(3): 265–273.

—— 1957. *Theory of Social Structure*. Glencoe, Ill.: Free Press.

Nader, Laura. 1965. "The Anthropological Study of Law." *American Anthropologist* 67 (6, 2): 3–32.

Nader, Laura, and Barbara Yngvesson. 1973. "On Studying the Ethnography of Law and Its Consequences." Pp. 883–921 in *Handbook of Social and Cultural Anthropology*, ed. John J. Honigmann. Chicago: Rand McNally.

Nagel, Ernest. 1961. *The Structure of Science: Problems in the Logic of Scientific Explanation*. New York: Harcourt, Brace, and World.

Nason, Alvin, and Robert L. Dehaan. 1973. *The Biological World*. New York: Wiley.

Navin, Thomas R. 1970. "The 500 Largest American Industrials in 1917." *Business History Review* 44(3): 360–386.

Nelson, Daniel. 1974. "Scientific Management, Systematic Management, and Labor, 1880–1915." *Business History Review* 48(4): 479–500.

———— 1975. *Managers and Workers: Origins of the New Factory System in the United States, 1880–1920.* Madison: University of Wisconsin Press.

Nelson, Phillip. 1975. "The Economic Consequences of Advertising." *Journal of Business* 48(2): 213–241.

Nelson, Ralph L. 1959. *Merger Movements in American Industry, 1895–1956.* Princeton, N.J.: Princeton University Press.

Nevins, Allan, and Frank Ernest Hill. 1954. *Ford: The Times, the Men, the Company.* New York: Scribner.

Newell, Allen, and Herbert A. Simon. 1972. *Human Problem Solving.* Englewood Cliffs, N.J.: Prentice-Hall.

Newton, Bernard. 1968. *The Economics of Francis Amasa Walker: American Economics in Transition.* New York: A. M. Kelley.

New York *Times*. 1983. "The Boxcar Follies." Editorial. *New York Times* 132 (August 31): A26.

Nora, Simon, and Alain Minc. 1978. *The Computerization of Society: A Report to the President of France.* Cambridge, Mass.: MIT Press, 1980.

North, Douglass C. 1966. *The Economic Growth of the United States, 1790–1860.* New York: Norton.

North, Douglass C., and Robert Paul Thomas. 1973. *The Rise of the Western World: A New Economic History.* Cambridge: Cambridge University Press.

Novick, David, Melvin Anshen, and William C. Trupper. 1949. *Wartime Production Controls.* New York: Columbia University Press.

Nyquist, Harry. 1924. "Certain Factors Affecting Telegraph Speed." *Bell System Technical Journal* 3(2): 324–346.

———— 1928. "Certain Topics in Telegraph Transmission Theory." *AIEE Transactions* 47 (April): 617 ff.

Odle, Thomas. 1964. "Entrepreneurial Cooperation on the Great Lakes: The Origin of the Methods of American Grain Marketing." *Business History Review* 38(4): 439–455.

Oettinger, Anthony G. 1962. "Retiring Computer Pioneer—Howard Aiken." *Communications of the ACM* 5(6): 298–299.

———— 1971. "Compunications in the National Decision-Making Process." Pp. 73–114 in *Computers, Communications, and the Public Interest,* ed. Martin Greenberger. Baltimore: Johns Hopkins University Press.

Oettinger, Anthony G., Paul J. Berman, and William H. Read. 1977. *High and Low Politics: Information Resources for the 80's.* Cambridge, Mass.: Ballinger.

Ogilvy, David. 1980. "Introduction." In Claude Hopkins, *Scientific Advertising: The Classic Book on the Fundamentals of Advertising.* New York: Chelsea House.

———— 1983. *Ogilvy on Advertising.* New York: Crown.

Osterweis, Rollin G. 1949. *Romanticism and Nationalism in the Old South.* New Haven, Conn.: Yale University Press.

Packer, C. 1977. "Reciprocal Altruism in *Papio anubis.*" *Nature* 265 (February 3): 441–443.

Parlin, Charles Coolidge. 1914. *Merchandising of Automobiles*. Philadelphia: Curtis Publishing.

Parsons, Talcott, 1937. *The Structure of Social Action*. New York: Free Press, 1968.

—— 1951. *The Social System*. New York: Free Press.

—— 1960. *Structure and Process in Modern Societies*. Glencoe, Ill.: Free Press.

—— 1961. "The General Interpretation of Action." Pp. 85–97 in *Theories of Society: Foundations of Modern Sociological Theory*, vol. 1, ed. Talcott Parsons, Edward Shils, Kaspar D. Naegele, and Jesse R. Pitts. New York: Free Press of Glencoe.

—— 1966. *Societies: Evolutionary and Comparative Perspectives*. Englewood Cliffs, N.J.: Prentice-Hall.

—— 1967. Review of *The Sociological Tradition*, Robert A. Nisbet. *American Sociological Review* 32(4): 640–643.

—— 1969. *Politics and Social Structure*. New York: Free Press.

—— 1971. *The System of Modern Societies*. Englewood Cliffs, N.J.: Prentice-Hall.

—— 1975. "Social Structure and the Symbolic Media of Interchange." Chap. 6, pp. 94–120 in *Approaches to the Study of Social Structure*, ed. Peter M. Blau. New York: Free Press.

Paullin, Charles O. 1932. *Atlas of the Historical Geography of the United States*, ed. John K. Wright. Washington: Carnegie Institute and the American Geographical Society of New York.

Peirce, Charles Sanders. 1931. *Collected Papers*, vol. 5, *Pragmatism and Pragmaticism*, ed. Charles Hartshorne and Paul Weiss. Cambridge, Mass.: Harvard University Press.

Perkins, Edwin J. 1975. *Financing Anglo-American Trade: The House of Brown, 1800–1880*. Cambridge, Mass.: Harvard University Press.

Phillips, Kevin P. 1975. *Mediacracy: American Parties and Politics in the Communications Age*. Garden City, N.Y.: Doubleday.

Piaget, Jean. 1970. *Structuralism*, trans. Chaninah Maschler. New York: Basic.

Piore, Michael J., and Charles F. Sabel. 1984. *The Second Industrial Divide: Possibilities for Prosperity*. New York: Basic.

Pittendrigh, Colin S. 1958. "Adaptation, Natural Selection, and Behavior." Pp. 390–416 in *Behavior and Evolution*, ed. Anne Roe and George Gaylord Simpson. New Haven, Conn.: Yale University Press.

—— 1970. Personal correspondence with Ernst Mayr, February 26. Pp. 391–392 in Ernst Mayr, *Evolution and the Diversity of Life: Selected Essays*. Cambridge, Mass.: Belknap Press of Harvard University Press, 1976.

Politz, Alfred. 1952. "Introduction and Commentary." In Claude C. Hopkins, *Scientific Advertising: Republished in the Interests of Better Advertising*. New York: Moore.

Pool, Ithiel de Sola, and Manfred Kochen. 1978. "Contacts and Influence." *Social Networks* 1(1): 5–51.

Pope, Daniel. 1983. *The Making of Modern Advertising*. New York: Basic Books.

Porat, Marc Uri. 1977. *The Information Economy: Definition and Measurement*. Washington: Office of Telecommunications, U.S. Department of Commerce.

Porter, Glenn, and Harold C. Livesay. 1971. *Merchants and Manufacturers: Studies in the Changing Structure of Nineteenth-Century Marketing*. Baltimore: Johns Hopkins University Press.

Porter, Michael E. 1980. *Competitive Strategy: Techniques for Analyzing Industries and Competitors*. New York: Free Press.

Post, Emil L. 1936. "Finite Combinatory Processes—Formulation, I." *Journal of Symbolic Logic* 1(3): 103–105.

Prague Academy. 1973. *Man, Science, and Technology: A Marxist Analysis of the Scientific Technological Revolution*. Prague: Academia Prague.

Pred, Allan R. 1973. *Urban Growth and the Circulation of Information: The United States System of Cities, 1790–1840*. Cambridge, Mass.: Harvard University Press.

——— 1980. *Urban Growth and City-Systems in the United States, 1840–1860*. Cambridge, Mass.: Harvard University Press.

Presbrey, Frank. 1929. *The History and Development of Advertising*. Garden City, N.Y.: Doubleday, Doran.

Quesnay, François. 1758. *Tableau Oeconomique*. Ann Arbor, Mich.: University Microfilms International, 1980.

Quine, Willard Van Orman. 1940. *Mathematical Logic*. Cambridge, Mass.: Harvard University Press, rev. ed., 1951.

——— 1945. "On the Logic of Quantification." *Journal of Symbolic Logic* 10(1): 1–12.

——— 1950. *Methods of Logic*. New York: Holt-Dryden, rev. ed., 1959.

Quinn, William G., and Yadin Dudai. 1976. "Memory Phases in *Drosophila*." *Nature* 262 (August 12): 576–577.

Radcliffe-Brown, A. R. 1933. "Primitive Law." Pp. 202–206 in *Encyclopedia of the Social Sciences*, vol. 9, ed. Edwin R. A. Seligman. New York: Macmillan.

——— 1952. *Structure and Function in Primitive Society: Essays and Addresses*. New York: Free Press, 1965.

Radford, George S. 1922. *The Control of Quality in Manufacturing*. New York: Ronald.

Rae, John B. 1965. *The American Automobile: A Brief History*. Chicago: University of Chicago Press.

Ramsey, Frank Plumpton. 1926. "Truth and Probability." Pp. 156–198 in *The Foundations of Mathematics and Other Logical Essays*, ed. R. B. Braithwaite. London: Kegan Paul, Trench, Trubner, 1931.

Randell, Brian, ed. 1982. *The Origins of Digital Computers: Selected Papers*, 3rd ed. New York: Springer-Verlag.

Rapoport, Anatol, and Albert Chammah. 1965. *Prisoner's Dilemma: A Study in Conflict and Cooperation*. Ann Arbor: University of Michigan Press.

Rappaport, Roy A. 1968. *Pigs for the Ancestors: Ritual in the Ecology of a*

New Guinea People. New Haven, Conn.: Yale University Press.

——— 1971a. "The Flow of Energy in an Agricultural Society." *Scientific American* 224(3): 116–132.

——— 1971b. "The Sacred in Human Evolution." *Annual Review of Ecology and Systematics* 2: 23–44.

Reach, Karl. 1939. "Some Comments on Grelling's Paper 'Zur Logik der Sollsatze'." *Unity of Science Forum* (April): 72.

Redlich, Fritz. 1951. *The Molding of American Banking, Men and Ideas.* New York: Johnson Reprint Corporation, 1968.

Rees, James Frederick. 1920. *A Social and Industrial History of England, 1815–1918.* New York: Dutton.

Rehm, Hans-Jürgen, and Gerald Reed, eds. 1981. *Biotechnology: A Comprehensive Treatise,* 8 vols. Weinheim, F.R.G.: Verlag Chemie.

Reilly, William J. 1929. *Marketing Investigations.* New York: Arno, 1978.

Rescher, Nicholas, ed. 1966. *The Logic of Decision and Action.* Pittsburgh: University of Pittsburgh Press.

Resor, Stanley. 1912. *Population and Its Distribution, Compiled from the United States Census Figures of 1910.* New York: J. Walter Thompson Company.

Resseguie, Harry E. 1965. "Alexander Turney Stewart and the Development of the Department Store, 1823–1876." *Business History Review* 39(3): 301-322.

Ricardo, David. 1817. *The Principles of Political Economy and Taxation.* London: J. M. Dent, 1911.

Richta, Radovan, ed. 1967. *Civilization at the Crossroads: Social and Human Implications of the Scientific and Technological Revolution.* White Plains, N.Y.: International Arts and Sciences Press.

Riesman, David. 1950. *The Lonely Crowd: A Study of the Changing American Character,* with Reuel Denney and Nathan Glazer. New Haven, Conn.: Yale University Press.

Rodman, Peter S. 1973. "Population Composition and Adaptive Organisation among Orang-utans of the Kutai Reserve." Pp. 171–209 in *Comparative Ecology and Behaviour of Primates,* ed. Richard P. Michael and John H. Crook. Proceedings of a conference held at the Zoological Society, London, November 1971. London: Academic Press.

Roe, Joseph Wickham. 1916. *English and American Tool Builders.* New Haven: Yale University Press.

Roethlisberger, Fritz J., and William J. Dickson. 1939. *Management and the Worker: An Account of a Research Program Conducted by the Western Electric Company, Hawthorne Works, Chicago.* Cambridge, Mass.: Harvard University Press.

Rogers, Everett M. 1962. *Diffusion of Innovations.* New York: Free Press, 3rd ed., 1983.

Rogers, Everett M., with F. Floyd Shoemaker. 1971. *Communication of Innovations: A Cross-Cultural Approach,* 2nd ed. New York: Free Press.

Ross, Dorothy G. 1972. *G. Stanley Hall: The Psychologist as Prophet.* Chicago: University of Chicago Press.

Rosser, J. Barkley. 1936. "Extensions of Some Theorems of Gödel and Church." *Journal of Symbolic Logic* 1: 87–91.

Rostow, Walt W. 1960. *The Stages of Economic Growth.* Cambridge: Cambridge University Press.

Rothstein, Morton. 1965. "The International Market for Agricultural Commodities, 1850–1873." Pp. 62–82 in *Economic Change in the Civil War Era,* ed. David T. Gilchrist and W. David Lewis. Greenville, Del.: Eleutherian Mills-Hagley Foundation.

Rousseau, Jean Jacques. 1762. *The Social Contract,* trans. Charles Frankel. New York: Hafner, 1947.

Rowell, George Presbury. 1906. *Forty Years an Advertising Agent, 1865–1905.* New York: Printers' Ink Publishing.

Rowell's American Newspaper Directory. 1869. New York: George P. Rowell & Company.

Russell, William Howard. 1866. *The Atlantic Telegraph.* London: Day.

Ryan, Daniel Edward. 1880. *Human Proportions in Growth.* New York: Griffith & Byrne.

Ryle, Gilbert. 1949. *The Concept of Mind.* New York: Barnes and Noble, 1959.

Salsbury, Stephen. 1967. *The State, the Investor, and the Railroad: The Boston and Albany, 1825–1867.* Cambridge, Mass.: Harvard University Press.

Sanderlin, Walter S. 1946. *The Great National Project: A History of the Chesapeake and Ohio Canal.* Johns Hopkins University Studies in Historical and Political Science, series 64, no. 1. Baltimore: Johns Hopkins Press.

Sapir, Edward. 1921. *Language: An Introduction to the Study of Speech.* New York: Harcourt, Brace Harvest, 1949.

—— 1949. *Selected Writings of Edward Sapir in Language, Culture and Personality,* ed. David G. Mandelbaum. Berkeley: University of California Press.

Saussure, Ferdinand de. 1916. *Course in General Linguistics,* trans. Wade Baskin. New York: McGraw-Hill, 1966.

Savage, Leonard J. 1951. "The Theory of Statistical Decision." *Journal of the American Statistical Association* 46 (March): 55–67.

—— 1954. *The Foundations of Statistics.* New York: Wiley.

Schaller, George B. 1972. *The Serengeti Lion: A Study in Predator-Prey Relations.* Chicago: University of Chicago Press.

Schapera, Isaac. 1938. *A Handbook of Tswana Law and Custom.* London: Oxford University Press.

Scheflen, Albert E. 1967. "On the Structuring of Human Communication." *American Behavioral Scientist* 10(8): 8–12.

Scheiber, Harry N. 1969. *Ohio Canal Era: A Case Study of Government and the Economy, 1820–1861.* Athens: Ohio University Press.

Schmalensee, Richard. 1974. "Brand Loyalty and Barriers to Entry." *Southern Economic Journal* 40(4): 579–588.

Schmoller, Gustav. 1884. *The Mercantile System and Its Historical Significance, Illustrated Chiefly from Prussian History.* New York: Macmillan, 1896.

Schmookler, Jacob. 1966. *Invention and Economic Growth*. Cambridge, Mass.: Harvard University Press.

Schneirla, Theodore C. 1953. "Modifiability in Insect Behavior." Pp. 723–747 in *Insect Physiology*, ed. Kenneth D. Roeder. New York: Wiley.

Schotter, Andrew. 1976. "Auctions and Economic Theory." Pp. 3–12 in *Bidding and Auctioning for Procurement and Allocation*, ed. Yakov Amihud. Proceedings of a conference at the Center for Applied Economics, New York University. New York: New York University Press.

Schreyer, Helmut. 1939. "Technical Computing Machines." Pp. 171–173 in *Origins of Digital Computers: Selected Papers*, 3rd ed., ed. Brian Randell. New York: Springer-Verlag, 1982.

Schrödinger, Erwin. 1944. *What Is Life? The Physical Aspect of the Living Cell*. Based on lectures at Trinity College, Dublin, February 1943. Cambridge: Cambridge University Press.

Schubert, Paul. 1928. *The Electric Word: The Rise of Radio*. New York: Macmillan.

Scott, John P. 1967. "The Evolution of Social Behavior in Dogs and Wolves." *American Zoologist* 7(2): 373–381.

Scott, John P., and John L. Fuller. 1965. *Genetics and the Social Behavior of the Dog*. Chicago: University of Chicago Press.

Scott, Walter Dill. 1903. *The Theory of Advertising: A Simple Exposition of the Principles of Psychology in Their Relation to Successful Advertising*. Boston: Small, Maynard.

——— 1908. *The Psychology of Advertising: A Simple Exposition of the Principles of Psychology in Their Relation to Successful Advertising*. Boston: Small, Maynard.

Seidenberg, Roderick. 1950. *Posthistoric Man: An Inquiry*. Chapel Hill: University of North Carolina Press.

Seldon, Arthur, and F. G. Pennance. 1976. *Everyman's Dictionary of Economics*, 2nd ed. London: J. M. Dent.

Selfridge, Oliver G. 1959. "Pandemonium: A Paradigm for Learning." Pp. 511–526 in *Mechanisation of Thought Processes*, vol. 1. National Physical Laboratory, Symposium no. 10 (November 24–27, 1958). London: Her Majesty's Stationery Office, 1962.

Shannon, Claude E. 1938. "A Symbolic Analysis of Relay and Switching Circuits." Pp. 3–24 in *Computer Design Development: Principal Papers*, ed. Earl E. Swartzlander, Jr. Rochelle Park, N.J.: Hayden, 1976.

——— 1948. "The Mathematical Theory of Communication." *Bell System Technical Journal* 27(3): 379–423; (4): 623–656.

Shannon, Claude E., and Warren Weaver. 1949. *The Mathematical Theory of Communication*. Urbana: University of Illinois Press.

Shaw, Arch Wilkinson. 1916. *An Approach to Business Problems*. Cambridge, Mass.: Harvard University Press, 1926.

Shepherd, James F., and Gary M. Walton. 1972. *Shipping, Maritime Trade, and the Economic Development of Colonial North America*. Cambridge: Cambridge University Press.

Shewhart, Walter A. 1931. *Economic Control of Quality of Manufactured Product.* New York: Van Nostrand.

—— 1939. "The Future of Statistics in Mass Production." *Annals of Mathematical Statistics* 10(1): 88–90.

Shryer, William A. 1912. *Analytical Advertising.* Detroit: Business Service Corporation.

Silverman, Michael, and Melvin Simon. 1974. "Flagellar Rotation and the Mechanism of Bacterial Motility." *Nature* 249 (May 3): 73–74.

Simon, Herbert A. 1965. "The Logic of Rational Decision." *British Journal for the Philosophy of Science* 16: 169–186.

—— 1966. "The Logic of Heuristic Decision Making." Pp. 1–35 in *The Logic of Decision and Action,* ed. Nicholas Rescher. Pittsburgh: University of Pittsburgh Press.

Simpson, Tracy L. 1973. "Coloniality Among the Porifera." Pp. 549–565 in *Animal Colonies: Development and Function Through Time,* ed. Richard S. Boardman, Alan H. Cheetham, and William A. Oliver, Jr. Stroudsburg, Pa.: Dowden, Hutchinson, and Ross.

Sinclair, Upton. 1906. *The Jungle.* New York: New American Library, 1973.

Skinner, B.F. 1938. *The Behavior of Organisms: An Experimental Analysis.* New York: Appleton-Century.

Skinner, Wickham. 1984. "The Taming of Lions: How Manufacturing Leadership Evolved, 1780–1984." Paper presented to the Harvard Business School 75th Anniversary Colloquium on Technology and Productivity.

Smith, Adam. 1776. *An Inquiry into the Nature and Causes of the Wealth of Nations,* ed. Edwin Cannan. Chicago: University of Chicago Press, 1977.

Smith, Clayton Lindsay. 1923. *The History of Trade Marks.* New York: Thomas H. Stuart.

Smith, John E. 1982. *Biotechnology.* Baltimore: Edward Arnold.

Smith, R. Elberton. 1959. *The Army and Economic Mobilization.* Washington: Office of the Chief of Military History, Department of the Army.

Smith, W. John. 1977. *The Behavior of Communicating: An Ethological Approach.* Cambridge, Mass.: Harvard University Press.

Speiser, Ambros P. 1980. "The Relay Calculator Z4." *Annals of the History of Computing* 2(3): 242–245.

Spencer, Herbert. 1864–1867. *The Principles of Biology,* 2 vols. New York: Appleton, 1881.

—— 1876–1896. *The Principles of Sociology,* 3 vols. Westport, Conn.: Greenwood, 1975.

Starch, Daniel. 1923. *Principles of Advertising.* Chicago: A. W. Shaw.

Stent, Gunther S. 1971. *Molecular Genetics: An Introductory Narrative.* San Francisco: Freeman.

—— 1972. "Cellular Communication." Pp. 16–25 in *Communication,* ed. *Scientific American* Magazine. San Francisco: Freeman.

Sterling, Christopher H. 1984. *Electronic Media: A Guide to Trends in Broadcasting and Newer Technologies, 1920–1983.* New York: Praeger.

Stibitz, George R. 1940. "Computer." Pp. 247–252 in *Origins of Digital Computers: Selected Papers,* 3rd ed., ed. Brian Randell. New York: Springer-Verlag, 1982.

—— 1980. "Early Computers." Pp. 479–483 in *A History of Computing in the Twentieth Century*, eds. Nicholas C. Metropolis, Jack Howlett, and Gian-Carlo Rota. International Research Conference on the History of Computing, Los Alamos Scientific Laboratory, 1976. New York: Academic.

Stigler, George J. 1951. "The Division of Labor is Limited by the Extent of the Market." *Journal of Political Economy* 59(3): 185–193.

Stine, G. Harry. 1975. *The Third Industrial Revolution.* New York: G. P. Putnam's Sons.

Stonier, Tom. 1979. "The Third Industrial Revolution—Microprocessors and Robots." In *Microprocessors and Robots: Effects of Modern Technology on Workers.* Vienna: International Metalworkers' Federation.

Storck, John, and Walter Dorwin Teague. 1952. *Flour for Man's Bread: A History of Milling.* Minneapolis: University of Minnesota Press.

Stover, John F. 1961. *American Railroads.* Chicago: University of Chicago Press.

Sullivan, Mark. 1933. *Our Times: The United States, 1900–1925*, vol. 5, *Over Here, 1914–1918.* New York: Scribner's.

Sumner, William Graham. 1906. *Folkways: A Study of the Sociological Importance of Usages, Manners, Customs, Mores, and Morals.* Boston: Ginn, 1940.

Suppes, Patrick. 1956. "The Role of Subjective Probability and Utility in Decision-Making." Pp. 61–73 in *Proceedings of the Third Berkeley Symposium on Mathematical Statistics and Probability*, vol. 5, ed. Jerzy Neyman. Berkeley: University of California Press.

——— 1957. *Two Formal Models for Moral Principles.* Technical Report No. 15 (November 1). Stanford, Calif.: Applied Mathematics and Statistics Laboratory, Stanford University.

Swanson, Guy E. 1968. "Symbolic Interaction." Pp. 441–445 in *International Encyclopedia of the Social Sciences*, vol. 7, ed. David L. Sills. New York: Macmillan Free Press.

Swartzlander, Earl E., Jr., ed. 1976. *Computer Design Development: Principal Papers.* Rochelle Park, N.J.: Hayden.

Sylvester, Edward J., and Lynn C. Klotz. 1983. *The Gene Age: Genetic Engineering and the Next Industrial Revolution.* New York: Scribner's.

Szilard, Leo. 1929. "On the Decrease of Entropy in a Thermodynamic System by the Intervention of Intelligent Beings," trans. Anatol Rapoport and Mechthilde Knoller. *Behavioral Science* 9(4): 301–310.

Tansley, Arthur G. 1935. "The Use and Abuse of Vegetational Concepts and Terms." *Ecology* 16 (July): 284–307.

Taylor, Frederick Winslow. 1911a. *The Principles of Scientific Management.* New York: Norton, 1967.

——— 1911b. *Shop Management.* New York: Harper.

Taylor, George Rogers. 1951. *The Transportation Revolution, 1815–1860.* Economic History of the United States, vol. 4. New York: Rinehart.

Taylor, George Rogers, and Irene D. Neu. 1956. *The American Railroad Network, 1861–1890.* Cambridge, Mass.: Harvard University Press.

Taylor, Keith, ed. 1975. *Henri Saint-Simon (1760–1825): Selected Writings on Science, Industry, and Social Organization.* New York: Holmes and Meier.

Taylor, Philip A. S. 1969. *A New Dictionary of Economics*, 2nd ed. London: Routledge and Kegan Paul.

Temin, Peter. 1964. *Iron and Steel in Nineteenth-Century America: An Economic Inquiry*. Cambridge, Mass.: MIT Press.

—— 1966. "Steam and Waterpower in the Early Nineteenth Century." *Journal of Economic History* 26(2): 187–205.

—— 1969. *The Jacksonian Economy*. New York: Norton.

Tennant, Richard B. 1950. *The American Cigarette Industry: A Study in Economic Analysis and Public Policy*. New Haven, Conn.: Yale University Press.

Thomas, J. A. C. 1957. "The Auction Sale in Roman Law." *Juridical Review*, pt. 1 (April): 42–66.

—— 1968. "Roman Law." Pp. 1–27 in *An Introduction to Legal Systems*, ed. J. Duncan M. Derrett. New York: Praeger.

Thompson, Robert Luther. 1947. *Wiring a Continent: The History of the Telegraph Industry in the United States, 1832–1866*. Princeton, N.J.: Princeton University Press.

Thorelli, Hans B. 1954. *The Federal Antitrust Policy: Origination of an American Tradition*. Stockholm: Kungl. Boktryckeriet P. A. Norstedt.

Tilley, Nannie May. 1948. *The Bright-Tobacco Industry, 1860–1929*. Chapel Hill: University of North Carolina Press.

Tipper, Harry. 1914. *The New Business*. Garden City, N.Y.: Doubleday, Page.

Tipper, Harry, Harry L. Hollingworth, George Button Hotchkiss, and Frank Alvah Parsons. 1915. *Advertising: Its Principles and Practice*. New York: Ronald Press.

Toennies, Ferdinand. 1887. *Fundamental Concepts of Sociology (Gemeinschaft und Gesellschaft)*, trans. Charles P. Loomis. New York: American Book, 1940.

Toffler, Alvin. 1971. *Future Shock*. New York: Bantam Books.

—— 1980. *The Third Wave*. New York: William Morrow.

Tomeski, Edward Alexander. 1970. *The Computer Revolution: The Executive and the New Information Technology*. New York: Macmillan.

Tooker, Elva. 1955. *Nathan Trotter, Philadelphia Merchant, 1787–1853*. Cambridge, Mass.: Harvard University Press.

Touraine, Alain. 1971. *The Post-Industrial Society*. New York: Random House.

Toynbee, Arnold. 1884. *Lectures on the Industrial Revolution of the Eighteenth Century in England*. London: Longmans, Green, 1920.

Transeau, Edgar N. 1926. "The Accumulation of Energy by Plants." *Ohio Journal of Science* 26: 1–10.

Travers, Jeffrey, and Stanley Milgram. 1969. "An Experimental Study of the Small World Problem." *Sociometry* 32(4): 425–443.

Trefethen, Florence N. 1954. "A History of Operations Research." Pp. 3–35 in *Operations Research for Management*, ed. Joseph F. McCloskey and Florence N. Trefethen. Baltimore: Johns Hopkins University Press.

Trivers, Robert L. 1971. "The Evolution of Reciprocal Altruism." *Quarterly Review of Biology* 46(1): 35–57.

Truesdell, Leon E. 1965. *The Development of Punch Card Tabulation in the*

Bureau of the Census, 1890–1940, with Outlines of Actual Tabulation Programs. Washington: U.S. Government Printing Office.

Turck, Joseph A. V. 1921. *Origin of Modern Calculating Machines: A Chronicle of the Evolution of the Principles that Form the Generic Make-Up of the Modern Calculating Machine*. Chicago: Western Society of Engineers.

Turing, Alan M. 1936. "On Computable Numbers, with an Application to the Entscheidungsproblem." *Proceedings of the London Mathematical Society*, series 2, 42: 230–265; 43 (1937): 544–546.

Twyman, Robert W. 1954. *History of Marshall Field and Company, 1852–1906*. Philadelphia: University of Pennsylvania Press.

Tyler, David Budlong. 1939. *Steam Conquers the Atlantic*. New York: Appleton.

U.S. Bureau of the Census. 1975. *Historical Statistics of the United States, Colonial Times to 1970*, 2 vols. Washington: U.S. Government Printing Office.

U.S. Navy Department. 1963. *History of Communications-Electronics in the United States Navy*. Washington: U.S. Government Printing Office.

U.S. War Department. 1921. The Medical Department of the United States Army in the World War, vol. 15, *Statistics*, pt. 1, "Army Anthropology." Washington: U.S. War Department.

van Gennep, Arnold. 1909. *Rites of Passage*, trans. Monika B. Vizedon and Gabrielle L. Caffee. Chicago: University of Chicago Press, 1961.

van Heijenoort, Jean, ed. 1967. *From Frege to Gödel: A Source Book in Mathematical Logic, 1879–1931*. Cambridge, Mass.: Harvard University Press.

Vickers, Geoffrey. 1970. *Freedom in a Rocking Boat: Changing Values in an Unstable Society*. London: Allen Lane, Penguin.

Villee, Claude A. 1972. *Biology*, 6th ed. Philadelphia: Saunders.

von Neumann, John. 1927. "Zur Hilbertschen Beweistheorie." Pp. 256–300 in *Collected Works*, vol. 1, *Logic, Theory of Sets, and Quantum Mechanics*, ed. Abraham H. Taub. New York: Pergamon, 1961.

——— 1928. "On the Theory of Games of Strategy." Pp. 13–42 in *Contributions to the Theory of Games*, vol. 4, ed. A. W. Tucker and R. Duncan Luce. Princeton, N.J.: Princeton University Press, 1959.

——— 1966. *Theory of Self-Reproducing Automata*, ed. and completed by Arthur W. Burks. Based on five lectures at the University of Illinois, December 1949. Urbana: University of Illinois Press.

von Neumann, John, and Oskar Morgenstern. 1944. *The Theory of Games and Economic Behavior*. Princeton, N.J.: Princeton University Press.

von Wright, Georg Henrik. 1951. "Deontic Logic." *Mind* 60: 1–15.

Wade, John. 1833. *History of the Middle and Working Classes*. London: E. Wilson, 3rd ed., 1835.

Wald, Abraham. 1939. "Contributions to the Theory of Statistical Estimation and Testing Hypotheses." *Annals of Mathematical Statistics* 10: 299–326.

Walker, Francis A. 1879. *Report to the Secretary of the Interior on the "Temporary Nature of Census Operations"* (November 15). Washington: U.S. Government Printing Office.

Wall, Joseph Frazier. 1970. *Andrew Carnegie.* New York: Oxford University Press.

Walras, Leon. 1874. *Elements of Pure Economics*, trans. William Jaffe. London: Allen and Unwin, 1954.

Walsh, Joseph Leigh. 1950. *Connecticut Pioneers in Telegraphy: The Origin and Growth of the Telephone Industry in Connecticut.* New Haven, Conn.: Morris F. Tyler Chapter, Telephone Pioneers of America.

Ware, Caroline F. 1931. *The Early New England Cotton Manufacture: A Study in Industrial Beginnings.* Boston: Houghton Mifflin.

Warner, Sam Bass, Jr. 1968. *The Private City: Philadelphia in Three Periods of Its Growth.* Philadelphia: University of Pennsylvania Press.

Watkins, James L. 1908. *King Cotton: A Historical and Statistical Review, 1798 to 1908.* New York: Watkins.

Watson, James D., and Francis H. C. Crick. 1953. "Molecular Structure of Nucleic Acids: A Structure for Deoxyribose Nucleic Acid." *Nature* 171 (April 25): 737–738.

Weber, Max. 1905. *The Protestant Ethic and the Spirit of Capitalism*, trans. Talcott Parsons. New York: Scribner's, 1958.

—— 1922. *Economy and Society: An Outline of Interpretive Sociology*, 3 vols., ed. Guenther Roth and Claus Wittich. New York: Bedminster Press, 1968.

—— 1923. *General Economic History*, trans. Frank H. Knight. Glencoe, Ill.: Free Press, 1927.

Weisz, Paul B., and Richard N. Keogh. 1982. *The Science of Biology*, 5th ed. New York: McGraw Hill.

Westerfield, Ray Bert. 1920. "Early History of American Auctions—A Chapter in Commercial History." *Transactions of the Connecticut Academy of Arts and Sciences* 23(4): 159–210.

White, Percival. 1931. *Marketing Research Technique.* New York: Harper.

Whitehead, Alfred North, and Bertrand Russell. 1910–1913. *Principia Mathematica*, 3 vols. Cambridge: Cambridge University Press.

Whorf, Benjamin Lee. 1956. *Language, Thought and Reality: Selected Writings*, ed. John B. Carroll. Cambridge, Mass.: MIT Press and Wiley.

Whyte, William H., Jr. 1956. *The Organization Man.* New York: Simon and Schuster.

Wiener, Norbert. 1948. *Cybernetics: or Control and Communication in the Animal and the Machine.* Cambridge, Mass.: MIT Press, 2nd ed., 1961.

—— 1950. *The Human Use of Human Beings: Cybernetics and Society.* Boston: Houghton Mifflin, Riverside.

Wilkins, Mira. 1970. *The Emergence of Multinational Enterprise: American Business Abroad from the Colonial Era to 1914.* Cambridge, Mass.: Harvard University Press.

Williams, D. M. 1969. "Liverpool Merchants and the Cotton Trade, 1820–1850." Pp. 182–211 in *Liverpool and Merseyside: Essays in the Economic and Social History of the Port and Its Hinterland*, ed. John R. Harris. London: Frank Cass.

Williams, Frederick. 1982. *The Communications Revolution.* Beverly Hills, Calif.: Sage.

Williams, George C. 1966. *Adaptation and Natural Selection: A Critique of Some Current Evolutionary Thought.* Princeton, N.J.: Princeton University Press.

Williams, Thomas Harry. 1969. *Huey Long.* New York: Knopf Borzoi.

Williamson, Harold F., ed. 1951. *The Growth of the American Economy,* 2nd ed. Englewood Cliffs, N.J.: Prentice-Hall.

—— 1952. *Winchester: The Gun that Won the West.* Washington: Combat Forces Press.

Williamson, Harold F., and Arnold R. Daum. 1959. *The American Petroleum Industry: The Age of Illumination, 1859–1899.* Evanston, Ill.: Northwestern University Press.

Williamson, Oliver E. 1970. *Corporate Control and Business Behavior: An Inquiry into the Effects of Organization Form on Enterprise Behavior.* Englewood Cliffs, N.J.: Prentice Hall.

Wilson, Edward O. 1975. *Sociobiology: The New Synthesis.* Cambridge, Mass.: Belknap Press of Harvard University Press.

Wilson, Edward O., Thomas Eisner, Winslow R. Briggs, Richard E. Dickerson, Robert L. Metzenberg, Richard D. O'Brien, Millard Susman, and William E. Boggs. 1978. *Life on Earth.* 2nd ed. Sunderland, Mass.: Sinauer.

Winner, Langdon. 1977. *Autonomous Technology: Technics-out-of-Control as a Theme in Political Thought.* Cambridge, Mass.: MIT Press.

Wiseman, Alan, ed. 1982. *Principles of Biotechnology.* Surrey, England: Surrey University Press, Methuen.

Woodcock, F. Huntly, and W. R. Lewis. 1938. *Canned Foods and the Canning Industry.* London: Sir Isaac Pitman.

Woodman, Harold D. 1968. *King Cotton and His Retainers: Financing and Marketing the Cotton Crop of the South, 1800–1925.* Lexington: University of Kentucky Press.

Wren, Daniel A. 1972. *The Evolution of Management Thought.* New York: Ronald Press.

Wright, Larry. 1976. *Teleological Explanations: An Etiological Analysis of Goals and Functions.* Berkeley: University of California Press.

Wright, Sewall. 1931. "Evolution in Mendelian Populations." *Genetics* 16(2): 97–158.

Wynne-Edwards, Vero Copner. 1962. *Animal Dispersion in Relation to Social Behaviour.* New York: Hafner.

Young, John Zachary. 1964. *A Model of the Brain.* William Withering Lectures, University of Birmingham, 1960. Oxford: Oxford University Press, Clarendon.

Young, Michael. 1958. *The Rise of the Meritocracy 1870–2033: An Essay on Education and Equality.* Harmondsworth, England: Penguin, 1961.

Zaborsky, Oskar R. 1983. "Biotechnology Patents of 1983: An International Perspective." *Biotechnology* 1(1): 33–36.

Zevin, Robert Brooke. 1971. "The Growth of Cotton Textile Production after

1815." Pp. 122–147 in *The Reinterpretation of American Economic History*, ed. Robert William Fogel and Stanley L. Engerman. New York: Harper and Row.

Zuse, Konrad. 1936. "Method for Automatic Execution of Calculations with the Aid of Computers." Pp. 163–170 in *Origins of Digital Computers: Selected Papers*, 3rd ed., ed. Brian Randell. New York: Springer-Verlag, 1982.

———— 1962. "The Outline of a Computer Development from Mechanics to Electronics." Pp. 175–190 in *Origins of Digital Computers: Selected Papers*, 3rd ed., ed. Brian Randell. New York: Springer-Verlag, 1982.

———— 1980. "Some Remarks on the History of Computing in Germany." Pp. 611–627 in *A History of Computing in the Twentieth Century: A Collection of Essays*. International Research Conference on the History of Computing, Los Alamos Scientific Laboratory, 1976. New York: Academic.

Index